Studies in Computational Intelligence

Volume 892

Series Editor

Janusz Kacprzyk, Polish Academy of Sciences, Warsaw, Poland

The series "Studies in Computational Intelligence" (SCI) publishes new developments and advances in the various areas of computational intelligence—quickly and with a high quality. The intent is to cover the theory, applications, and design methods of computational intelligence, as embedded in the fields of engineering, computer science, physics and life sciences, as well as the methodologies behind them. The series contains monographs, lecture notes and edited volumes in computational intelligence spanning the areas of neural networks, connectionist systems, genetic algorithms, evolutionary computation, artificial intelligence, cellular automata, self-organizing systems, soft computing, fuzzy systems, and hybrid intelligent systems. Of particular value to both the contributors and the readership are the short publication timeframe and the world-wide distribution, which enable both wide and rapid dissemination of research output.

The books of this series are submitted to indexing to Web of Science, EI-Compendex, DBLP, SCOPUS, Google Scholar and Springerlink.

More information about this series at http://www.springer.com/series/7092

Vladik Kreinovich
Editor

Statistical and Fuzzy Approaches to Data Processing, with Applications to Econometrics and Other Areas

In Honor of Hung T. Nguyen's 75th Birthday

 Springer

Editor
Vladik Kreinovich
Department of Computer Science
University of Texas at El Paso
El Paso, TX, USA

ISSN 1860-949X   ISSN 1860-9503   (electronic)
Studies in Computational Intelligence
ISBN 978-3-030-45621-4   ISBN 978-3-030-45619-1   (eBook)
https://doi.org/10.1007/978-3-030-45619-1

This Springer imprint is published by the registered company Springer Nature Switzerland AG
The registered company address is: Gewerbestrasse 11, 6330 Cham, Switzerland

Foreword

It is always with great pleasure that we celebrate someone's birthday and it is, indeed, all the more pleasant for me to celebrate Hung T. Nguyen's 75th birthday, a colleague and dear friend I have known for almost 50 years.

In 1970, Hung was appointed at the University of Lille (France) where I was an associate professor. Both of us had already presented a (Third Cycle) doctoral thesis, Hung defending his in Paris while I defended mine in Lille.

That year, we had the chance to meet Joseph Kampé de Fériet, a university professor from Lille who had taken retirement in 1964 and whose research work on turbulence had been internationally known and appreciated.

He mentioned to us a new theory of information whose foundations he had rethought in 1966 with two Italian mathematicians, Pietro Benvenuti from the Institute of Applied Mathematics in Rome, and Bruno Forte from the University of Pavia. Considering that the notion of information is more primitive than that of probability, the new theory challenged the probability basis and suggested a more general version.

The research topic was new. Professor Kampé de Fériet felt very enthusiastic to work with two young researchers in Lille and offered to develop that new theory under his direction.

We thus spent five years working together, with this research making it possible for us to obtain the degrees of Doctorat d'État es Sciences Mathématiques.

Our careers then followed different paths, Hung went to the USA, namely to California (Berkeley), Massachusetts, and New Mexico, and later to Chiang Mai in Thailand, while I remained in Lille, developing cooperations in the field of statistics, first with Brazil and later with all South American states.

However, we have had many opportunities to meet each other, notably while attending congresses and conferences exemplified by: Bangkok (2000, 2001), Hanoi (2001), Chiang Mai (2003), Houston (2004), Kunming (2005), Chengdu (2006), Taipei (2005, 2006), Guillin (2009), and Saigon (2012, 2013). What is more, apart from these scientific events, we have often shared friendly family meetings, at home (Lille, Las Cruces, Chiang Mai) but also in Japan, Vietnam, Thailand, in northern Italy (Lake Como and Venice), in Prague, on a cruise down the Nile and also in Paris where Hung, a talented tennis player himself, attended the Roland Garros tournaments.

In light of all these landmarks, I therefore wish a memorable birthday to Prof. Hung T. Nguyen, a distinguished colleague and a great friend, hoping that he will remain in good health and will keep developing his numerous research ideas.

Claude Langrand
Professor Emeritus
University of Lille
Lille, France

Bibliography

1. J. Kampé De Fériet, B. Forte, Information et probabilité. C. R. Acad. Sci. Paris **265**, Sér.A, 110–114, 142–146, 350–353 (1967)
2. J. Kampé De Fériet, P. Benvenuti, Sur une classe d'informations. C. R. Acad. Sci. Paris **269**, Sér.A, 97–101 (1969)
3. J. Kampé De Fériet, B. Forte, P. Benvenuti, Forme générale de l'opération continue d'une information, C. R. Acad. Sci. Paris **269**, Sér.A, 529–534 (1969)
4. J. Kampé De Fériet, T. H. Nguyen, Temps d'entrée d'un processus stochastique et mesure de l'information. C. R. Acad. Sci. Paris **275**, Sér.A, 721–725 (1972)
5. C. Langrand, Constructions de m-précapacités. C. R. Acad. Sci. Paris **275**, Sér.A, 1243–1246 (1972)
6. N.-T. Hung, Mesures d'informations, capacités positives et sous-mesures. C. R. Acad. Sci. Paris **275**, Sér.A, 441–443 (1972)
7. C. Langrand, N.-T. Hung, Sur les mesures intérieures de l'information et les σ-précapacités. C. R. Acad. Sci. Paris **276**, Sér.A, 927–930 (1972)
8. C. Langrand, Mesures extérieures d'information. C. R. Acad. Sc. Paris **276**, Sér.A, 703–706 (1973)
9. J. Kampé De Fériet, N.T. Hung, Mesure de l'information, temps d'entrée et dimension de Hausdorff. C. R. Acad. Sci. Paris **276**, Sér.A, 807–811 (1973)
10. C. Langrand, Information généralisée; estimation et sélection, Thèse de doctorat es sciences mathématiques, Lille, 1973
11. J. Kampé De Fériet, La théorie généralisée de l'Information et la mesure subjective de l'Information Actes des Rencontres de Marseille-Luminy, in *Théories de l'information, Lectures Notes in Mathematics*, vol. 398 (Springer-Verlag, 1974), pp. 1–35, 5–7 juin 1973
12. C. Langrand, Précapacités fortes et mesures d'information, in *Lectures Notes in Mathematics*, vol. 398, (Springer-Verlag, 1974), pp. 36–48
13. N. T. Hung, Sur les mesures d'information de type Inf, in *Lectures Notes in Mathematics*, vol. 398 (Springer-Verlag, 1974), pp. 62–75

14. N.T. Hung, Mesures d'information sur les ensembles ordonnés. C. R. Acad. Sci. Paris **278**, Sér.A, 1139–1142 (1974)
15. C. Langrand, Composabilité d'une mesure d'information limite de mesures d'informations composables, C. R. Acad. Sci. Paris **279**, Sér.A, 727–730 (1974)
16. N.T. Hung, Mesures d'information, ensembles flous et espaces topologiques aléatoires, Thèse de doctorat es sciences mathématiques, Lille, 1975

Preface

It is my privilege and my honor to start this volume devoted to our good friend, great colleague, and wonderful teacher Hung T. Nguyen.

I have first met Professor Hung T. Nguyen twenty years ago, in 1990, shortly after my coming to the USA. Since then, we have published many joint papers, two joint books, and numerous joint edited books. During these twenty years, I continue to be impressed by his enthusiasm, by his productivity, and by his ability to transition to completely research areas.

His contributions are numerous, let me mention a few major ones. Papers co-authored by Hung T. Nguyen revolutionized our understanding of conditional probabilities—by showing that these probabilities can be interpreted as probabilities of naturally defined events. He showed that both Dempster–Shafer belief theory and fuzzy techniques, two of the most successful post-probabilistic approaches to uncertainty, can be naturally interpreted in probabilistic terms—namely in terms of random sets. He showed that a natural way to extend functions (and algorithms in general) to fuzzy data can be expressed in terms of the ranges of these functions over naturally defined sets. From statistics and fuzzy, he moved to econometrics, where he is now developing and actively promoting quantum ideas.

In many of these areas, not only he got interesting results, he also wrote textbooks and monographs and lecture notes that describe these—and other—results in a very clear and convincing way. These books helped several generations of researchers understand complex topics and complex results.

In this volume, we, his friends and colleagues, present examples of our results motivated by Hung's results and ideas. Happy birthday to Hung T. Nguyen. We are looking forward toward many more years of his ideas and his successes.

HUNG T. NGUYEN

A. Awards

(1) *Chair of Fuzzy Theory* (1992–1993): Tokyo Institute of Technology, Japan
(2) *Westhafer Award for Excellence in Research and Creativity* (1999–2000): New Mexico State University, USA
(3) *Distinguished Lukacs Professor of Statistics* (Spring 2002): Bowling Green State University, Ohio, USA
(4) *Distinguished Faculty Fellow* (Summers 2002&2003): American Association of Engineering Education/Summer Faculty Research Program, USA.

B. Selected Publication

1. A note on the extension principle for fuzzy sets. *J. Math.Anal. and Appl. (64) 369–380 (1978)*
2. On random sets and belief functions. *J. Math. Anal. Appl. (65), 541–542 (1978)*
3. Density estimation in a continuous-time, stationary markov process. *Annals of Statistics 7(2), 341–348 (1979)*
4. Sur l'utilisation du temps local en statistique des processus (co-authored with T. D. Pham). *Comptes Rendus Acad. Sci. Paris A-290, 165–168 (1980)*
5. Recursive estimation in diffusion model (co-authored with G. Banon). *SIAM J. Control and Optimization 19(5), 676–685 (1981)*
6. Identification of non-stationary diffusion model by the method od sieves (co-authored with T.D. Pham). *SIAM J. Control and Optimization 20(5), 603–611 (1982)*
7. Estimation on change-point hazard rate model (co-authored with G.S. Rogers and E.A. Walker). *Biometrika 71(2), 299–304 (1984)*
8. Strong consistency of maximum likelihood estimator in a change-point hazard rate model (co-authored with T.D. Pham). *J. Statistics 2(2), 203–216 (1990)*
9. On the scoring approach to admissibility of uncertainty measures in expert systems (co-authored with I.R. Goodman and G.S. Rogers). *J. Math. Anal. and Appl. 159(2), 550–594 (1991)*
10. Intervals in boolean rings: approximation and logic. *J. Foundations of Computing and Decision Sciences 17(3), 131–138 (1992)*
11. Bootstrapping the change-point in a hazard rate model (co-authored with T.D. Pham). *J. Inst. Statist. Math. 45(2) 331-340 (1993)*
12. A history and introduction to the algebra of conditional events and probability logic (co-authored with E.A. Walker). *IEEE Trans. Man Systems and Cyyernetics 24(2) 1671–1675 (1994)*

13. Robust reasoning with rules that have exceptions (co-authored with D. Bamber and I.R. Goodman). *Annals of Math. and Artifical Intell.(45), 83–171 (2005)*

14. Random and fuzzy sets in coarse data analysis (co-authored with B. Wu). *Comp.Statist. and Data Anal. (51) 70–85 (2006)*

15. On some claims related to Choquet integral risk measures (co-authored with U. Pham and H. Tran). *Annals of Operations Research 195(1) 5–31 (2012)*

16. Nonparametric estimation of a scalar diffusion model from discrete data: A survey (co-authored with C. Gourieroux and S, Sriboonchitta). *An- nals of Operations Research 256(2), 203–219 (2017)*

C. Books

1. *Uncertainty Models for knowledge-Based Systems* (co-authored with I.R. Goodman). North Holland (1985)

2. *Fundamentals of Mathematical Statistics Vol.I and Vol II* (co-authored with G. S. Rogers). Springer -Verlag University Text Book Series (1989)

3. *Conditioning Inference and Logic for Intelligent Systems A Theory of Measure-Free Conditioning* (co-authored with I.R. Goodman and E.A. Walker). North-Holland (1991)

4. *Fundamental of Uncertainty Calculi with Applications to Fuzzy Inference* (co-authored with M. Grabisch and E.A. Walker). Kluwer Academic (1994)

5. *Les Incertitudes dans les Systemes Intelligents (co-authored with B. Bouchon-Meunier).* Collection "Que Sais-Je?", Presses Universitaires de France (1996)

6. *Mathematics of Data Fusion* (co-authored with I.R. Goodman and R. Mahler). Kluwer Academic (1997)

7. *Applications of Continuous Mathematics to Computer Science* (co-authored with V. Kreinovich). Kluwer Academic (1997)

8. *A First Course in Fuzzy and Neural Control* (co-authored with N.R. Prasad, C. L. Walker and E.A. Walker). Chapman and Hall/CRC Press (2000)

9. *An Introduction to Random Sets.* Chapman and Hall/CRC Press (2006)

10. *Stochastic Dominance and Applications to Finance, Risk and Economics* (co-authored with S. Sriboonchitta, W.K. Wong and S. Dhompongsa). Chapman and Hall/CRC Press (2009)

11. *A First Course in Fuzzy Logic/Fourth Edition* (co-authored with C.L. and E.A. Walker). Chapman and Hall/CRC Press (2018)

El Paso, TX, USA

Vladik Kreinovich
e-mail: vladik@utep.edu

Contents

An Enjoyable Research Journey
on Uncertainty

Hung T. Nguyen

Abstract This is a story of research on uncertainty modeling from information measures without probability to Choquet capacities, possibility measures, fuzzy measures, imprecise probabilities, belief functions and finally, quantum probability. The main part of the paper is devoted, almost entirely, to an invitation to quantum probability for behavioral economics.

Keywords Behavioral economics · Decision theory · Quantum behavioral economics · Quantum entropy · Quantum finance · Quantum mechanics · Quantum probability · Uncertainty modeling

1 Introduction

It is all about uncertainty in its various faces we face in various aspects of life, either in natural or social sciences. Of course, we start out with the notion of chance as uncertainty. Recognizing that chance can be measured, various "ideas" about chance surfaced (see [7]) together with associated quantitative modeling of chance, called probability (although "Probability is the most important concept in modern science, especially as nobody has the slighest notion what it means", Bertrand Russell, 1929). And of course we talk about Kolmogorov's book, written in 1933, which provides the foundations for statistics. As mentioned in [17], Kolmogorov may write another kind of book if he was aware of a book written just a year earlier ([36], but written in German, in 1932) by von Neumann in which quantum probability was formulated, since quantum probability is just "a bit" more general than "standard" probability that Kolmogorov formulated in his 1933 book.

H. T. Nguyen (✉)
Department of Mathematical Sciences, New Mexico State University,
Las Cruces, NM 88003, USA
e-mail: hunguyen@nmsu.edu

Faculty of Economics, Chiang Mai University, Chiang Mai, Thailand

© The Editor(s) (if applicable) and The Author(s), under exclusive license
to Springer Nature Switzerland AG 2021
V. Kreinovich (ed.), *Statistical and Fuzzy Approaches to Data Processing, with Applications to Econometrics and Other Areas*, Studies in Computational Intelligence 892,
https://doi.org/10.1007/978-3-030-45619-1_1

Our journey will be from Kolmogorov to von Neumann (a strange one!). While the main purpose of this paper is providing a friendly invitation to quantum probability for applied econometricians (since other "introductions" to quantum probability are only written in journals of physics, for physicists or mathematicians), we will, however, trace back the journey we went through from Kolmogorov in the 1970s, passing by various kinds of uncertainty modelings (e.g., capacities, fuzzy measures, and belief functions), before, finally reaching quantum probability!

But let us elaborate a bit on why the jump from Kolmogorov to von Neumann? As stated clearly in [18]: "The proposed modification of the probability calculus appears more natural if we distinguish between *probability calculus* and *probability theory*. With *calculus* we denote the mathematical formalism devoid of any interpretation of this formalism. With *theory* we refer to the application of this calculus to various situations involving the occurrences of observed phenomena". This spirit is the cause of various proposals concerning non-additive uncertainty measures since von Neumann's work (in 1944) on behavioral economics (more specifically in decision analysis). In the context of quantum mechanics, as Feynman [8] spelled out clearly, while the meaning of probability is the same, nature computes probabilities of observed events differently, i.e., its probability calculus is different than that of Kolmogorov. A theory and calculus of probability is "suitable" for a given context (e.g., a sub-atomic level, or social sciences) if the application of the calculus to that context is "consistent" with observations, such as effectivity in predicting the probabilities of actually occuring events. As we will spell out in Sect. 4, this is precisely why we borrow quantum probability *calculus* from quantum mechanics to model uncertainty in human decision-making.

At the end of the journey (!), we arrive at quantum probability (a noncommutative generalization of Kolmogorov probability theory), the main topic of this paper. Considering it as a new topic, at least for applied econometricians (and not for probabilists since they have [22]!), I will proceed, in subsequent sections, to elaborate, in a "pedalogical" order, on the "what", "why" and "How". In doing so, we keep in mind that while there are compelling reasons why quantum probability calculus should replace Kolmogorov probability calculus in behavioral economics (as exposed in the literature since many years), we still see little efforts from econometricians to apply it to real-world problems. The reason for this phenomenon is this. Almost all efforts on promoting quantum probability seem to be for "econophysics" (where physics is the main theme), and not for Economics! For example, a nice article like [17] was not published in a economic (or statistics) journal. Research efforts seem to flourish outside of the economic community, such as [5, 16, 34, 37], while others seem not attract attention of econometricians because of "unclear" mathematical presentations (in fact, "complications deter") such as [1, 3, 14, 32, 33], recalling that serious statisticians and econometricians have no problems with Kolmogorov probability theory as the foundation of statistical science! If quantum probability is to be useful and appropriate outside of physics, e.g., in behavioral economics, the theory and calculus of quantum probability should be made as simple as possible (but not simpler!), at least at the introductory level. What seems to be really needed is an introductory text in the form of "A First Course in Quantum Probability for Economists"!

In the sequent, first we elaborate on some "popular" uncertainty measures in Sect. 2. The subsequent Sects. 3, 4, and 5 are devoted to an "invitation to quantum probability for applied statisticians and econometricians".

2 Some Uncertainty Measures

Measuring and computing probabilities (of uncertain events) provide information. Without asking (like Joseph Kampe' de Feriet) whether information comes before or after probability, let's concentrate on the theory and calculus of probability as formulated by Kolmogorov in 1933. Among tremendous impacts on sciences from this formalism, let's focus on the birth of quantitative economics in 1944 with von Neumann's classical work on game theory and economic behavior, in which, under the assumption of rationality of economic agents, Kolmogorov probability is used to model the uncertainty of decision-makers. Of course that assumption is false, as later, Stephen Hawking said [15]: "Economics is also an effective theory, based on the notion of free will plus the assumption that people evaluate their possible alternative courses of action and choose the best. That effective theory is only moderately successful in predicting behavior because, as we all know, decisions are often not rational or are based on a defective analysis of the consequences of the choice. That is why the world is in such a mess". The argument for assuming rationality is this. Rational behavior can be predicted, irrational behavior cannot!

In fact, if von Neumann's work is about "economic behavior", it should describe economic agents' behavior in making decisions under uncertainty. However, it is surprizing that he used Kolmogorov probability as behavioral probability rather than his own quantum probability. If he used quantum probability as behavioral probability instead, based on the "analogy" of intrinsic randomness of particles' motion with human's free will (rational or not), we could have then behavioral economics ever since! As a technical note, the main ingredient in von Neumann's decision theory is the concept of expected utility, an integral operator which operates on functions which form a commutative algebra, whereas in quantum probability, it is the trace operator, replacing integral, which operates on linear operators (of a Hilbert space) which forms a noncommutative algebra. As we will see, this extension from commutative algebras to noncommutative algebras is essential for behavioral economics. In other words, we could have the most appropriate behavioral economics without going through all efforts since M. Allais in 1953, and the work [19] which opened the doors to the application of empirical psychology to economics flourishing the actual behavioral economics. However, such efforts are not in vain, since, as a "side effect" (!), they revealed different useful uncertainty measures as we proceed to mention next.

Kampe' de Feriet's information measures without probability led to a formal connection with (Choquet) capacities in potential theory. By themselves, capacities are not uncertainty measures per se. However, they can be used, as generalized probability measures in, say, financial risk management. This can be seen as follows.

Since a financial (loss) variable X, defined on a probability space $(\Omega, \mathscr{A}, P)$, can be heavy-tailed, the notion of value-at-risk is more realistic to use rather than the variance. If $F(.)$ is the distribution function of X, then its value-at-risk, at level $\alpha \in (0, 1)$ is its quantile: $VaR_\alpha(X) = VaR_\alpha(F) = F^{-1}(\alpha)$. But,

$$F^{-1}(\alpha) = \int_0^\infty 1_{(1-\alpha,1]}(1 - F(t))dt =$$

$$\int_0^\infty g_\alpha(1 - F(t))dt = \int_0^\infty (g_\alpha \circ P)(X > t)dt$$

where $g_\alpha(.) : [0, 1] \to [0, 1]$, $g_\alpha(x) = 1_{(1-\alpha]}(x)$. The non-additive set-function $\nu(.) = g_\alpha \circ P(.) : \mathscr{A} \to [0, 1]$ is non-decreasing ($A \subseteq B \Longrightarrow \nu(A) \le \nu(B)$), and $\nu(\varnothing) = 0$, $\nu(\Omega) = 1$ which is called a capacity. Thus, $VaR_\alpha(F)$ has a Choquet integral representation $\int_0^\infty \nu(X > t)dt$. Likewise, the coherent risk measure "tail value-at-risk" $TVaR_\alpha(F) = \frac{1}{1-\alpha} \int_\alpha^1 F^{-1}(t)dt$ has the Choquet integral representation with capacity $\nu(.) = g_\alpha \circ P(.)$, with $g_\alpha(x) = \min\{1, \frac{x}{1-\alpha}\}$.

In 1965, L. A. Zadeh considered a type of "uncertain evidence" (called fuzzy data). In general, as spelled out in [7], an uncertain evidence is an observation which could be imprecise, ambiguous, or fuzzy (i.e., no sharply defined boundaries). For example, an uncertain evidence could be the realization of a random set (see [28]) as a coarse data. Note that if we have an "evidence", we are going to use it to make some decisions (based on it), although, as we all know, an evidence alone is not sufficient to reach a reasonable decision, exemplified by the use of P-value alone in frequentist testing of hypotheses. The specific kind of uncertain evidence, so to speak, that Zadeh considered is fuzziness in observations, mostly expressed in a natural language. Viewing events as subsets of some set Ω, a fuzzy subset of Ω (a fuzzy event) is a generalization of an ordinary (crisp) subset of Ω. Whereas an ordinary subset $A \subseteq \Omega$ has members all of full membership (1), a fuzzy subset of Ω could contain elements with partial membership (between 0 and 1). A direct extension is not possible. An indirect one is "simple": Each subset A is equivalent to its membership (indicator) function 1_A in the sense that $\omega \in A \Longleftrightarrow 1_A(\omega) = 1$. Now $1_A(.) : \Omega \to \{0, 1\}$. To define fuzzy subsets of Ω, it suffices to extend the range of indicator functions from $\{0, 1\}$ to the whole unit interval $[0, 1]$, so that a function $\mu(.) : \Omega \to [0, 1]$ represents (or in fact, defines) a fuzzy subset of Ω with the interpretation that the value $\mu(\omega) \in [0, 1]$ is the membership degree (full or partial) of $\omega \in \Omega$ in the fuzzy set under consideration. With its membership function specified, a "fuzzy evidence" provides some information. Clearly, fuzziness is different than randomness. The membership function is not a probability distribution! Fuzziness is another kind of uncertainty (lack of total knowledge). In fact, fuzziness and randomness can coexist.

At a practical level, how to "manipulate" fuzzy data? i.e., how to carry out computations with functions defined on fuzzy sets? It turns out that such computations can be carried out by using interval mathematics, via a result in [23], known as Nguyen's theorem, see also [10, 11].

In the context of fuzzy theory (see e.g., [29]), there are various associated uncertainty measures. Zadeh's possibility measure [40] is a special capacity. As a generalization of probability measure, a first attempt to define conditional possibility distribution was [24]. See also [4, 26]. More generally, Sugeno's fuzzy measures were considered for subjective evaluation problems. A fuzzy measure $\mu(.)$ is a capacity with the interpretation that the value $\mu(A)$ of a crisp subset $A \subseteq \Omega$, being a number in $[0, 1]$, is the subjective evaluation expressing, say, the degree of matching of the observed A with some intended target. The measure is called "fuzzy" by analogy with membership function of a fuzzy set (viewing the set-function $\mu(.)$ as a fuzzy subset of 2^{Ω}).

With the interpretation that fuzziness in observed data is a type of "uncertain evidence", this is no wonder why Zadeh was immediately interested in Shafer's "mathematical theory of evidence" in 1976 [35], inspired from Demspter's upper and lower probabilities [6] (and robust Bayesian statistics, imprecise probabilities, see e.g. [38]). The framework in which the so-called belief functions were developed is this. Let U be a finite set, a function $F(.) : 2^{U} \to [0, 1]$ is called a belief function when it has the following properties: $F(\emptyset) = 0$, $F(U) = 1$, and, for $k \geq 2$, and $A_j, j = 1, 2, \ldots, k$ subsets of U:

$$F(\cup_{j=1}^{k} A_j) \geq \sum_{\emptyset \neq I \subseteq \{1,2,\ldots,k\}} (-1)^{|I|+1} F(\cap_{i \in I} A_i)$$

where $|I|$ denotes the cardinality of the set I.

Note that if $F(.)$ is a probability measure on 2^{U}, then the above inequality is an (H. Poincaré) equality, so that belief functions are simply slightly generalized probability measures.

The intent of the theory was to provide another kind of uncertainty measure different than randomness. It was quickly pointed out in [25] that it is not so, i.e., a belief function $F(.)$ is, in fact, a bona fide probability distribution function, not of a random variable, but of a *random set* (i.e., a random element whose values are subsets of some set), so that randomness is around!

This can be seen simply as follows. Given a belief function $F(.)$, consider the set-function $f(.) : 2^{U} \to [0, 1]$ defined by

$$f(A) = \sum_{B \subseteq A} (-1)^{|A \setminus B|} F(B)$$

then $f(.)$ is a bona fide probability density function on 2^{U}, and we have $F(A) = \sum_{B \subseteq A} f(B)$, i.e., $F(.)$ acts as the probability distribution function with density $f(.)$.

More specifically, given $F(.)$, there exists a random set $S : (\Omega, \mathscr{A}, P) \to 2^{U}$, such that $F(A) = P(S \subseteq A)$. Note that in random set theory (see [28]), the partial order $\subseteq$ on subsets replaces $\leq$ on numbers. Thus, like fuzzy set theory which can be studied via interval mathematics, belief function theory can be studied in the context of random set theory (which is based on Kolmogorov probability theory).

In summary, we got quite a number of non-additive uncertainty measures, generalizing Kolmogorov probability measures, say, by the 90s. In closing his text on the theory of choice in 1988, Kreps [20] had this to say "These data provide a continuing challenge to the theorist, a challenge to develop and adapt the standard models so that they are more descriptive of what we see. It will be interesting to see what will be in a course on choice theory in ten or twenty years time". Now, over 30 years, what do we see as possible advances on choice theory? Since choice theory is clearly part of "economic behavior", and in view of the current flourishing approach to economics, namely behavioral economics, we could answer Kreps as: An update course on choice theory, for the 21st century, should contain quantum probability since only quantum probability captures the noncommutativity of "observables" that all the other uncertainty measures did not.

But how to formalize a noncommutative probability theory? Well, as an old saying goes "There is nothing new under the sun", we got it for free from quantum mechanics!

3 What Is Quantum Probability

Upfront, quantum probability (borrowed from quantum mechanics, including the term "quantum" which is used here simply to distinguish it from conventional notion of probability) is simply a generalization of Kolmogorov probability theory. We spell out next what to be "generalized" (whereas "why" such a generalization will be explained in the next Sect. 4).

Following David Hilbert's advice ("what is clear and easy to grasp attracts us, complications deter"), here is a "clear and easy" exposition to grasp! Essentially, quantum probability refers to a noncommutative generalization of Kolmogorov theory. Specifically, random variables in Kolmogorov's framework are functions so that they are commutative with respect to multiplication. We wish to extend random variables to a noncommutative setting. For "easy to grasp", it suffices to consider the simplest case, namely a finite probability space $(\Omega, \mathscr{A}, P)$ where Ω is a finite set, say, $\Omega = \{1, 2, \ldots, n\}$, $\mathscr{A}$ is its power set, and $P(.) : \mathscr{A} \rightarrow [0, 1]$ is a probability measure with probability density function $\rho(.) : \Omega \rightarrow [0, 1]$, $\rho(\omega) = P(\{\omega\})$.

Objects related to $(\Omega, \mathscr{A}, \rho)$ are functions $f(.) : \Omega \rightarrow \mathbb{R}$, such as random variables, the probability density $\rho(.)$, and events $A \in \mathscr{A}$ via their indicator functions $1_A(.) : \Omega \rightarrow \{0, 1\}$.

We wish to extend the commutative algebra of functions to a noncommutative algebra.

For those familiar with fuzzy set theory, remember how Zadeh generalized crisp sets to fuzzy sets? A subset A of Ω cannot be extended to a fuzzy subset of Ω directly. However, it can be if we do it indirectly, i.e., looking for an equivalent representation which can be extended. $A \subseteq \Omega$ is equivalent to its indicator function $1_A(.) : \Omega \to \{0, 1\}$. Extending the range $\{0, 1\}$ to $[0, 1]$ we obtain generalized membership functions which are used as definition of fuzzy subsets. We face the same situation here in extending Kolmogorov probability space to quantum probability space. We seek a feasible indirect way to achieve this.

First, when $\Omega = \{1, 2, \ldots, n\}$, such functions $f(.)$ are identified with the transpose vector $(f(1), f(2), \ldots, f(n))' \in \mathbb{R}^n$. In turn, each vector $X = (X(1), X(2), \ldots, X(n))' \in \mathbb{R}^n$ is equivalent to the diagonal $n \times n$ matrix $[X]$ whose diagonal terms are $X(1), X(2), \ldots, X(n)$. We have transformed functions to (special) matrices. Now a $n \times n$ matrix is a linear map (called operator) from $\mathbb{R}^n$ to $\mathbb{R}^n$.

If X is a random variable, then we say that X is "represented" by $[X]$. In particular, an event $A \in \mathscr{A}$, is represented by $[1_A]$ whose diagonal terms are 0 or 1. Such a diagonal matrix is of course symmetric, i.e., $[1_A] = [1_A]'$, but also idempotent, i.e., $[1_A]^2 = [1_A]$, so that $[1_A]$ is a (special) projection (operator) onto some closed subspace of the (Hilbert) space $\mathbb{R}^n$.

The probability density function $\rho(.)$ is represented by the diagonal matrix $[\rho]$ whose diagonal terms are nonnegative and sum up to one (its trace is one). Such an operator is "positive" in the sense that for any $x \in \mathbb{R}^n$, the scalar product $\langle x, [\rho]x \rangle \geq 0$ noting that arbitrary positive operators are symmetric.

All these diagonal matrices are of course symmetric: random variables "are" (diagonal) symmetric matrices, events are (special) projections, and probability densities are (special) positive operators with unit trace. Note that in this equivalent representation of standard (Kolmogorov) probabilistic objects, we obtain also equivalent computational procedures such as: for $A \in \mathscr{A}$,

$$P(A) = \sum_{\omega \in A} \rho(\omega) = tr([\rho][1_A])$$

where $tr(.)$ denotes the trace operator operating on matrices, and

$$EX = \sum_{\omega \in \Omega} X(\omega)\rho(\omega) = tr([\rho][X])$$

i.e., the trace operator replaces expectation operator (thus replacing "integral" in continuous case).

As observations on random variables are necessary for predictions (say, in physics or in psychological experiments to discover "laws"), as well as for statistics in general (!), we see that, when a random variable X is represented by $[X]$, its range (i.e., its possible values) is the set of its diagonal entries of $[X]$, which is the set of its eigenvalues (its spectrum), recalling that the spectrum on any symmetric (or more generally, self adjoint) matrix (or operator) lies in $\mathbb{R}$ (eigenvalues of self adjoint operators are real-valued). These observations motivate what we continue now to arrive as quantum counterpart of Kolmogorov probability.

Now diagonal matrices (as special case of symmetric matrices) form a commutative subagebra of the noncommutative algebra of all arbitrary symmetric matrices.

It is this noncommutative algebra of (arbitrary) symmetric matrices which extends Kolmogorov probability space to quantum probability space. Specifically, a finite quantum probability space is a triple $(\mathbb{R}^n, \mathscr{P}(\mathbb{R}^n), \rho)$, where $\mathscr{P}(\mathbb{R}^n)$ denotes the space of all (orthogonal) projections on $\mathbb{R}^n$ (i.e., projections p onto closed subspaces of $\mathbb{R}^n$, characterized by $p = p^2 = p^*$(transpose)), and $\rho(.)$ is a "density matrix", i.e., an arbitrary positive operator with unit trace. Of course, any Kolmogorov probability spaces are examples of quantum probability spaces! And of course, we leave the general quantum probability spaces for "advanced readers" to consult texts such as [30].

After spelling out "why quantum probability spaces are appropriate for behavioral economics?" in the next Sect. 4, we will elaborate on quantum probability *calculus* in Sect. 5.

4 Why Do We Need Quantum Probability in Economics?

Upfront, we are going to motivate (in fact, to explain) the need to use quantum probability as *behavioral probability* to model decision-making behavior in social sciences, especially in economics which, in turn, will suggest that the actual flourishing field of behavioral economics should be *quantum behavioral economics.*

In neoclassical economics (or more generally, in conventional social sciences) the uncertainty involved (in human decision-making) is measured and manipulated by standard probability (SP) of Kolmogorov. As stated in Sect. 2, various modifications of SP were proposed based on empirical facts that indicated that SP cannot explain certain behavior of decision-makers. This is consistent with the spirit of natural sciences (e.g., physics) where a good model should be able to predict well what are to be observed, exemplified by the replacement of classical mechanics by quantum mechanics in the first quarter of the last century. There are empirical facts in human decision context which call for further extension of SP to obtain a better behavioral probability formalism (see e.g., [3, 19]). These empirical facts were known for quite some time, but we reproduce them here to give a clear motivation. Specifically, these are empirical facts related to the so-called "conjunction fallacy", "disjunction fallacy", "order effect" and "preference reversal" which cannot be "explained" if SP is used to model uncertainty, i.e., the associated empirical facts are not consistent with SP. Just like in physics—when a model (e.g., a suggestive "law") does not predict well the observations, it should be not only abandoned, but also replaced by some other model which can explain what we see—it seems natural to do the same thing in social sciences, including economics.

Roughly speaking, experiments on human decision-making provide empirical probabilities such as $\Pr(A\&B) > \Pr(A)$ (probability measures are not monotone increasing), $\Pr(A or B) \neq \Pr(A) + \Pr(B)$ (the additivity, or the "law of total probability" fails), $\Pr(A\&B) \neq \Pr(B\&A)$ (the commutativity fails). Thus, since economic decisions involve humans, these facts should ring the bell! Moreover, as far as modeling of economic/financial data is concerned, should we take seriously the investigation in [34]?:

A natural explanation of extreme irregularities in the evolution of prices in financial markets is provided by quantum effects

And, as far as statistics with quantum probability is concerned, see e.g., [12, 27, 31].

It turns out that the mentioned fallacies are very similar to what physicists see in quantum mechanics (which is intrinsically random), exemplified by the well-known two-slit experiment (see, e.g., [8]), in which the observed probabilities do not obey the calculus of SP. Since quantum mechanics is intrinsically random, physicists need to figure out ways to compute probabilities of quantum events (e.g., measurements). With the knowledge of the Schrodinger equation (counterpart of Newtonian law of motion in classical mechanics), a "quantum probability calculus" was discovered and applied successfully in explaining (and predicting) quantum phenomena, as we know today. In fact, psychologists have shown that quantum probability theory (as we outlined in Sect. 3, and will be elaborating more in the next Sect. 5) can explain the mentioned fallacies. Moreover, it seems that the "order effect" could be a testimony for the *quantum nature of human judgement*. All the above form the basis for considering seriously the use of quantum probability as behavioral probability, to improve behavioral economics. In a sense, such a shift from neoclassical economics to quantum behavioral economics could be compared with the extension from Newtonian mechanics to quantum mechanics.

5   How to Use Quantum Probability in Economics?

Just like with Kolmogorov probability, we need to learn how to carry out quantum probability calculus in order to investigate economic issues! However, we do not have (yet) anything similar to standard probability calculus from which statistical applications follow, although quantum probability is as old as Kolmogorov probability. Perhaps the reason is that most, if not all, writings on quantum probability are for physics and cognitive science, rather for economics. Hopefully, with the flourishing field of behavioral economics, and the aggressive push in favor of quantum probability in it by a large community of researchers, we will see soon textbooks on *quantum probability calculus for economic applications*. Until then, a flavor of it could be useful as a starting point.

Recall that standard probability calculus is built on a probability space $(\Omega, \mathscr{A}, P)$. Similarly, quantum probability calculus is based upon a quantum probability space $(H, \mathscr{P}(H), \rho)$ where H is a separable, complex Hilbert space, $\mathscr{P}(H)$ is the set of all projectors on H, and ρ is a positive operator, with unit trace, on H. To be simple (avoiding unnecessary mathematical complications at this stage), as in

Sect. 2, it suffices to consider a finite probability space $(\Omega, \mathscr{A}, P)$, with, say, $\Omega = \{1, 2, \ldots, n\}$, so that its extension to quantum probability space $(H, \mathscr{P}(H), \rho)$ is simply $(\mathbb{R}^n, \mathscr{P}(\mathbb{R}^n), \rho)$.

A random variable X (called also an "observable") is "represented" by a symmetric matrix $[X]$ (diagonal or not) whose spectrum $\sigma(X)$ (set of its real eigenvalues) is the range of X.

As such, we need to specify its probability distribution on $\sigma(X)$ or $\sigma([X])$. For this, we will rely upon the spectral decomposition of symmetric matrices. Let $\sigma([X]) = \{\lambda_1, \lambda_2, \ldots, \lambda_n\} \subseteq \mathbb{R}$, and $u_1, u_2, \ldots, u_n$ (in $\mathbb{R}^n$) be its corresponding eigenvectors. Let $P_j = u_j u_j^*$ (the "outer product" on the finitely dimensional Hilbert space $\mathbb{R}^n$, where u_j^* is the transpose on u_j). Each P_j is an operator (a matrix) on $\mathbb{R}^n$, orthogonal projection onto the closed subspace of $\mathbb{R}^n$ spaned by the eigenvector u_j, which is in fact a (orthogonal) projector since $P_j = P_j^2 = P_j^*$. The spectral decomposition of $[X]$ is this.

$$[X] = [u_1, u_2, \ldots, u_n] \begin{bmatrix} \lambda_1 & & \\ & \cdot & \\ & & \lambda_n \end{bmatrix} \begin{bmatrix} u_1^* \\ \cdot \\ u_n^* \end{bmatrix} = \sum_{j=1}^{n} \lambda_j u_j u_j^* = \sum_{j=1}^{n} \lambda_j P_j$$

noting that this decomposition is the counterpart of the "decomposition" of a random variable (in classical setting) as a linear combination of indicator (elementary) variables.

As stated in Sect. 3, quantum events, i.e., $A \subseteq \Omega$, are represented by projectors $[1_A]$, so that $P(A)$ in $(\mathbb{R}^n, \mathscr{P}(\mathbb{R}^n), \rho)$ means $tr(\rho[1_A])$, noting that the trace operator $tr(.)$ is commutative. More generally, the expected value, under ρ (a density matrix, i.e., a positive operator with unit trace) of an observable $[X]$ is $E_\rho[X] = tr(\rho[X])$. Note however that quantum events, as projectors, might not be commutative in general: for $p, q \in \mathscr{P}(\mathbb{R}^n)$, pq is an event (i.e., a projector) if and only if they commute since $(pq)^* = pq = q^* p^* = qp$, so that the problem of "joint distribution" arises (see later).

Now for $B \in \mathscr{B}(\mathbb{R})$, the event $(X \in B)$ is the projector $\zeta(B) = \sum_{\lambda \in B} P_\lambda$, where P_λ is the event "X takes the value λ". This set-function $\zeta(.) : \mathscr{B}(\mathbb{R}) \to \mathscr{P}(\mathbb{R}^n)$ takes values in $\mathscr{P}(\mathbb{R}^n)$, so that it is a "projection-valued" mapping. In fact, it is a projection-valued measure, called a spectral measure. With the above spectral decomposition of a symmetric matrix $[X]$, we have a spectral measure associated with each observable X, denoted as $\zeta_X(.)$. Thus, the probability distribution of $[X]$ on $(\mathbb{R}, \mathscr{B}(\mathbb{R}))$ is $\mu_X(.) = tr(\rho \zeta_X(.))$, where $\mu_X(B) = tr(\rho \zeta_X(B))$ is the probability that X takes a value in B, when the density matrix is ρ.

Remark. In general, i.e., for general quantum probability spaces, $\mu_X(.)$ plays the role of a probability distribution for the observable X, so that

$$E_\rho(X) = E_\rho([X]) = \int_{\mathbb{R}} \lambda \mu_X(d\lambda) = \int_{\mathbb{R}} \lambda tr(\rho \zeta_X(d\lambda) = tr(\rho[X])$$

Indeed, $\rho = \sum_j a_i (u_j u_j^*)$ for any orthonormal basis $\{u_j\}$ of an arbitrary, separable, complex Hilbert space H, with $a_j > 0$, $\sum_j a_i = 1$. For $B \in \mathcal{B}(\mathbb{R})$, we have

$$\mu_{\rho, X}(B) = tr(\rho \zeta_X(B)) = \sum_j \langle \rho \zeta_X(B) u_j, u_j \rangle = \sum_j a_j \langle \zeta_X(B) u_j, u_j \rangle$$

Thus, since $[X] = \int x \zeta_X(dx)$, the spectral decomposition of $[X]$, we have

$$E_\rho(X) = \int x \mu_{\rho, x}(dx) = \sum_j a_j \int_{\mathbb{R}} x \langle \zeta_X(dx) u_j, u_j \rangle =$$

$$\sum_j a_j \langle [X] u_j, u_j \rangle = \sum_j a_j \langle \rho[X] u_j, u_j \rangle = tr(\rho[X])$$

When the density matrix ρ is induced by a normalized "wave function" $\psi \in H$ (in the Schrodinger's equation) on $\mathbb{R}^3$ as $\rho = \sum_j a_j |u_j \rangle \langle u_j|$, with $a_j = |c_j|^2$, $c_j = \langle \psi, u_j \rangle$, we have

$$E_\psi(X) = \int_{\mathbb{R}^3} \psi^*(x)[X]\psi(x)dx = \langle [X]\psi, \psi \rangle$$

We use *Dirac's notations* in the above analysis just to have an occasion to explain now what they are and help the readers when they read the literature!

The case where $H = \mathbb{R}^n$ makes things clear. For $x \in H$, we rewrite it as $|x\rangle$ (called a "ket"). It is a "column vector", such as $x = (x_1, x_2, \ldots, x_n)^*$, whereas we write $< x|$ (a "bra") to designate an element of the dual of H, i.e., a map from H to $\mathbb{R}$, so that $\langle x|(y) = \langle x|(|y >) = \langle x|y \rangle$ is the "inner (scalar) product" on H, such as $\langle x|y \rangle = \langle x, y \rangle$ (getting a "bracket") $= \sum_{j=1}^n x_j y_i$. On the other hand, the "*outer product*" on H is a "multiplication" of a column vector $x \in \mathbb{R}^n$ with a "row vector" y (i.e., the transpose y^* of the column vector $y \in \mathbb{R}^n$) resulting in a matrix (i.e., an operator on $\mathbb{R}^n$) is denoted as $|x\rangle\langle y|$ which acts on $\mathbb{R}^n$ as, for $z \in \mathbb{R}^n$, $(|x\rangle\langle y|)(z) = \langle y, z\langle x \in \mathbb{R}^n$. Of course, on $\mathbb{R}^n$, $|x\rangle\langle y|$ is simply xy^* (multiplication of two matrices: x is a $n \times 1$ matrix, and y^* is a $1 \times n$ matrix).

With respect to obtaining the amplitude $|\psi|^2$ of the wave function ψ in the Schrodinger's equation, a "practical" method is using Feynman's path integral (see [1, 9]).

Recall that the standard interpretation of quantum mechanics is this. The dynamics of a particle with mass m is "described" by a wave function $\psi(x, t)$, where $x \in \mathbb{R}^3$ is the particle position at time t, which is the solution of the Schrodinger's equation (counterpart of Newton's law of motion of macroobjects):

$$ih\frac{\partial \psi(x, t)}{\partial t} = -\frac{h^2}{2m}\Delta_x \psi(x, t) + V(x)\psi(x, t)$$

As such, particles in motion do not have trajectories (in their phase space), or put it more specifically, their motion cannot be described (mathematically) by trajectories (because of the Heisenberg's uncertainty principle). The probability amplitude $|\psi(x, t)|^2$ is used to make probabilistic statements about the particle motion. There is another, not "official", interpretation of quantum mechanics in which it is possible to consider trajectories for particles, called Bohmian mechanics. This mechanics formulation is suitable to use as models for, say, financial data (where time series data are like "trajectories" of moving objects). In fact, they could be considered as "useful models" in G. Box's sense since they display an extra term which can be used to represent the missing human factor.

Now as the "density matrix" ρ on a general quantum probability space plays the role of an ordinary probability density function f (whose ordinary entropy is $-\int f(x) \log f(x) dx$), its *quantum entropy* (as defined by von Neumann [36]) is $-tr(\rho \log \rho)$. As maximum (Kolmogorov) entropy principle provides equilibrium models in statistical mechanics or other stochastic systems, it also enters financial econometrics as the most diversified portfolio selection, see e.g. [2, 13].

Let's specify von Neunman's quantum entropy in a simple case, e.g., when $H = \mathbb{C}^n$.

For a density matrix ρ (extension of a probability density function) on $(\mathbb{C}^n, \mathscr{P}(\mathbb{C}^n))$, $\rho \log \rho$ is a $n \times n$ self adjoint matrix (operator) which is defined as follows (by using spectral theorem). The spectral theorem says this. Since ρ is a self adjoint operator on $\mathbb{C}^n$, there exists an orthonormal basis of $\mathbb{C}^n$, $\{u_1, u_2, \ldots, u_n\}$ consisting of eigenvectors of ρ, with associated eigenvalues $\{\lambda_1, \lambda_2, \ldots, \lambda_n\}$ (the spectrum of ρ). If we let P_j be the projector onto the closed subspace spanned by u_j, then $\rho = \sum_{j=1}^n \lambda_j P_j$.

For $g(.) : \mathbb{R} \to \mathbb{R}$, $g(x) = x \log x$, the (self adjoint) operator $g(\rho) = \rho \log \rho$ is defined by $\sum_{j=1}^n g(\lambda_j) P_j$ whose trace is

$$tr(\rho \log \rho) = \sum_{j=1}^n \langle u_j, (\rho \log \rho) u_j \rangle = \sum_{j=1}^n \lambda_j \log \lambda_j$$

so that the quantum entropy of ρ is $-tr(\rho \log \rho) = -\sum_{j=1}^n \lambda_j \log \lambda_j$ which depends only on the eigenvalues of ρ.

One more thing we like to mention about quantum probability calculus is the noncommutativity of observables. When events $p, q \in \mathscr{P}(H)$ do not commute, pq is not an event, so that the problem of "joint distribution" of non commuting observables X and Y, arises. By considering them separately, we can, in principle get their "marginal distributions". In classical SP, models for joint distribution of (X, Y) can be obtained by using *copulas*. The situation in quantum probability theory is different: there is no notion of copulas in it. For this issue, see e.g., [21, 39].

References

1. B.E. Baaquie, *Quantum Finance: Path Integrals and Hamiltonians for Options and Interest Rates* (Cambridge University Press, Cambridge, 2007)
2. A. Bera, S. Park, Optimal portfolio diversification using maximum entropy principle. Econ. Rev. **27**, 484–512 (2008)
3. J.R. Busemeyer, P.D. Bruza, *Quantum Models of Cognition and Decision* (Cambridge University Press, Cambridge, 2012)
4. L. Coroianu, R. Fuller, Nguyen type theorem for extension principle based on a joint possibility distribution. Intern. J. Approx. Reason. **95**, 22–35 (2018)
5. P. Darbyshire, Quantum physics meets classical finance. Phys. World **25–29** (2005)
6. A. Dempster, Upper and lower probabilities induced by a multivalued mapping. Ann. Math. Statist. **38**, 325–339 (1967)
7. P. Diaconis, B. Skyrms, *Ten Great Ideas About Chance* (Princeton University Press, Princeton and Oxford, 2018)
8. R. Feynman, The concept of probability in quantum mechanics, in *Berkeley Symposium on Mathematical Statistics* (1951), pp. 533–541
9. R. Feynman, A. Hibbs, *Quantum Mechanics and Path Integrals* (Dover, New York, 1965)
10. R. Fuller, T. Keresztfalvi, On generalization of Nguyen's theorem. Fuzzy Sets Syst. **41**(3), 371–374 (1991)
11. R. Fuller, On generalization of Nguyen's theorem: a short survey of recent developments. Adv. Soft Comput. Intell. Robot. Control., 183–190 (2014)
12. A. Gelman, M. Bethancourt, Does quantum uncertainty have a place in everyday applied statistics. Behav. Brain Sci. **36**(3), 285 (2013)
13. A. Golan, G. Judge, D. Miller, *Maximum Entropy Econometrics* (Wiley, New York, 1996)
14. E. Haven, A. Khrennikov, *Quantum Social Science* (Cambridge University Press, Cambridge, 2013)
15. S. Hawking, L. Mlodinow, *The Grand Design* (Bantam Books, London, 2010)
16. B. Herzog, Quantum models of decision-making in economics. J. Quantum Inf. Sci. **5**, 1–5 (2015)
17. R. Hudson, A short walk in quantum probability. Phil. Trans. R. Soc. A **376**, 20170226 (2017)
18. J.M. Jauch, The quantum probability calculus. Synthese **29**, 131–154 (1974)
19. D. Kahneman, A. Tversky, Prospect theory: an analysis of decision under risk. Econometrica **47**, 263–292 (1979)
20. D.M. Kreps, *Notes on the Theory of Choice* (Westview Press, 1988)
21. J.M. Malley, A. Fletcher, Joint distribution and quantum nonlocal models. Axioms **3**, 166–176 (2014)
22. P.A. Meyer, *Quantum Probability for Probabilists* (Springer, Berlin, 1995)
23. H.T. Nguyen, A note on the extension principle for fuzzy sets. J. Math. Anal. Appl. **64**, 369–380 (1978)
24. H.T. Nguyen, On conditional possibility distributions. Fuzzy Sets Syst. **1**(4), 299–309 (1978)
25. H.T. Nguyen, On random sets and belief functions. J. Math. Anal. Appl. **65**, 531–542 (1978)
26. H.T. Nguyen, On the entropy of random sets and possibility distributions, in *The Analysis of Fuzzy Information*, ed. by J. Bezdek (CRC Press, 1987), pp. 45–156
27. H.T. Nguyen, Toward improving models for decision making in economics. Asian J. Econ. Bank. **3**(01), 1–19 (2019)
28. H.T. Nguyen, *An Introduction to Random Sets* (Chapman and Hall/CRC Press, Boca Raton, 2006)
29. H.T. Nguyen, C.L. Walker, E.A. Walker, *A First Course in Fuzzy Logic*, 4th edn. (Chapman and Hall/CRC Press, Boca Raton, 2019)
30. K.R. Parthasarathy, *An Introduction to Quantum Stochastic Calculus* (Springer, Basel, 1992)
31. L. Pasca, A critical review of the main approaches on financial market dynamics modeling. J. Heterodox Econ. **2**(2), 151–167 (2015)

32. S. Patra, A quantum framework for economic science: new directions. Economics 2019–20 (2019)
33. E.W. Piotrowski, J. Sladkowski, Quantum games in finance. Quant. Financ. **4**(6), 61–67 (2004)
34. S. Segal, I.E. Segal, The Black-Scholes pricing formula in the quantum context. Proc. Nat. Acad. Sci. USA **95**, 4072–4075 (1998)
35. G. Shafer, *A Mathematical Theory of Evidence* (Princeton University, University Press, Princeton, 1976)
36. J. von Neumann, *Mathematical Foundations of Quantum Mechanics*, New edn. (Princeton University Press, Princeton, 2018)
37. V. Vukotic, Quantum economics. Panoeconomicus **2**, 267–276 (2011)
38. P. Walley, *Statistical Reasoning with Imprecise Probabilities* (Chapman and Hall, London, 1991)
39. V.I. Yukalov, D. Sornette, Quantum probabilities of composite events in quantum measurements with multimode states. Lazer Phys. **23**, 105502 (2013)
40. L.A. Zadeh, Fuzzy sets as a basis for a theory of possibility. Fuzzy Sets Syst. **1**, 3–28 (1978)

A Bayesian Dilemma

Donald Bamber

Abstract An example is given in which a Bayesian reasoner fails to learn the obvious. That failure calls into question whether Bayesian epistemology is complete and correct as it stands.

Keywords Bayesian epistemology · Bayesian reasoning · Intuitive reasoning · Rationality · Degree of belief · Epistemic Bayesian · Pragmatic Bayesian

1 Intuitive Rationality and Bayesian Rationality

Usually the conclusions that people arrive at intuitively agree approximately with conclusions arrived at by Bayesian reasoning.[1] Thus, intuitive rationality and Bayesian rationality tend to agree.

However, there are circumstances where conclusions that are judged rational by intuitive standards are judged grossly irrational by Bayesian standards and, conversely, conclusions judged rational by Bayesian standards are judged grossly irrational by intuitive standards.

For example, suppose that a large number N trials of an experiment have been run. On each trial, it is observed whether a particular event A occurred or not. Across all N trials, the event A never occurred. After observing such empirical results, most intuitive reasoners would assert that they had only a very small degree of belief—close to zero—that A would occur on Trial $N + 1$. Moreover, they would regard it to

---

D. Bamber (✉)
Cognitive Sciences Department, University of California, Irvine, USA
e-mail: dbamber@uci.edu

[1]By "Bayesian reasoning" is meant the type of reasoning advocated by *epistemic* Bayesians, that is by adherents of Bayesian epistemology. (As discussed in [3], epistemic Bayesians are much different from pragmatic Bayesians.) In Bayesian epistemology, one should update one's beliefs by conditioning them on observations. For more detailed information, see [8, Chap. 2] and [10, Chap. 4], the latter book having been reviewed by me in [2]. For a briefer, less complete account, see my paper [3].

© The Editor(s) (if applicable) and The Author(s), under exclusive license
to Springer Nature Switzerland AG 2021
V. Kreinovich (ed.), *Statistical and Fuzzy Approaches to Data Processing, with Applications to Econometrics and Other Areas*, Studies in Computational Intelligence 892,
https://doi.org/10.1007/978-3-030-45619-1_2

be grossly irrational to have a degree of belief close to one-half that A would occur on Trial $N + 1$.

However, there are circumstances where, after observing no occurrences of A in N trials (N large), a Bayesian reasoner would have degree of belief around one-half that A would occur on Trial $N + 1$. Moreover, the Bayesian reasoner would regard as grossly irrational a near-zero belief that A would occur on Trial $N + 1$.

How can this happen? The next section will explain.

2 A Case of Counterintuitive Bayesian Reasoning

2.1 The Robotic Black Box

Suppose that B.R., a Bayesian reasoner, happens upon a black box. Although nothing inside the box can be seen, an informant provides the following information:

- On the exterior of the black box, there is pushbutton and two side-by-side display windows.
- Inside the black box, there is a robot having one arm, a hand, and an eye.
- On the floor of the black box lie a gold coin and a silver coin.
- When the pushbutton on the outside of the box is pressed, the robot picks up the two coins in its hand and tosses them in the air so that they tumble about as they fall.
- After the coins land on the floor, the robot observes whether the gold coin landed heads or tails. Then the robot puts either the printed word "heads" or the word "tails", as appropriate, in the left display window.
- Next the robot examines the silver coin on the floor and puts either the printed word "heads" or the word "tails", as appropriate, in the right display window.

2.2 Two-Coins Model

B.R. decides to perform an experiment on the black box. A single trial of the experiment will consist of pressing the pushbutton on the black box and observing the results in the display window. The experiment will consist of running many such trials.

After hearing the informant's description of the inner workings of the black box, B.R formulates his/her beliefs about the results of the experiment. To describe those beliefs, we will need some notation.

2.2.1 Notation

Define the following events:

$$\mathbf{H}_G^i = \text{gold coin lands heads on trial } i. \tag{1}$$
$$\mathbf{T}_G^i = \text{gold coin lands tails on trial } i. \tag{2}$$
$$\mathbf{H}_S^i = \text{silver coin lands heads on trial } i. \tag{3}$$
$$\mathbf{T}_S^i = \text{silver coin lands tails on trial } i. \tag{4}$$

Also, define the random variable

$$Res^i = \text{results of Trial } i. \tag{5}$$

In other words, Res^i specifies whether the gold coin landed heads or tails on Trial i and whether the silver coin landed heads or tails on Trial i. To be specific,

$$Res^i \in \{(\text{heads, heads}), (\text{heads, tails}), (\text{tails, heads}), (\text{tails, tails})\}, \tag{6}$$

where the first member of each ordered pair indicates how the gold coin landed and the second member indicates how the silver coin landed.

2.2.2 B.R.'s Beliefs

Given the facts provided by the informant, B.R. formulates his/her beliefs as listed below. (Bear in mind that, because B.R. is a Bayesian, all probabilities are degrees of belief.) To emphasize that the probability measure employed here represents B.R.'s degrees of belief, it will be denoted $P_{\text{B.R.}}(\cdot)$. So, B.R.'s beliefs may be stated as follows.

- There is an ordered pair of parameters (Θ_G, Θ_S) that, in B.R.'s belief, have a distribution that is uniform on the unit square $[0, 1]^2$. In other words, Θ_G and Θ_S are each distributed uniformly on the interval $[0, 1]$ and are independent of each other.
- On each Trial i, B.R.'s degree of belief that the gold coin lands heads is determined by the value of Θ_G. And B.R.'s degree of belief that the silver coin lands heads is determined by the value of Θ_S. To be more precise:

$$P_{\text{B.R.}}[\mathbf{H}_G^i \mid (\Theta_G, \Theta_S) = (\theta_g, \theta_s)] = \theta_g; \tag{7}$$
$$P_{\text{B.R.}}[\mathbf{H}_S^i \mid (\Theta_G, \Theta_S) = (\theta_g, \theta_s)] = \theta_s. \tag{8}$$

- On each trial, in B.R.'s belief, the landings of the gold coin and the silver coin are conditionally independent given (Θ_G, Θ_S). To be precise:

$$P_{\text{B.R.}}[\mathbf{H}_G^i \,\&\, \mathbf{H}_S^i \mid (\Theta_G, \Theta_S) = (\theta_g, \theta_s)]$$
$$= P_{\text{B.R.}}[\mathbf{H}_G^i \mid (\Theta_G, \Theta_S) = (\theta_g, \theta_s)] \cdot P_{\text{B.R.}}[\mathbf{H}_S^i \mid (\Theta_G, \Theta_S) = (\theta_g, \theta_s)]. \qquad (9)$$

- The results across trials Res^1, Res^2, Res^3, ... are conditionally independent given (Θ_G, Θ_S).

Given what B.R. was told about the black box by the informant, these beliefs seem quite reasonable.

2.3   Results of the First N Trials of the Experiment

B.R. runs an experiment with N trials, where N is large, perhaps a thousand or a million.

Let N_{HT} denote the number of trials in the experiment where the gold coin landed heads and the silver coin landed tails. Let N_{HH}, N_{TH}, and N_{TT} be defined analogously. In addition, let

$$N_{H\bullet} = N_{HH} + N_{HT}; \qquad (10)$$
$$N_{T\bullet} = N_{TH} + N_{TT}; \qquad (11)$$
$$N_{\bullet H} = N_{HH} + N_{TH}; \qquad (12)$$
$$N_{\bullet T} = N_{HT} + N_{TT}. \qquad (13)$$

Then, for example, $N_{H\bullet}$ is the number of trials where the gold coin landed heads, and $N_{\bullet H}$ is the number of trials where the silver coin landed heads.

The results of the experiment may be summarized by a table of the form:

	silver head	silver tail
gold head	N_{HH}	N_{HT}
gold tail	N_{TH}	N_{TT}

(14)

Let $\mathcal{TBL}_N$ denote the set of possible tables (14) for an N-trial experiment. There is one such table for every choice of nonnegative integers N_{HH}, N_{HT}, N_{TH}, and N_{TT} that sum to N. Thus, every $Table \in \mathcal{TBL}_N$ is a specification of the values of N_{HH}, N_{HT}, N_{TH}, and N_{TT} plus, via Eqs. 10–13, a specification of the values of $N_{H\bullet}$, $N_{T\bullet}$, $N_{\bullet H}$, and $N_{\bullet T}$ as well.

2.4   B.R.'s Beliefs About Trial $N + 1$

Having observed the results of the first N trials of the experiment, what will B.R.'s degrees of belief concerning Trial $N + 1$ be? Because B.R. is a Bayesian reasoner, his/her belief will be conditioned on the observed $Table$. It is shown in the Appendix (Sect. 6, Eqs. 62 and 64) that, for all possible values of $Table$,

$$P_{\text{B.R.}}(\mathbf{H}_G^{N+1} \mid Table) = \frac{N_{H\bullet} + 1}{N + 2}; \tag{15}$$

$$P_{\text{B.R.}}(\mathbf{H}_S^{N+1} \mid Table) = \frac{N_{\bullet H} + 1}{N + 2}. \tag{16}$$

Equation 15 shows that B.R.'s belief about the behavior of the gold coin on Trial $N + 1$ conforms to Laplace's Rule of Succession [8, p. 111], [10, p. 72]. Likewise, Eq. 16 shows that B.R.'s belief about the silver coin's behavior will also conform to Laplace's Rule. Furthermore, Eq. 66 of the Appendix implies that, for all possible values of $Table$,

$$P_{\text{B.R.}}(\mathbf{H}_G^{N+1} \ \& \ \mathbf{H}_S^{N+1} \mid Table)$$
$$= P_{\text{B.R.}}(\mathbf{H}_G^{N+1} \mid Table) \cdot P_{\text{B.R.}}(\mathbf{H}_S^{N+1} \mid Table). \tag{17}$$

Thus, Eq. 17 shows that B.R.'s belief about the gold coin's behavior on Trial $N + 1$—conditioned on the results of the first N trials—will be independent of his/her belief about the silver coin's behavior on Trial $(N + 1)$.

All this (Eqs. 15, 16, and 17) seems very reasonable. But, as will be seen, there are circumstances where intuitively it would seem that belief about the gold coin on Trial N+1 should be influenced by belief about the silver coin on that trial.

2.5   What B.R. Wasn't Told About the Black Box

Imagine that there is an important fact[2] about the black box that B.R. was never told. (Perhaps B.R.'s informant was unaware of the fact.) Specifically, B.R. was not told the following:

- The gold coin and the silver coin have both been embedded side by side in a thin slab of transparent plastic.
- As a result, when the robot picks up the coins and flips them, either they will both land heads or they will both land tails—no exceptions.

Consequently, over all N trials, no matter how large N is, B.R. will never observe a trial where one coin lands heads and the other tails. Thus the results of the first N trials of the experiment will have the following form:

---

[2]Which we will call the *Plastic-Yoking Mechanism*.

	silver head	silver tail
gold head	N_{HH}	0
gold tail	0	N_{TT}

$$(18)$$

Let the set of tables for an N-trial experiment that have the diagonal form above be denoted $\Delta \mathscr{TBL}_N$.

2.6 B.R.'s Counterintuitive Reasoning

Let **Diff** ^{i} denote the event that the two coins land *differently* on Trial i. That is: The gold coin lands heads and the silver coin lands tails or vice versa. If the observed $Table$ has the diagonal form (18), then the event **Diff** ^{i} did not occur on any Trial i, for $i = 1, \ldots, N$.

Intuitive reasoning. Suppose then N is large and that **Diff** ^{i} did not occur on any Trial i for $i = 1, \ldots, N$. Then, for most intuitive reasoners, their degree of belief in **Diff** $^{N+1}$ would be close to zero.

B.R.'s reasoning. It is shown in the Appendix (Eq. 73) that, for all $Table \in \Delta \mathscr{TBL}_N$,

$$P_{\text{B.R.}}(\textbf{Diff}^{N+1} \mid Table)$$

$$= 2 \left(\frac{N_{H\bullet} + 1}{N + 2} \right) \left[1 - \left(\frac{N_{H\bullet} + 1}{N + 2} \right) \right]. \tag{19}$$

Thus, if $Table$ has the diagonal form (18) and if N_{HH}/N is close to one half, then $P_{\text{B.R.}}(\textbf{Diff}^{N+1} \mid Table)$ will be close to one half. To see why this is so, note that $(N_{HH} + 1)/(N + 2)$ always lies closer to $1/2$ than N_{HH}/N does. Thus, if the latter is close to $1/2$, then the former will be closer yet. For example:

$$\text{If } 0.4 \leq \frac{N_{HH}}{N} \leq 0.6, \text{ then } 0.4 < \frac{N_{HH} + 1}{N + 2} < 0.6. \tag{20}$$

So, if $Table$ has the diagonal form (18) and if (20) holds, then, by (19),

$$0.48 < P_{\text{B.R.}}(\textbf{Diff}^{N+1} \mid Table) \leq 0.50. \tag{21}$$

In fact, even if N_{HH}/N is not close to $1/2$, as long as it is not close to either zero or one, then $P_{\text{B.R.}}(\textbf{Diff}^{N+1} \mid Table)$ will not be close to zero. For example, if

$$0.1 \leq \frac{N_{HH}}{N} \leq 0.9, \tag{22}$$

then

$$0.18 < P_{\text{B.R.}}(\textbf{Diff}^{N+1} \mid Table) \leq 0.50. \tag{23}$$

2.7 Who Will Think B.R. to Be Irrational?

So, as we have seen, most intuitive reasoners would come to a different conclusion than B.R. Over N trials, where N is large, if $\textbf{Diff}^i$ never occurred, most intuitive reasoners would have only a very small degree of belief in $\textbf{Diff}^{N+1}$. But, if N_{HH}/N is not close to either zero or one, then B.R.'s degree of belief in $\textbf{Diff}^{N+1}$ would not be small. And, if $N_{HH}/N \approx 1/2$, then B.R.'s degree of belief in $\textbf{Diff}^{N+1}$ would be near $1/2$. This would hold even if N is *very large*, e.g., $10^{100} = $ a googol [11]. So, to most intuitive reasoners, B.R.'s degree of belief in $\textbf{Diff}^{N+1}$ would seem absurd.

2.7.1 B.R.'s Dilemma

If B.R. sticks to his/her opinion and has a degree of belief in $\textbf{Diff}^{N+1}$ close to $1/2$, most intuitive reasoners (i.e., most people) will think B.R. to be irrational. On the other hand, if B.R. modifies his/her opinion so that his/her degree of belief in $\textbf{Diff}^{N+1}$ is near zero, then B.R. will judge him/herself to be irrational.

B.R. cannot win!

3   No Way Out of the Dilemma

Should B.R. drop the Two-Coins Model and start over with a new model?[3] This might seem like a sensible response to B.R.'s dilemma.

Some people would advise B.R. that the Two-Coins Model is too inflexible and that he/she should have adopted a model that was more flexible. Thus, the Two-Coins Model does not have enough parameters to be flexible. In addition to the Θ_G and Θ_S, a third parameter could have been added: the so-called "log odds ratio" LOR[4] that expresses the strength of association (conditional on Θ_G and Θ_S) between the gold coin landing heads and the silver coin landing heads. In the Two-Coins Model, only the explicit parameters Θ_G and Θ_S were allowed to vary and the *potential* parameter LOR was held fixed at zero. If the model were modified to have three parameters, Θ_G, Θ_S, and LOR, and the parameters were not restricted by their prior distribution, then the modified model would give a more reasonable value for $P_{\text{B.R.}}(\textbf{Diff}^{N+1} \mid Table)$

---

[3]Many pragmatic Bayesians would advise this [9, Chap. 6]. For a discussion of the difference between pragmatic Bayesians and epistemic Bayesians, see [3].

[4]For explanation and discussion of the log odds ratio, also called the "log cross-product ratio", see [5, pp. 13–18] or [4, pp. 26–28].

when confronted with a diagonal table of the form (18). (Call this modified model the *Cooperating-Coins Model*.)

Because B.R.'s beliefs are expressed by the Two-Coins Model, then $P_{B.R.}(\textbf{Diff}^{N+1} \mid Table)$ has an unreasonable value when $Table \in \Delta \mathcal{TBL}_N$. Should B.R. start over with a new model, say the Cooperating-Coins Model?

The problem with this "move" is that it undermines the whole rationale for Bayesian epistemology. According to that rationale, there is only one rational way to update one's beliefs. That is: conditioning those beliefs on observations. In the "move" just mentioned, one (a) starts with beliefs expressed by a model, (b) observes data, (c) updates one's beliefs by conditioning on the observed data, (d) notes whether the updated beliefs seem intuitively reasonable, (e) possibly replaces the original model with a modified model, (f) updates beliefs by conditioning the *modified* model on the observed data and (g) retains the new updated beliefs if they seem intuitively reasonable. But, this procedure is simply not the same as updating by conditioning the *original* model on observations. Therefore, according to Bayesian epistemology, following steps (a)–(g) is not a rational way to update one's beliefs.

It seems B.R. has no way out of his/her dilemma.

4 A Limitation of Bayesian Inference?

In Bayesian epistemology, probabilities express degrees of belief. For example, the probability measure $P_{B.R.}(\cdot)$ expresses B.R.'s beliefs.

In contrast, in large parts of science, a common form of modeling employs probabilities to describe variability in the behavior of Nature [1, Sect. 1.3]. The idea is that Nature is not deterministic, but instead has a *propensity* described by probability to behave in different ways on different occasions.[5] Many models in science are formulated as hypotheses about *mechanisms*, described in terms of probabilistic propensities, underlying the behavior of Nature.[6] Classical, non-Bayesian, statistics is typically used to make inferences about such models by estimating parameter values and evaluating goodness of fit.

Bruno de Finetti, one of the "Fathers" of Bayesian epistemology, was scornful of the use of probability to describe Nature *itself*, rather than *beliefs* about Nature. He wrote [7, Preface]: PROBABILITY DOES NOT EXIST, meaning that probabilities do not exist in Nature, but only in the mind. He likened the concept of natural probability to the concept of phlogiston, which had to be disposed of in order to make scientific progress.

---

[5]Philosophers of science have developed multiple theories of propensity: see [10, Chaps. 6 and 7] and [8, pp. 76–77].

[6]An example of a simple model that describes hypothesized natural mechanisms and employs probabilistic propensities is the All-or-None Model of Paired-Associate Learning [6], [1, Chap. 3] that I discussed in [3].

Naturally de Finetti thought it absurd to use Bayesian methods to make inferences about hypothesized physical mechanisms expressed in terms of natural probabilities. But, what about physical mechanisms whose description does not involve the use of natural probabilities? Should Bayesian methods be applicable to making inferences about such mechanisms?

The Plastic-Yoking Mechanism (Sect. 2.5) is a physical mechanism that explains why the two coins always land either both heads or both tails. The explanation for this yoking of the two coins is that they are embedded side-by-side in a clear slab of plastic. In the description of the Plastic-Yoking Mechanism, no natural probabilities were employed whatsoever. Thus, de Finetti could not complain that the Plastic-Yoking Mechanism is conceptually absurd. Should Bayesian methods be capable of inferring whether some yoking mechanism, such as the Plastic-Yoking Mechanism, exists or probably exists?

Clearly B.R. failed to infer the existence of any kind of yoking mechanism. Should this non-inference be regarded as a failure of Bayesian methodology? Or should the non-inference be regarded as an instance of Bayesian epistemology not doing something that it was never intended to do? If the latter, does this indicate that there are important kinds of inference that Bayesian epistemology is incapable of making?

5 Conclusions

The Bayesian reasoner B.R. was given *partial* information about the inner workings of a black box. Based on that information, B.R. formulated his/her beliefs as a model of the black box. B.R. then performed N trials (N large) of an experiment on the black box and collected data from those trials. Applying the precepts of Bayesian epistemology, B.R. arrived at an absurd conclusion.

It certainly is true that, if B.R. had been given more complete information about the inner workings of the black box, then B.R.'s initial model would have been different and his/her final conclusion would have been reasonable rather than absurd.

Nevertheless, the above counterfactual does not mitigate B.R.'s failure to learn the obvious: Given that something has not occurred on each of Trials 1 to N, where N is large, then almost certainly that something will not occur on Trial $(N + 1)$.

The failure to learn the obvious calls into question whether Bayesian epistemology is complete and correct as it stands. Or do additional principles need to be incorporated?

6 Appendix: Derivation of Mathematical Results

6.1 B.R.'s Beliefs About Trials 1 to N

The set of possible outcomes of a single trial is denoted:

$$O = \{(\text{heads, heads}), (\text{heads, tails}), (\text{tails, heads}), (\text{tails, tails})\}. \tag{24}$$

In each of the ordered pairs above, the left member indicates whether the gold coin landed head or tails, and right member indicates whether the silver coin landed heads or tails. The set of possible outcomes of $N \geq 1$ trials is the set of N-tuples:

$$O^N = \overbrace{O \times \cdots \times O}^{N \text{ factors}}. \tag{25}$$

Let $\mathbf{S}_N$ denote a random element of O^N whose probabilities are governed by the *Two-Coins Model*.

Let $\text{count}_N(\cdot)$ denote a function that takes any N-tuple in O^N and counts the number of (heads, heads), (heads, tails), (tails, heads) and (tails, tails) in that N-tuple to produce a table of the form (14). Thus, $\text{count}_N(\cdot)$ maps O^N onto $\mathcal{TBL}_N$.

Three useful functions. Recall that every $Table \in \mathcal{TBL}_N$ specifies values for N_{HH}, N_{HT}, N_{TH}, and N_{TT} via (14) and also values for $N_{H\bullet}$, $N_{T\bullet}$, $N_{\bullet H}$, and $N_{\bullet T}$ via Eqs. 10–13. For every $Table \in \mathcal{TBL}_N$, define the functions:

$$F_{Table}^{\text{G \& S}}(\theta_g, \theta_s)$$
$$= [\theta_g \theta_s]^{N_{HH}} [\theta_g (1 - \theta_s)]^{N_{HT}} [(1 - \theta_g)\theta_s]^{N_{TH}} [(1 - \theta_g)(1 - \theta_s)]^{N_{TT}}; \tag{26}$$
$$F_{Table}^{\text{G}}(\theta_g)$$
$$= \theta_g^{N_{H\bullet}} (1 - \theta_g)^{N_{T\bullet}}; \tag{27}$$
$$F_{Table}^{\text{S}}(\theta_s)$$
$$= \theta_s^{N_{\bullet H}} (1 - \theta_s)^{N_{\bullet T}}. \tag{28}$$

For every Trial i,

$$Res^i = (\text{heads, heads}) \ \text{iff} \ \mathbf{H}_G^i \& \mathbf{H}_S^i \ \text{happens}; \tag{29}$$
$$Res^i = (\text{heads, tails}) \ \text{iff} \ \mathbf{H}_G^i \& \mathbf{T}_S^i \ \text{happens}; \tag{30}$$
$$Res^i = (\text{tails, heads}) \ \text{iff} \ \mathbf{T}_G^i \& \mathbf{H}_S^i \ \text{happens}; \tag{31}$$
$$Res^i = (\text{tails, tails}) \ \text{iff} \ \mathbf{T}_G^i \& \mathbf{T}_S^i \ \text{happens}. \tag{32}$$

Therefore, applying (7), (8), and (9) yields:

$$P_{\text{B.R.}}[Res^i = (\text{heads, heads}) \mid (\Theta_G, \Theta_S) = (\theta_g, \theta_s)] = \theta_g \theta_s; \qquad (33)$$

$$P_{\text{B.R.}}[Res^i = (\text{heads, tails}) \mid (\Theta_G, \Theta_S) = (\theta_g, \theta_s)] = \theta_g (1 - \theta_s); \qquad (34)$$

$$P_{\text{B.R.}}[Res^i = (\text{tails, heads}) \mid (\Theta_G, \Theta_S) = (\theta_g, \theta_s)] = (1 - \theta_g)\theta_s; \qquad (35)$$

$$P_{\text{B.R.}}[Res^i = (\text{tails, tails}) \mid (\Theta_G, \Theta_S) = (\theta_g, \theta_s)] = (1 - \theta_g)(1 - \theta_s). \qquad (36)$$

Possible values of Table and seq. From here through the end of Sect. 6.2, let $Table$ be any table in $\mathscr{TBL}_N$ and set seq be any N-tuple in O^N such that $\text{count}_N(seq) = Table$. In other words, the results derived below will hold for every $Table$ and seq such that $Table \in \mathscr{TBL}_N$ and $\text{count}_N(seq) = Table$.

Recall that, in the Two-Coins Model, $Res^1, Res^2, Res^3, \ldots$ are conditionally independent given (Θ_G, Θ_S). Therefore, from Eqs. 33–36, it follows that

$$P_{\text{B.R.}}(\mathbf{S}_N = seq \mid \Theta_G = \theta_g \ \& \ \Theta_S = \theta_s) = F_{Table}^{G\,\&\,S}(\theta_g, \theta_s). \qquad (37)$$

Then, since (Θ_G, Θ_S) is uniformly distributed on the unit square $[0, 1] \times [0, 1]$,

$$P_{\text{B.R.}}(\mathbf{S}_N = seq) = \int_0^1 \int_0^1 F_{Table}^{G\,\&\,S}(\theta_g, \theta_s) \, d\theta_g \, d\theta_s. \qquad (38)$$

But

$$F_{Table}^{G\,\&\,S}(\theta_g, \theta_s) = F_{Table}^{G}(\theta_g) \cdot F_{Table}^{S}(\theta_s). \qquad (39)$$

Therefore,

$$P_{\text{B.R.}}(\mathbf{S}_N = seq) = \int_0^1 \int_0^1 F_{Table}^{G}(\theta_g) \cdot F_{Table}^{S}(\theta_s) \, d\theta_g \, d\theta_s \qquad (40)$$

$$= \left[\int_0^1 F_{Table}^{G}(\theta_g) \, d\theta_g \right] \cdot \left[\int_0^1 F_{Table}^{S}(\theta_s) \, d\theta_s \right]. \qquad (41)$$

But it is well known[7] that, for any integer k, with $0 \leq k \leq N$,

$$\int_0^1 x^k (1 - x)^{N-k} \, dx = \left[(N + 1) \binom{N}{k \ \ N-k} \right]^{-1}. \qquad (42)$$

Hence

---

[7] https://math.stackexchange.com/questions/122296/how-to-evaluate-this-integral-relating-to-binomial.

$$\int_0^1 F_{Table}^G(\theta_g)\, d\theta_g = \left[(N+1)\binom{N}{N_{H\bullet}\ \ N_{T\bullet}}\right]^{-1};$$ (43)

$$\int_0^1 F_{Table}^S(\theta_s)\, d\theta_s = \left[(N+1)\binom{N}{N_{\bullet H}\ \ N_{\bullet T}}\right]^{-1}.$$ (44)

Thus, for *every seq* such that $\text{count}_N(seq) = Table$,

$$P_{\text{B.R.}}(\mathbf{S}_N = seq) = \left[(N+1)^2\binom{N}{N_{H\bullet}\ \ N_{T\bullet}}\binom{N}{N_{\bullet H}\ \ N_{\bullet T}}\right]^{-1}.$$ (45)

6.2 B.R.'s Beliefs About Trial $(N+1)$

Let O^{N+1} and $\mathbf{S}_{N+1}$ have meanings analogous to O^N and $\mathbf{S}_N$. If *seq* is an N-tuple, let the $(N+1)$-tuple consisting of the concatenation of *seq* and (heads, heads) be denoted $seq\frown$(heads, heads). If $\text{count}_N(seq)$ is the table (14), then

$$\text{count}_{N+1}[seq\frown(\text{heads, heads})]$$

is the table

	silver head	silver tail
gold head	$N_{HH}+1$	N_{HT}
gold tail	N_{TH}	N_{TT}

(46)

Then, by a derivation analogous to the derivation of (45), we have:

$$P_{\text{B.R.}}[\mathbf{S}_{N+1} = seq\frown(\text{heads, heads})]$$
$$= \left[(N+2)^2\binom{N+1}{N_{H\bullet}+1\ \ N_{T\bullet}}\binom{N+1}{N_{\bullet H}+1\ \ N_{\bullet T}}\right]^{-1}.$$ (47)

Note that

$$P_{\text{B.R.}}[(\mathbf{S}_N = seq)\ \&\ \mathbf{H}_G^{N+1}\ \&\ \mathbf{H}_S^{N+1}] = P_{\text{B.R.}}[\mathbf{S}_{N+1} = seq\frown(\text{heads, heads})].$$ (48)

Therefore,

$$P_{\text{B.R.}}[\mathbf{H}_G^{N+1}\ \&\ \mathbf{H}_S^{N+1}\,|\,\mathbf{S}_N = seq] = \frac{P_{\text{B.R.}}[\mathbf{S}_{N+1} = seq\frown(\text{heads, heads})]}{P_{\text{B.R.}}(\mathbf{S}_N = seq)}$$ (49)

Then, dividing the right side of (47) by the right side of (45) yields:

$$P_{\text{B.R.}}[\mathbf{H}_G^{N+1} \ \& \ \mathbf{H}_S^{N+1} \mid \mathbf{S}_N = seq] = \left(\frac{N_{H\bullet} + 1}{N + 2}\right) \cdot \left(\frac{N_{\bullet H} + 1}{N + 2}\right). \tag{50}$$

Analogous to the derivation of (50), we may also derive the following:

$$P_{\text{B.R.}}[\mathbf{H}_G^{N+1} \ \& \ \mathbf{T}_S^{N+1} \mid \mathbf{S}_N = seq] = \left(\frac{N_{H\bullet} + 1}{N + 2}\right) \cdot \left(\frac{N_{\bullet T} + 1}{N + 2}\right). \tag{51}$$

$$P_{\text{B.R.}}[\mathbf{T}_G^{N+1} \ \& \ \mathbf{H}_S^{N+1} \mid \mathbf{S}_N = seq] = \left(\frac{N_{T\bullet} + 1}{N + 2}\right) \cdot \left(\frac{N_{\bullet H} + 1}{N + 2}\right). \tag{52}$$

$$P_{\text{B.R.}}[\mathbf{T}_G^{N+1} \ \& \ \mathbf{T}_S^{N+1} \mid \mathbf{S}_N = seq] = \left(\frac{N_{T\bullet} + 1}{N + 2}\right) \cdot \left(\frac{N_{\bullet T} + 1}{N + 2}\right). \tag{53}$$

Summing Eqs. 50 and 51 yields Eq. 54 below. And Eqs. 55–57 may be derived analogously.

$$P_{\text{B.R.}}[\mathbf{H}_G^{N+1} \mid \mathbf{S}_N = seq] = (N_{H\bullet} + 1)/(N + 2). \tag{54}$$

$$P_{\text{B.R.}}[\mathbf{T}_G^{N+1} \mid \mathbf{S}_N = seq] = (N_{T\bullet} + 1)/(N + 2). \tag{55}$$

$$P_{\text{B.R.}}[\mathbf{H}_S^{N+1} \mid \mathbf{S}_N = seq] = (N_{\bullet H} + 1)/(N + 2). \tag{56}$$

$$P_{\text{B.R.}}[\mathbf{T}_S^{N+1} \mid \mathbf{S}_N = seq] = (N_{\bullet T} + 1)/(N + 2). \tag{57}$$

Comparing Eqs. 50–53 with Eqs. 54–57, we see that:

$\mathbf{H}_G^{N+1}$ and $\mathbf{H}_S^{N+1}$ are conditionally independent given $\mathbf{S}_N = seq$. (58)

$\mathbf{H}_G^{N+1}$ and $\mathbf{T}_S^{N+1}$ are conditionally independent given $\mathbf{S}_N = seq$. (59)

$\mathbf{T}_G^{N+1}$ and $\mathbf{H}_S^{N+1}$ are conditionally independent given $\mathbf{S}_N = seq$. (60)

$\mathbf{T}_G^{N+1}$ and $\mathbf{T}_S^{N+1}$ are conditionally independent given $\mathbf{S}_N = seq$. (61)

Conditioning on Table rather than on $\mathbf{S}_N = seq$. Recall that Eqs. 54–61 hold for all $seq \in O^N$ such that $\text{count}_N(seq) = Table$. Therefore, those equations hold when the conditioning on $\mathbf{S}_N = seq$ is replaced by conditioning on $Table$. In other words:

$$P_{\text{B.R.}}[\mathbf{H}_G^{N+1} \mid Table] = (N_{H\bullet} + 1)/(N + 2). \tag{62}$$

$$P_{\text{B.R.}}[\mathbf{T}_G^{N+1} \mid Table] = (N_{T\bullet} + 1)/(N + 2). \tag{63}$$

$$P_{\text{B.R.}}[\mathbf{H}_S^{N+1} \mid Table] = (N_{\bullet H} + 1)/(N + 2). \tag{64}$$

$$P_{\text{B.R.}}[\mathbf{T}_S^{N+1} \mid Table] = (N_{\bullet T} + 1)/(N + 2). \tag{65}$$

And:

$\mathbf{H}_G^{N+1}$ and $\mathbf{H}_S^{N+1}$ are conditionally independent given $Table$.   (66)

$\mathbf{H}_G^{N+1}$ and $\mathbf{T}_S^{N+1}$ are conditionally independent given $Table$.   (67)

$\mathbf{T}_G^{N+1}$ and $\mathbf{H}_S^{N+1}$ are conditionally independent given $Table$.   (68)

$\mathbf{T}_G^{N+1}$ and $\mathbf{T}_S^{N+1}$ are conditionally independent given $Table$.   (69)

Equations 62–69 hold for all $Table \in \mathscr{TBL}_N$.

6.3 B.R.'s Beliefs About Whether the Coins Will Land Differently on Trial $(N + 1)$

Recall that **Diff** $^{N+1}$ denotes the event that the two coins land *differently* on Trial $N + 1$. That is: The gold coin lands heads and the silver coin lands tails or vice versa. So, for all $Table \in \mathscr{TBL}_N$,

$$P_{\text{B.R.}}(\textbf{Diff}^{N+1} \mid Table) \tag{70}$$
$$= P_{\text{B.R.}}(\mathbf{H}_G^{N+1} \,\&\, \mathbf{T}_S^{N+1} \mid Table) + P_{\text{B.R.}}(\mathbf{T}_G^{N+1} \,\&\, \mathbf{H}_S^{N+1} \mid Table).$$

Then, from Eqs. 62–69, it follows that

$$P_{\text{B.R.}}(\textbf{Diff}^{N+1} \mid Table)$$
$$= \left(\frac{N_{H\bullet} + 1}{N + 2}\right)\left(\frac{N_{\bullet T} + 1}{N + 2}\right) + \left(\frac{N_{T\bullet} + 1}{N + 2}\right)\left(\frac{N_{\bullet H} + 1}{N + 2}\right). \tag{71}$$

Suppose, however, $Table$ has the diagonal form (18). In other words, suppose that $Table \in \Delta\mathscr{TBL}_N$. Then

$$N_{\bullet H} = N_{H\bullet} \text{ and } N_{T\bullet} = N_{\bullet T} = N - N_{H\bullet}. \tag{72}$$

Hence, for all $Table \in \Delta\mathscr{TBL}_N$,

$$P_{\text{B.R.}}(\textbf{Diff}^{N+1} \mid Table)$$
$$= 2\left(\frac{N_{H\bullet} + 1}{N + 2}\right)\left[1 - \left(\frac{N_{H\bullet} + 1}{N + 2}\right)\right]. \tag{73}$$

References

1. R.C. Atkinson, G.H. Bower, E.J. Crothers, *An Introduction to Mathematical Learning Theory* (Wiley, New York, 1965)
2. D. Bamber, What is probability? (Review of the book [10].) J. Math. Psychol. **47**, 377–382 (2003)
3. D. Bamber, Statisticians should not tell scientists what to think, in *Structural Changes and Their Econometric Modeling*, ed. by V. Kreinovich, S. Sriboonchitta (Proceedings of the meeting of the Thailand Econometric Society, January 2019, Chiang Mai.), pp. 63–82. Studies in Computational Intelligence, vol. 808 (Springer, Cham, Switzerland, 2019)
4. D. Bamber, I.R. Goodman, H.T. Nguyen, High-probability logic and inheritance, in *Mathematical Models of Perception and Cognition: A Festschrift for James T. Townsend*, ed. by J.W. Houpt, L.M. Blaha, vol. 1 (Psychology Press, New York, 2016), pp. 13–36
5. Y.M.M. Bishop, S.E. Fienberg, P.W. Holland, *Discrete Multivariate Analysis: Theory and Practice* (MIT Press, Cambridge, MA, 1975)
6. G.H. Bower, Application of a model to paired-associate learning. Psychometrika **26**, 255–280 (1961)
7. B. de Finetti, *Theory of Probability: A Critical Introductory Treatment*, vol. 1 (Wiley, New York, 1974)
8. P. Diaconis, B. Skyrms, *Ten Great Ideas About Chance* (Princeton University Press, Princeton, New Jersey, 2018)
9. A. Gelman, J.B. Carlin, H.S. Stern, D.B. Dunson, A. Vehtari, D.B. Rubin, *Bayesian Data Analysis*, 3rd edn. (CRC Press, Boca Raton, Florida, 2013)
10. D. Gillies, *Philosophical Theories of Probability* (Routledge, London, 2000)
11. E. Kasner, J.R. Newman, *Mathematics and the Imagination* (Simon and Schuster, New York, 1940)

The Fell Compactification of a Poset

G. Bezhanishvili and J. Harding

For Hung Nguyen on his 75th birthday

Abstract A poset P forms a locally compact T_0-space in its Alexandroff topology. We consider the hit-or-miss topology on the closed sets of P and the associated Fell compactification of P. We show that the closed sets of P with the hit-or-miss topology is the Priestley space of the bounded distributive lattice freely generated by the order dual of P. The Fell compactification of $H(P)$ is shown to be the Priestley space of a sublattice of the upsets of P consisting of what we call Fell upsets. These are upsets that are finite unions of those obtained as upper bounds of finite subsets of P. The restriction of the hit topology to $H(P)$ is a stable compactification of P. When P is a chain, we show that this is the least stable compactification of P.

Keywords Hit-or-miss topology · Fell compactification · Stable compactification · Order-compactification · Poset · Free lattice · Distributive lattice · Priestley space

MSC: 54B20, 54D35, 54D45, 06E15, 06B25

G. Bezhanishvili (✉) · J. Harding
Department of Mathematical Sciences, New Mexico State University, Las Cruces, NM 88003, USA
e-mail: guram@nmsu.edu

J. Harding
e-mail: hardingj@nmsu.edu

V. Kreinovich (ed.), *Statistical and Fuzzy Approaches to Data Processing, with Applications to Econometrics and Other Areas*, Studies in Computational Intelligence 892, https://doi.org/10.1007/978-3-030-45619-1_3

1 Introduction

For a metric space X, the Hausdorff metric associates a distance between compact subsets of X. It is the first of a number of examples of topologies placed on some collection of subsets of a topological space. Other examples include the Vietoris topology and the hit-or-miss topology. Our focus here is the hit-or-miss topology on the closed sets $\mathscr{F}(X)$ of a topological space X, and the associated Fell compactification of a locally compact T_0-space, which is the closure of its image in its hit-or-miss topology.

Hyperspace topologies, such as the Vietoris and hit-or-miss topology, feature prominently in branches of logic, lattice theory, and theoretical computer science. The Vietoris topology is a primary feature in co-algebraic logic (see, e.g., [16]), while the hit-or-miss topology features in the study of continuous lattices and is closely tied to the Lawson topology [7, Sect. III-1]. Hyperspace topologies also feature prominently in the powerdomain constructions [7, Sect IV-8].

The authors of this note work in logic and lattice theory. It came as a surprise when the statistician in the office next door, Hung Nguyen, came asking pointed questions about continuous lattices and Lawson topologies.

His interests at the time lay in the area of random sets, and their extension to the fuzzy setting. In defining random sets it is customary to consider only closed sets. One then requires a topology on the collection of closed sets of a topological space to generate a Borel measure. Such is provided by the hit-or-miss topology.

In extending to the fuzzy random setting, one notes that the indicator function of a closed set is upper semicontinuous. Hung sought a topology on the collection $\mathrm{USC}(X)$ of $[0, 1]$-valued upper semicontinuous functions on a locally compact Hausdorff space X. His approach was through domain theory. The key facts are as follows. The closed sets $\mathscr{F}(X)$ under $\supseteq$ form a continuous lattice whose Lawson topology is the hit-or-miss topology; the interval $[0, 1]$ is a continuous lattice; and under the pointwise $\geq$-order, $\mathrm{USC}(X)$ is a continuous lattice that is isomorphic to the continuous lattice $\mathscr{F}(X)^{[0,1]}$ of Scott-continuous functions from $[0, 1]$ to $\mathscr{F}(X)$. The Lawson topology on $\mathrm{USC}(X)$ provides a compact Hausdorff topology whose Borel algebra allows to define random fuzzy sets. Hung's work in this area is summarized in [12].

In this note we discuss the hit-or-miss topology, and associated Fell compactification, when it is applied to a poset P with its Alexandroff topology. From a lattice-theoretic standpoint this provides a pleasing theory with connections to free lattices and the theory of stable compactifications that originated in theoretical computer science. We do not investigate direct applications in the study of random sets, but are happy to provide a paper that recalls an unexpected collaboration with our colleague and friend of many years.

2 Hit-or-Miss Topology and Fell Compactification

Here we consider topological spaces that are not necessarily Hausdorff. A *compact space* is the one in which every open cover has a finite subcover, and a *locally compact space* is one in which compact sets form a neighborhood base. For a topological space X, let $\mathcal{O}(X)$ be the set of open sets, $\mathcal{F}(X)$ the set of closed sets, and $\mathcal{K}(X)$ the set of compact sets of X.

Definition 2.1 Let X be a topological space.

(1) For $S \subseteq X$, define

$$\Diamond_S = \{F \in \mathcal{F}(X) \mid F \cap S \neq \varnothing\} \text{ and } \Box_S = \{F \in \mathcal{F}(X) \mid F \cap S = \varnothing\}.$$

(2) Let $\tau_\Diamond$ be the topology on $\mathcal{F}(X)$ given by the subbasis $\{\Diamond_U \mid U \in \mathcal{O}(X)\}$.
(3) Let $\tau_\Box$ be the topology on $\mathcal{F}(X)$ given by the subbasis $\{\Box_K \mid K \in \mathcal{K}(X)\}$.
(4) Let $\pi = \tau_\Diamond \vee \tau_\Box$.

We call $\tau_\Diamond$ the *hit topology*, $\tau_\Box$ the *miss topology*, and π the *hit-or-miss topology*.

The next lemma is easily seen.

Lemma 2.2 *Let X be a topological space. For any collection $\{S_i \mid i \in I\}$ of subsets of X, we have*

$$\bigcap_{i \in I} \Box_{S_i} = \Box_{\bigcup_{i \in I} S_i} \text{ and } \bigcup_{i \in I} \Diamond_{S_i} = \Diamond_{\bigcup_{i \in I} S_i}.$$

Therefore, the subbasis for $\tau_\Box$ is actually a basis, and the hit-or-miss topology has a basis of sets of the form

$$\{\Box_K \cap \Diamond_{U_1} \cap \cdots \cap \Diamond_{U_n} \mid K \in \mathcal{K}(X) \text{ and } U_1, \ldots, U_n \in \mathcal{O}(X)\}.$$

If X is locally compact, then the hit-or-miss topology π on $\mathcal{F}(X)$ is compact Hausdorff [6, Theorem 1]. Moreover, if X is compact Hausdorff, then it is easy to see that the hit-or-miss topology coincides with the Vietoris topology [10, Sect. III-4]. The next result is well known.

Proposition 2.3 *The map $e : X \to \mathcal{F}(X)$ that sends x to its closure $\overline{\{x\}}$ has the following properties.*

(1) *e is 1–1 iff X is T_0.*
(2) *If $U \in \mathcal{O}(X)$, then $e^{-1}(\Diamond_U) = U$; hence e is continuous with respect to $\tau_\Diamond$.*
(3) *If X is T_1 and $K \in \mathcal{K}(X)$, then $e^{-1}(\Box_K) = X \setminus K$; hence if X is Hausdorff, then e is continuous with respect to $\tau_\Box$.*
(4) *If X is Hausdorff, then e is continuous with respect to π.*

An *embedding* of a space X into a space Y is a 1-1 map $e : X \to Y$ that is a homeomorphism from X to the image $e(X)$ given the subspace topology from Y. Classically, a *compactification* of a space X is an embedding of X into a compact Hausdorff space Y where the image of X is dense in Y.

Definition 2.4 (see, e.g., [9, p. 57]) For X a locally compact T_0-space, its *Fell compactification* $H(X)$ is the closure of the image of X in the hit-or-miss topology of $\mathscr{F}(X)$.

Since X is locally compact, the hit-or-miss topology is compact Hausdorff, so the closed subset $H(X)$ of $\mathscr{F}(X)$ is a compact Hausdorff space. When X is a non-compact locally compact Hausdorff space, $e : X \to H(X)$ is an embedding, and the Fell compactification is the one-point compactification of X (see [6, p. 475]). When X is non-Hausdorff, $e : X \to H(X)$ is no longer an embedding. In the next section we show that there are two ways in which we can consider $e : X \to H(X)$ to be a form of compactification. By enriching the topology of X we can view e as a classical compactification, and by decreasing the topology of $H(X)$ we can view e as what is known as a stable compactification.

3   Stable and Order-Compactifications

In a topological space, a closed set is *irreducible* if it cannot be written as the union of two proper closed subsets. A space X is *sober* if each irreducible closed set is the closure of a unique singleton. Sober spaces are T_0. A set that is an intersection of open sets is *saturated*.

Definition 3.1 (see, e.g., [7, Definition VI-6.7]) A space is *stably compact* if it is compact, locally compact, sober, and the intersection of any family of compact saturated sets is compact.

For a stably compact space (X, τ), the *co-compact topology* τ^k on X has as opens the complements of compact saturated sets, and the *patch topology* $\pi = \tau \vee \tau^k$ is the smallest topology on X containing the original and co-compact topologies.

Definition 3.2 An *ordered topological space* $(X, \pi, \leq)$ is a set X with a partial ordering $\leq$ and topology π. It is a *Nachbin space* if π is compact and $\leq$ is closed in the product topology.

Remark 3.3 The study of ordered topological spaces in general, and of Nachbin spaces in particular, was pioneered by Nachbin in the 1940s (see [11]); the name Nachbin space appears to originate from [1, Definition 2.5].

We recall that a subset S of a poset P is an *upset* if $x \in S$ and $x \leq y$ imply $y \in S$, and is a *downset* if $x \in S$ and $y \leq x$ imply $y \in S$. We use $\uparrow S$ for the smallest upset containing S, $\downarrow S$ for the smallest downset containing S, and for $x \in P$ we use $\uparrow x$ for $\uparrow\{x\}$ and $\downarrow x$ for $\downarrow\{x\}$. Finally, we use $S^u = \{y \in P \mid x \leq y \text{ for all } x \in S\}$ for the set of upper bounds of S. We will use repeatedly that in a Nachbin space, if S is closed then $\uparrow S$ and $\downarrow S$ are closed.

Definition 3.4 For a Nachbin space $(X, \pi, \leq)$, the *upper topology* π_u is defined as the open upsets and the *lower topology* π_ℓ as the open downsets.

It is known that every Nacbin space is Hausdorff, and the upper and lower topologies give rise to stably compact spaces (X, π_u) and (X, π_ℓ) (see, e.g., [7, Proposition VI-6.11]). Conversely, every stably compact space gives rise to a Nachbin space. To see this, we require the following.

Definition 3.5 (see, e.g., [7, Definition O-5.2]) The *specialization order* of a topological space is defined by setting $x \leq y$ if $x \in \overline{\{y\}}$.

The specialization order is a partial order on X iff X is T_0. For a stably compact space (X, τ) with specialization order $\leq$ and patch topology $\pi = \tau \vee \tau^k$, we have that $(X, \pi, \leq)$ is a Nachbin space with upper topology τ and lower topology τ^k. This provides a 1-1 correspondence between stably compact spaces and Nachbin spaces (see, e.g., [7, Sect. VI-6]).

Definition 3.6 A *stable compactification* of a T_0-space X is an embedding of X into a stably compact space Y where the image of X is dense in the patch topology of Y.

Smyth [14] introduced stable compactifications to generalize the classical theory of compactifications to the setting of T_0-spaces. The definition given above is a simplification that was provided in [1, Theorem 3.5]. of his original definition. A related notion is the following.

Definition 3.7 An *order-compactification* of an ordered topological space $(X, \pi, \leq)$ consists of a Nachbin space $(Y, \pi, \leq)$ and a map $e : X \to Y$ that is both a topological embedding and an order embedding.

Our connection to Nachbin spaces stems from the fact that for a locally compact T_0-space X, if π is the hit-or-miss topology on $\mathscr{F}(X)$, then $(\mathscr{F}(X), \pi, \subseteq)$ is a Nachbin space [9, p. 57]. The lower topology π_ℓ of this Nachbin space is the hit topology $\tau_\Diamond$, and the upper topology π_u is the miss topology $\tau_\Box$. We use these facts to address the problem raised at the end of the previous section, that the map $e : X \to H(X)$ of the Fell compactification need not be an embedding. We now provide two ways to alter the topologies of X or $H(X)$ to provide an embedding.

Theorem 3.8 *If X is locally compact T_0, then the Fell compactification $H(X)$ with the restriction of the hit topology is a stable compactification of X.*

Proof Since X is a locally compact T_0-space, $(\mathscr{F}(X), \pi, \subseteq)$ is a Nachbin space, and since $H(X)$ is a closed subset, it naturally forms a Nachbin space as well. The upper topology of $\mathscr{F}(X)$ is the hit topology $\tau_\Diamond$, and it follows that the restriction of $\tau_\Diamond$ to $H(X)$ is its upper topology. So under the restriction of $\tau_\Diamond$ we have that $H(X)$ is a stably compact space. By definition, $H(X)$ is the closure of the image of X under the topology π, hence this image is dense in the patch topology of the stably compact space $H(X)$. $\qquad \square$

Theorem 3.9 *Let (X, τ) be a locally compact T_0-space, $\leq$ its specialization order, and σ the smallest topology on X making $e : X \to \mathscr{F}(X)$ continuous with respect to the hit-or-miss topology. Then $\tau \subseteq \sigma$ and $e : (X, \sigma, \leq) \to (H(X), \pi, \subseteq)$ is an order-compactification of $(X, \sigma, \leq)$.*

Proof By Proposition 2.3(2), $\tau \subseteq \sigma$. Also, since e is 1-1 by Proposition 2.3(1), e is a topological embedding of (X, σ) into $(H(X), \pi)$. Therefore, $e : (X, \sigma) \to (H(X), \pi)$ is a compactification of (X, σ). To see that it is an order-compactification, note the τ-closure $\overline{\{x\}}$ is equal to $\downarrow x$, so $e(x) \subseteq e(y)$ iff $\overline{\{x\}} \subseteq \overline{\{y\}}$ iff $\downarrow x \subseteq \downarrow y$ iff $x \leq y$. $\qquad\qquad\square$

4   Priestley Spaces and Order-Compactifications

We recall that a subset of a topological space is *clopen* if it is both closed and open. A topological space is *zero-dimensional* if it has a basis of clopen sets, and a *Stone space* if it is zero-dimensional compact Hausdorff.

Definition 4.1 [13] A *Priestley space* is an ordered topological space $(X, \pi, \leq)$ that is compact and for each $x \nleq y$ there is a clopen upset U with $x \in U$ and $y \notin U$.

Every Priestley space is both a Stone space and a Nachbin space. We next describe briefly Priestley duality which establishes close connection between Priestley spaces and bounded distributive lattices. For convenience, we will often write an ordered topological space $(X, \pi, \leq)$ simply as X.

Definition 4.2 For an ordered topological space X, let X^* be the collection of clopen upsets of X.

The collection of clopen upsets of an an ordered topological space X is a bounded distributive lattice. The following result is the object level part of a categorical duality between bounded distributive lattices and Priestley spaces.

Theorem 4.3 [13] *For each bounded distributive lattice D there is a Priestley space D_* such that $(D_*)^*$ is isomorphic as a distributive lattice to D. Further, for each Priestley space X we have $(X^*)_*$ is order-isomorphic and homeomorphic to X.*

By Priestley duality, homomorphic images of a bounded distributive lattice D correspond to closed subsets of its Priestley space D_*. As a result, we have:

Lemma 4.4 *Let X be a Priestley space and Y a closed subspace of X. Then Y is a Priestley space and for each clopen upset U of Y there is a clopen upset V of X such that $U = V \cap Y$.*

Definition 4.5 [3] A *Priestley order-compactification* of an ordered topological space is an order-compactification that is a Priestley space.

Remark 4.6 Suppose (X, τ) is a stably compact space and $(X, \pi, \leq)$ the corresponding Nachbins space. Then the sets that are the clopen upsets of the Nachbin space are exactly the compact open sets of the stably compact space. Therefore, the Nachbin space is a Priestley space iff the compact open sets in (X, τ) form a basis which is closed under finite intersections. Such stably compact spaces (X, τ) whose compact opens form a basis and are closed under finite intersections are known as *spectral spaces*. They are exactly the Zariski spectra of commutative rings [8]. Thus, under the above correspondence between stably compact spaces and Nachbin spaces, spectral spaces correspond exactly to Priestly spaces.

5 The Hit-or-Miss Topology of a Poset

Throughout this section P is a poset with partial ordering $\leq$. The collection of upsets of P is closed under arbitrary unions and arbitrary intersections, hence forms a topology on P called the *Alexandroff topology*. We denote it τ_A. Clearly the closed sets of τ_A are the downsets of P. It is easy to see that for each $x \in P$, its upset $\uparrow x$ is compact. It follows that τ_A is locally compact.

Definition 5.1 For a poset P we use $(\mathscr{F}(P), \pi, \subseteq)$, or simply $\mathscr{F}(P)$, for the downsets of P partially ordered by set inclusion with the topology π being the hit-or-miss topology derived from the Alexandroff topology of P.

The hit-or-miss topology will simplify in the setting of $\mathscr{F}(P)$. To describe this, we first recall from Definition 2.1(1) that for S a subset of P, we have the following subsets of $\mathscr{F}(P)$:

$$\lozenge_S = \{D \mid S \cap D \neq \varnothing\} \text{ and } \square_S = \{D \mid S \cap D = \varnothing\}.$$

In the case that $S = \{x\}$ is a singleton, we write these simply as $\lozenge_x$ and $\square_x$. We need a new notion to better describe the topology of $\mathscr{F}(P)$.

Definition 5.2 For S a subset of P set $\triangle_S = \{D \mid S \subseteq D\}$.

Again, when $S = \{x\}$ is a singleton, we simply write $\triangle_x$. Since $\triangle_x = \lozenge_x$, in the sequel we will use $\lozenge_x$. The following lemma follows easily from the observation that a downset D intersects a principal upset $\uparrow y$ nontrivially iff $y \in D$.

Lemma 5.3 *If $G = \{y_1, \ldots, y_n\}$ is a finite subset of P, then*

$$\lozenge_{\uparrow y_1} \cap \cdots \cap \lozenge_{\uparrow y_n} = \triangle_G.$$

Proposition 5.4 *For a poset P, the hit-or-miss topology of $\mathscr{F}(P)$ has a subbasis of sets of the form $\square_x \cap \lozenge_y$ where $x, y \in P$ and a basis of sets of the form $\square_F \cap \triangle_G$ where F and G are finite subsets of P.*

Proof The hit-or-miss topology on the closed sets of an arbitrary space has as a basis all sets of the form $\Box_K \cap \Diamond_{U_1} \cap \cdots \cap \Diamond_{U_n}$ where K is compact and $U_1, \ldots, U_n$ are open. For any set U, trivially $\Diamond_U = \bigcup\{\Diamond_y \mid y \in U\}$. Using the distributive property of union and intersection, we then have

$$\Box_K \cap \Diamond_{U_1} \cap \cdots \cap \Diamond_{U_n} = \bigcup\{\Box_K \cap \Diamond_{y_1} \cap \cdots \cap \Diamond_{y_n} \mid y_1 \in U_1, \ldots, y_n \in U_n\}.$$

In general, sets of the form $\Diamond_y$ need not be open since the singleton $\{y\}$ need not be open. So this identity as it is cannot be used to improve our description of a basis of the hit-or-miss topology.

In the poset P, we have that y belongs to a downset D iff $D \cap \uparrow y \neq \emptyset$. Therefore, $\Diamond_y = \Diamond_{\uparrow y}$, and since $\uparrow y$ is open in τ_A we have $\Diamond_y$ is an open set of the hit-or-miss topology of $\mathcal{F}(P)$. Thus, the hit-or-miss topology of $\mathcal{F}(P)$ has a basis of sets of the form $\Box_K \cap \Diamond_{y_1} \cap \cdots \cap \Diamond_{y_n}$ where K is compact and $y_1, \ldots, y_n \in P$. Equivalently, as follows from Lemma 5.3, there is a basis of sets of the form $\Box_K \cap \triangle_G$ where K is compact and G is a finite set.

It is easily seen that $K \subseteq P$ is compact in τ_A iff there is a finite set $F \subseteq K$ with $\uparrow K = \uparrow F$. For a downset D and any set S, we have $D \cap S = \emptyset$ iff $D \cap \uparrow S = \emptyset$. It follows that if K is compact and F is finite with $\uparrow K = \uparrow F$, then $\Box_K = \Box_F$. Therefore, the sets of the form $\Box_K$ for K compact are exactly the sets of the form $\Box_F$ for F finite. So the hit-or-miss topology of $\mathcal{F}(P)$ has a basis of sets of the form $\Box_F \cap \triangle_G$ where F and G are finite subsets of P.

For $F = \{x_1, \ldots, x_m\}$ and $G = \{y_1, \ldots, y_n\}$ we have

$$\Box_F = \Box_{x_1} \cap \cdots \cap \Box_{x_m} \text{ and } \triangle_G = \Diamond_{y_1} \cap \cdots \cap \Diamond_{y_n}.$$

Thus, the sets of the form $\Box_x \cap \Diamond_y$ are a subbasis.   □

Remark 5.5 In view of Proposition 5.4, for a poset P with its Alexandroff topology, the hit-or-miss topology on the closed sets $\mathcal{F}(P)$ becomes what might be called an *all-or-nothing* topology. Here we have a basis of sets $\Box_F \cap \triangle_G$ for finite subsets F and G of P. The set $\triangle_G$ is all closed sets D that contain G, and the topology they generate could be called the *all* topology. The set $\Box_F$ is all closed sets that are disjoint from F, and the topology they generate could be called the *nothing* topology.

Note that for $x \in P$, by definition $\Diamond_x$ and $\Box_x$ are complementary sets, and as both are open in the hit-or-miss topology, they are clopen. Further, $\Diamond_x$ is clearly an upset and $\Box_x$ a downset. As noted earlier, for a locally compact space X, we have $(\mathcal{F}(X), \pi, \subseteq)$ is a Nachbin space.

Theorem 5.6 *For P a poset, $(\mathcal{F}(P), \pi, \subseteq)$ is a Priestley space.*

Proof It is sufficient to verify the Priestley separation axiom of Definition 4.1. Let $D, E \in \mathcal{F}(P)$ with $D \nsubseteq E$, so there is $x \in D$ with $x \notin E$. Then $\Diamond_x$ is a clopen upset containing D and not E.   □

For a poset P, we next describe the distributive lattice of clopen upsets of the Priestley space $(\mathscr{F}(P), \pi, \subseteq)$. In Theorem 5.12 we describe this same lattice in terms more directly related to the poset P.

Proposition 5.7 *For a poset P, the clopen upsets of $(\mathscr{F}(P), \pi, \subseteq)$ are exactly finite unions of sets $\triangle_G$, and clopen downsets are exactly finite unions of sets $\square_F$ where F, G are finite subsets of P.*

Proof Since $\triangle_G = \bigcap\{\lozenge_x \mid x \in G\}$ and $\square_F = \bigcap\{\square_x \mid x \in F\}$ and F, G are finite, we have $\triangle_G$ is a clopen upset and $\square_F$ is a clopen downset. Let $\mathscr{U}$ be a clopen upset and $D \in \mathscr{U}$. For $E \notin \mathscr{U}$, there is $x \in P$ with $D \in \lozenge_x$ and $E \notin \lozenge_x$, so $E \in \square_x$. The $\square_x$ cover the complement of $\mathscr{U}$, which is a closed downset. By compactness, there are $x_1, \ldots, x_n \in P$ such that $\square_{x_1}, \ldots, \square_{x_n}$ cover the complement of $\mathscr{U}$. Let $G = \{x_1, \ldots, x_n\}$. Then $D \in \triangle_G \subseteq \mathscr{U}$. Therefore, $\mathscr{U}$ is a union of sets $\triangle_G$ for G a finite subset of P. Since $\mathscr{U}$ is compact, it is a finite union. That clopen downsets are exactly finite unions of sets $\square_F$ for F a finite subset of P is proved similarly. □

Lemma 5.8 *For each clopen subset $\mathscr{U}$ of $\mathscr{F}(P)$ we have $\downarrow\mathscr{U}$ is clopen.*

Proof By Proposition 5.4 and compactness, clopen sets are finite unions of sets of the form $\square_F \cap \triangle_G$ where F, G are finite subsets of P. Since $\downarrow$ commutes with unions, it is sufficient to show that for $\square_F \cap \triangle_G$ nonempty, its downset is clopen. Clearly $\square_F$ is a downset that contains $\square_F \cap \triangle_G$. For the converse, since $\square_F \cap \triangle_G$ is nonempty, $\downarrow G$ is disjoint from F. So if $D \in \square_F$, then $D \cup \downarrow G$ is disjoint from F, hence belongs to $\square_F \cap \triangle_G$. So $\square_F$ is contained in the downset of $\square_F \cap \triangle_G$, and therefore is equal to it. □

In the following, we provide an alternate way to look at the hit-or-miss topology of $\mathscr{F}(P)$ that is more algebraic in nature. Let 2 be the 2-element lattice $\{0, 1\}$ with the discrete topology, and consider the power 2^P which consists of all functions $\alpha : P \to 2$. This is both a lattice under the product order, and a topological space under the product topology. We recall that the product topology has a basis of sets of the form $B(F, G)$ for finite subsets F, G of P where

$$B(F, G) = \{\alpha \mid \alpha(x) = 0 \text{ for } x \in F \text{ and } \alpha(y) = 1 \text{ for } y \in G\}.$$

In fact, 2^P is a Priestley space since it is a product of Priestly spaces.

Theorem 5.9 *Let P be a poset with its Alexandroff topology. Then the Priestley space $(\mathscr{F}(P), \pi, \subseteq)$ of closed subsets of P with the hit-or-miss topology is order-isomorphic and homeomorphic to the closed subspace of the Priestley space 2^P of all order-reversing maps via the mapping $\Phi : \mathscr{F}(P) \to 2^P$ where*

$$\Phi(D)(p) = \begin{cases} 1 & \text{if } p \in D \\ 0 & \text{if } p \notin D \end{cases}$$

Further, for each finite $F, G \subseteq P$ we have $\Phi^{-1}(B(F, G)) = \square_F \cap \triangle_G$.

Proof It is easily seen that Φ is an order-embedding and that each $\Phi(D)$ is an order-reversing map. Conversely, if $\alpha : P \to 2$ is an order-reversing map, then $D = \alpha^{-1}(1)$ is a downset and $\Phi(D) = \alpha$. So Φ is an isomorphism to the lattice of order-reversing maps. For F, G finite subsets of P, the definitions provide directly that the inverse image under Φ of $B(F, G)$ is $\square_F \cap \triangle_G$. Since the $B(F, G)$ and $\square_F \cap \triangle_G$ form bases of the respective spaces, Φ is a homeomorphism to its image. $\qquad\square$

We next give a more insightful description of the distributive lattice of clopen upsets of $(\mathscr{F}(P), \pi, \subseteq)$, but this requires a new notion [15, p. 267].

Definition 5.10 For a poset P, a bounded distributive lattice L is said to be *freely generated by* P if P is a subposet of L, L is generated as a bounded lattice by P, and for any order-preserving map $f : P \to M$ of P into a bounded distributive lattice M there is a (unique) bounded lattice homomorphism $\overline{f} : L \to M$ that extends f.

Care is required when discussing this notion since related definitions with similar names also require all existing finite joins and meets in P to agree with those in F, as is the case with Dilworth's original definition of a lattice freely generated by a poset [4]. The following is a small modification of [15, Lemma 2.4] that accommodates the case of the empty poset and is easier to state.

Lemma 5.11 *For a poset P, a bounded distributive lattice L that contains P as a subposet is freely generated by P iff for any finite subsets F and G of P we have that $\bigwedge F \leq \bigvee G$ implies there exist $x \in F$ and $y \in G$ with $x \leq y$.*

Theorem 5.12 *Let P be a poset and L the distributive lattice of clopen upsets of the Priestley space $(\mathscr{F}(P), \pi, \subseteq)$. Then L is a bounded distributive lattice freely generated by a poset that is dually isomorphic to P.*

Proof For $x \in P$ we have $\Diamond_x$ is a clopen upset of $\mathscr{F}(P)$. Let $P' = \{\Diamond_x \mid x \in P\}$. It is easily seen that $y \leq x$ iff $\Diamond_x \subseteq \Diamond_y$, so P' is a subposet of L that is dually isomorphic to P.

Let F, G be finite subsets of P and set $F' = \{\Diamond_x \mid x \in F\}$ and $G' = \{\Diamond_y \mid y \in G\}$. All finite subsets of P' arise in this way. Suppose that in L we have $\bigwedge F' \leq \bigvee G'$. This means that
$$\bigcap \{\Diamond_x \mid x \in F\} \subseteq \bigcup \{\Diamond_y \mid y \in G\}.$$

Let $D = \mathord{\downarrow} F$. Then $x \in D$ for each $x \in F$, so D belongs to $\bigcap \{\Diamond_x \mid x \in F\}$. Therefore, D belongs to $\bigcup \{\Diamond_y \mid y \in G\}$, so there is $y \in G$ with $D \in \Diamond_y$. This means that $y \in D$, and since $D = \mathord{\downarrow} F$, there is $x \in F$ with $y \leq x$. So there are $x \in F$ and $y \in G$ with $y \leq x$, and hence there are $\Diamond_x \in F'$ and $\Diamond_y \in G'$ with $\Diamond_x \subseteq \Diamond_y$. By Lemma 5.11, L is freely generated by P'. $\qquad\square$

Corollary 5.13 *For a poset P, the Priestly space $(\mathscr{F}(P), \pi, \subseteq)$ of downsets of P with the hit-or-miss topology is the dual space of the free bounded distributive lattice generated by the order-dual of P.*

A *Heyting algebra* is a bounded distributive lattice D such that for each $a, b \in D$ the set $\{x \in D \mid a \wedge x \leq b\}$ has the largest element denoted $a \rightarrow b$. Priestley spaces dual to Heyting algebras were studied by Esakia [5] and are known as Esakia spaces. An *Esakia space* is a Priestley space in which the downset of each clopen is clopen. Applying Lemma 5.8 and Corollary 5.13, we obtain:

Corollary 5.14 *For P a poset, the Priestly space $(\mathscr{F}(P), \pi, \subseteq)$ is an Esakia space. Thus, the free bounded distributive lattice generated by the order-dual of P is a Heyting algebra.*

6 The Fell Compactification of a Poset

Definition 6.1 By the Fell compactification $H(P)$ of a poset P, we mean the Fell compactification of P with its Alexandroff topology.

For a general locally compact T_0-space X, its Fell compactification $H(X)$ is the closure of the image of $e : X \rightarrow \mathscr{F}(X)$ in the hit-or-miss topology of $\mathscr{F}(X)$. We have $e(x) = \downarrow x$ since the closure $\overline{\{x\}}$ is equal to $\downarrow x$ where $\leq$ is the specialization order. For a poset P, the specialization order of its Alexandroff topology is the order of the poset. The following is then immediate.

Proposition 6.2 *The Fell compactification of a poset P is the closure of the set of principal downsets of P in the Priestly space $(\mathscr{F}(P), \pi, \subseteq)$ of downsets of P with the hit-or-miss topology. Thus, the Fell compactification is a Priestley space.*

We next give an explicit description of the Fell compactification as a subspace of the space $\mathscr{F}(P)$ of all downsets of P.

Theorem 6.3 *The Fell compactification of P consists of the set of downsets D of P with the property that for each finite $F \subseteq D$ at least one of the following holds:*

(1) *F has an upper bound in D.*
(2) *The set of upper bounds of F is not compact.*

We call such downsets Fell downsets and let $\mathscr{D}_F(P)$ be the set of Fell downsets of P.

Proof The image of $e : P \rightarrow \mathscr{F}(P)$ is the set of all principal downsets $\downarrow x$. So the Fell compactification is the closure of the set of all principal downsets in the hit-or-miss topology. We show this closure consists exactly of the Fell downsets.

Suppose D is a downset that is not Fell. Then there is a finite subset G of D such that G has no upper bound in D and the set G^u of upper bounds of G is compact. So there is a finite set F such that $G^u = \uparrow F$. Then $\square_F \cap \triangle_G$ is a neighborhood of D. We claim that this neighborhood does not contain any principal downsets. If $\downarrow x \in \triangle_G$, then $G \subseteq \downarrow x$. This implies that x is an upper bound of G, and since $G^u = \uparrow F$, there is some $y \in F$ with $y \leq x$. This shows that $\downarrow x \notin \square_F$.

Conversely, suppose D is a Fell downset. We show it is in the closure of the set of principal downsets. Consider a basic open neighborhood $\Box_F \cap \triangle_G$ of D where F and G are finite subsets of P. Then $F \cap D = \varnothing$ and $G \subseteq D$. Since D is a Fell downset, either G has an upper bound x that belongs to D or the set G^u of upper bounds of G is not compact. In the first case, x being an upper bound of G implies that $\downarrow x$ contains G so belongs to $\triangle_G$; and since $\downarrow x \subseteq D$ and $F \cap D = \varnothing$ we have $\downarrow x \in \Box_F$. In the second case, since G^u is not compact, there is an upper bound x of G with $x \notin \uparrow F$. Since x is an upper bound of G we have $G \subseteq \downarrow x$, giving $\downarrow x \in \triangle_G$; and since $x \notin \uparrow F$ we have $F \cap \downarrow x = \varnothing$, hence $\downarrow x \in \Box_F$. In either case, $\Box_F \cap \triangle_G$ intersects the principal downsets non-trivially, so D belongs to the closure of the principal downsets. $\qquad\square$

For a poset P we have seen in Proposition 6.2 that its Fell compactification $H(P)$ is a Priestly space. We now determine its lattice of clopen upsets.

Definition 6.4 For a poset P, we say U is a *basic Fell upset* if there is a finite $G \subseteq P$ with $U = G^u$ where G^u is the set of upper bounds of G. A *Fell upset* is one that is a finite union of basic Fell upsets. Let $\mathscr{U}_F(P)$ be the set of Fell upsets of P partially ordered by set inclusion.

Note that P is the set of upper bounds of the empty set, so it is a Fell upset. Since the union of the empty set of basic Fell upsets is empty, the empty set is a Fell upset. Since $G_1^u \cap G_2^u = (G_1 \cup G_2)^u$ the collection of basic Fell upsets is closed under finite intersections. Therefore, the collection of Fell upsets is closed under finite unions and finite intersections. So the Fell upsets are a bounded sublattice of the lattice of upsets of P.

Theorem 6.5 *For a poset P, the distributive lattice of clopen upsets of its Fell compactification is isomorphic to the lattice $\mathscr{U}_F(P)$ of Fell upsets of P.*

Proof By Proposition 5.7, the clopen upsets of $(\mathscr{F}(P), \pi, \subseteq)$ are the finite unions of sets of the form $\triangle_G$ where G is a finite subset of P. So by Lemma 4.4, the clopen upsets of the Fell compactification $H(P)$ are finite unions of sets $\triangle_G \cap H(P)$. Define a map Γ from the lattice of Fell upsets $\mathscr{U}_F(P)$ to the lattice of clopen upsets of the Fell compactification $H(P)$ by setting

$$\Gamma(G_1^u \cup \cdots \cup G_n^u) = (\triangle_{G_1} \cup \cdots \cup \triangle_{G_n}) \cap H(P).$$

To show Γ is well defined, we first note that a principal downset $\downarrow x$ belongs to $\triangle_{G_1} \cup \cdots \cup \triangle_{G_n}$ iff x belongs to $G_1^u \cup \cdots \cup G_n^u$. If $(\triangle_{G_1} \cup \cdots \cup \triangle_{G_n}) \cap H(P)$ is not equal to $(\triangle_{F_1} \cup \cdots \cup \triangle_{F_m}) \cap H(P)$, then since these sets are clopen, there is a nonempty clopen set $\mathscr{U}$ contained in one and disjoint from the other. Since the principal downsets are by definition dense in $H(P)$, the clopen $\mathscr{U}$ contains a principal downset belonging to one of these sets and not the other. This shows $G_1^u \cup \cdots \cup G_n^u$ is not equal to $F_1^u \cup \cdots \cup F_m^u$. So Γ is well defined.

Let $S = G_1^u \cup \cdots \cup G_n^u$ and $T = F_1^u \cup \cdots \cup F_m^u$. If $S \subseteq T$, then $T = S \cup T$, and $\Gamma(T) = \Gamma(S \cup T)$. It follows from the definition of Γ that $\Gamma(S) \subseteq \Gamma(T)$. So Γ

is order preserving. For the converse, suppose $S \not\subseteq T$, so there is $x \in S \setminus T$. Then there is $i \leq n$ with $x \in G_i^u$ and $x \notin F_1^u, \ldots, F_m^u$. Therefore, $\downarrow x \in \triangle_{G_i}$ and $\downarrow x \notin \triangle_{F_1} \cup \cdots \cup \triangle_{F_m}$. Thus, $\Gamma(S) \not\subseteq \Gamma(T)$. So Γ is an order-embedding. By Proposition 5.7 and Lemma 4.4, Γ maps onto the lattice of clopen upsets of $H(P)$, and hence is an isomorphism. $\qquad\square$

Corollary 6.6 *The Fell compactification of a poset P is the Priestley space of the distributive lattice of Fell upsets of P.*

Remark 6.7 In contrast to Corollary 5.14, the distributive lattice of clopen upsets of the Fell compactification of a poset need not be a Heyting algebra; see Corollary 7.3. It is however a bounded sublattice of the Heyting algebra of the clopen upsets of $\mathscr{F}(P)$.

We recall that the Fell compactification can be thought of as a stable compactification of (P, τ_A) (Theorem 3.8) or as an order-compactification of $(P, \sigma, \leq)$ where σ is the smallest topology on P making e continuous (Theorem 3.9). Indeed, this is even a Priestley order-compactification of $(P, \sigma, \leq)$ (Proposition 6.2). The topology σ arises in a natural intrinsic way.

Proposition 6.8 *For P a poset, σ is the smallest topology containing τ_A and with the property that $(P, \sigma, \leq)$ has an order-compactification.*

Proof By Theorem 3.9, $\tau_A \subseteq \sigma$ and $(P, \sigma, \leq)$ has an order-compactification. Suppose Δ is another topology on P that contains τ_A and is such that $(P, \Delta, \leq)$ has an order-compactification. Then $(P, \Delta, \leq)$ is homeomorphic and order-isomorphic to a subspace of a Nachbin space, so each principal upset $\uparrow x$ is closed in Δ.

The topology σ is generated by τ_A and the collection of sets of the form $e^{-1}(\square_K)$ where K is compact in τ_A. As we have seen in the proof of Proposition 5.4, for a compact set K there is a finite set $F = \{x_1, \ldots, x_n\}$ with $\square_K = \square_{x_1} \cap \cdots \cap \square_{x_n}$. Thus,

$$e^{-1}(\square_K) = e^{-1}(\square_{x_1}) \cap \cdots \cap e^{-1}(\square_{x_n}).$$

For $x \in P$ we have $e^{-1}(\square_x) = \{y \mid x \notin \downarrow y\} = \{y \mid y \notin \uparrow x\}$. Therefore, $e^{-1}(\square_x)$ is the complement of the principal upset $\uparrow x$. Since principal upsets are closed in Δ, each $e^{-1}(\square_x)$ is open in Δ. So for K compact, we have $e^{-1}(\square_K)$ is open in Δ. Thus, $\sigma \subseteq \Delta$. $\qquad\square$

Remark 6.9 It is a classical result that a topological space has a compactification iff it is completely regular. There is a corresponding result for ordered spaces. An ordered topological space $(P, \pi, \leq)$ has an order-compactification iff it is completely order-regular. The definition of completely order-regular involves the existence of certain order-preserving and order-reversing continuous maps to the unit interval to separate points and closed sets. Thus, σ is the smallest completely order-regular topology on P containing the Alexandroff topology.

Fig. 1 The Fell
compactification of an
antichain

7 Examples

In this final section we consider some simple examples. We begin with the case
when P is an antichain, meaning that distinct $x, y \in P$ are incomparable. In this
case the Alexandroff topology is discrete. The following is well known and is seen
as a consequence of Theorem 5.9.

Proposition 7.1 *If P is an antichain, then the set of closed sets $\mathscr{F}(P)$ is the powerset
$\mathscr{P}(P)$. The bijection between $\mathscr{P}(P)$ and 2^P then yields that $(\mathscr{F}(P), \pi, \subseteq)$ is homeomorphic and order-isomorphic to 2^P with the product topology and order.*

We next describe the Fell compactification of the antichain P. This is the closure in
$\mathscr{P}(P)$ of the set $S = \{\{p\} \mid p \in P\}$ of singletons of P. Since the Alexandroff topology of P is discrete and hence Hausdorff, by [6, p. 475], the Fell compactification
is the one-point compactification of P.

Proposition 7.2 *For an antichain P, the Fell compactification $(H(P), \pi, \subseteq)$ is the
Priestley space in Fig. 1 where the topology is the one-point compactification of the
discrete space S.*

Proof If $D \subseteq P$ contains distinct elements y_1, y_2, then for $G = \{y_1, y_2\}$, we have
$D \in \triangle_G$ and $\triangle_G$ is disjoint from S. Therefore, $D \notin H(P)$. Any open neighborhood
of $\varnothing$ contains one of the form $\square_F$ for $F \subseteq P$ finite, hence intersects S nontrivially.
So $\varnothing \in H(P)$.   □

Corollary 7.3 *The Fell compactification of a poset need not be an Esakia space.*

Proof It is well known that the Priestley space of Fig. 1 is not an Esakia space since
each singleton $\{p\}$ is clopen, but $\downarrow\{p\} = \{\{p\}, \varnothing\}$ is not open.   □

We next consider the case when P is a chain.

Proposition 7.4 *For a chain P its Fell compactification $H(P)$ is an Esakia space
that consists of all downsets of P if P has no least element, and of all nonempty
downsets of P if P has a least element. Its distributive lattice of clopen upsets is
dually isomorphic to the chain P with a new top added, and with a bottom added
if P has no bottom.*

Proof The description of the Fell compactification $H(P)$ of a chain P follows
directly from Theorem 6.3: Every nonempty downset D is a Fell downset since

every finite subset of D has an upper bound in D, and the empty set is a Fell downset iff its set of upper bounds is not compact, which means that P has no least element.

The description of the lattice of clopen upsets of $\mathscr{F}(P)$ follows from the description of Fell upsets in Theorem 6.5: The basic Fell upsets are ones of the form G^u where G is a finite subset of P. If G is nonempty, then G^u is a principle upset of P, and if G is empty then G^u is P. Finite unions of basic Fell upsets are the empty set, and the basic Fell upsets. So the lattice $\mathscr{U}_F(P)$ of Fell upsets consists of the principal upsets, $\varnothing$, and P. Noting that P is a principal upset iff P has a bottom, the description of the clopen upsets of $H(P)$ follows.

Since these clopen upsets form a chain, and hence a Heyting algebra, $H(P)$ is an Esakia space.   □

For a chain P and $x \in P$, the *strict principal downset* of x is $\Downarrow x = \{p \mid p < x\}$. The *half-open topology* of P is that generated by the collection of all principal upsets and all strict principal downsets. The name arises as $\uparrow x \cap \Downarrow y$ is the *half-open interval*

$$[x, y) = \{p \mid x \le p < y\}.$$

The real numbers with the half-open topology is also known as the Sorgenfrey line.

Proposition 7.5 *Let P be a chain. The topology σ of Theorem 3.9 is its half-open topology, and $e : (P, \sigma, \le) \to (H(P), \pi, \subseteq)$ is the smallest order-compactification of the chain P with the half-open topology.*

Proof The topology σ has as a subbasis all $e^{-1}(\Box_x)$ and $e^{-1}(\Diamond_y)$ where $x, y \in P$. These are the sets $\Downarrow x$ and $\uparrow y$ respectively. So $(P, \sigma, \le)$ is the chain P with its half-open topology. That e is the least order-compactification follows from the description of all order-compactifications of an ordered topological space whose underlying poset is a chain given in [2, Theorem 2.13].   □

Acknowledgements We would like to thank Vladik Kreinovich for the opportunity to contribute to this volume.

References

1. G. Bezhanishvili, J. Harding, Stable compactifications of frames. Cah. Topol. Géom. Différ. Catég. **55**(1), 37–65 (2014)
2. G. Bezhanishvili, P.J. Morandi, Order-compactifications of totally ordered spaces: revisited. Order **28**(3), 577–592 (2011)
3. G. Bezhanishvili, P.J. Morandi, Priestley rings and Priestley order-compactifications. Order **28**(3), 399–413 (2011)
4. R.P. Dilworth, Lattices with unique complements. Trans. Amer. Math. Soc. **57**, 123–154 (1945)
5. L. Esakia, Topological Kripke models. Soviet Math. Dokl. **15**, 147–151 (1974)
6. J.M.G. Fell, A Hausdorff topology for the closed subsets of a locally compact non-Hausdorff space. Proc. Amer. Math. Soc. **13**, 472–476 (1962)

7. G. Gierz, K.H. Hofmann, K. Keimel, J.D. Lawson, M. Mislove, D.S. Scott, *Continuous Lattices and Domains* (Cambridge University Press, Cambridge, 2003)
8. M. Hochster, Prime ideal structure in commutative rings. Trans. Amer. Math. Soc. **142**, 43–60 (1969)
9. R.E. Hoffmann, The Fell compactification revisited, in *Continuous Lattices and Their Applications* (Bremen, 1982), Lecture Notes in Pure and Applied Mathematics, vol. 101 (Dekker, New York, 1985), pp. 57–116
10. P.T. Johnstone, *Stone Spaces* (Cambridge University Press, Cambridge, 1982)
11. L. Nachbin, *Topology and Order* (D. Van Nostrand Co., Inc., Princeton, N.J.-Toronto, Ont.-London, 1965)
12. H.T. Nguyen, A continuous lattice approach to random sets. Thai J. Math. **5**(1), 137–142 (2007)
13. H.A. Priestley, Representation of distributive lattices by means of ordered Stone spaces. Bull. Lond. Math. Soc. **2**, 186–190 (1970)
14. M.B. Smyth, Stable compactification. I. J. Lond. Math. Soc. **45**(2), 321–340 (1992)
15. C. Tsinakis, Free objects and free extensions in the category of frames. Math. Slovaca **65**(2), 265–272 (2015)
16. Y. Venema, J. Vosmaer, Modal logic and the Vietoris functor, in *Leo Esakia on Duality in Modal and Intuitionistic Logics*. Outstanding Contributions to Logic, vol. 4 (Springer, Dordrecht, 2014), pp. 119–153

Is Time Fuzzy?

Bernadette Bouchon-Meunier

Abstract Imprecision of time measurements, subjective perception of time, flexible management of time, are examples of reasons to use a fuzzy modeling of time. Although all fuzzy set-based knowledge representations can be applied to time, its particular nature leads to specific treatments. We give examples of fuzzy methods to deal with time, in temporal reasoning, linguistic summarization of data, forecasting and scoring and also in spatio-temporal reasoning.

Keywords Time · Fuzzy set · Possibility theory · Temporal reasoning · Linguistic summarization · Time series · Forecasting

1 Introduction

Is time fuzzy? The answer depends on the point of view. In his famous painting "The Persistence of Memory" shown at the Museum of Modern Art (MoMA) in New York City, Salvador Dalí represents melting watches which can be regarded as the symbol of the fuzziness of time.

Time is not unique and it is necessary to distinguish between the absolute time of the universe and its measurement, representation, perception, utilisation by human beings.

The absolute time is usually evaluated in a very precise way. The most recent definition of a second is "exactly 9,192,631,770 times the period of the radiation corresponding to the transition between the two hyperfine levels of the ground state of the caesium-133 atom" [1] and there is absolutely no fuzziness in it. When we measure the performance of an athlete, we can count in fractions of seconds and a 110 m hurdles champion can win with a two thousandths of a second difference after a race of thirteen seconds. Would not we call this preciseness?

B. Bouchon-Meunier (✉)
CNRS, Laboratoire d'Informatique de Paris 6, LIP6, Sorbonne Université, 75005 Paris, France
e-mail: Bernadette.Bouchon-Meunier@lip6.fr

© The Editor(s) (if applicable) and The Author(s), under exclusive license
to Springer Nature Switzerland AG 2021
V. Kreinovich (ed.), *Statistical and Fuzzy Approaches to Data Processing, with Applications to Econometrics and Other Areas*, Studies in Computational Intelligence 892,
https://doi.org/10.1007/978-3-030-45619-1_4

But if the absolute time is supposed to be the time of the universe, we can quote Stephen Hawking [2] saying that "the universe had a beginning about 15 billion years ago". There is no possible preciseness in such estimation. So even the absolute time underlying the universe can be regarded as precise or fuzzy, because we cannot handle it without using a form of representation and measurement, and we are submitted to an uncertainty due to the complexity of the universe.

As opposed to the absolute time of the universe, the human time is the "local" time, the time in which human beings are embedded and it is obviously depending on the individual, the environment, the culture. We can distinguish the objective time, running our life and activities, from the subjective time depending of our perception. To catch a plane at 3:35 pm is very precise and should not be considered as fuzzy if we want to avoid any problem. Nevertheless, we all know that the plane will not necessarily leave at 3:35 pm and the actual time of departure is rather fuzzy.

If you are asked "*Can you do this task for tomorrow?*", you may answer "Impossible, I am really too busy" and, after a second question of your interlocutor such as "*Are you sure? you would help me a lot in doing this*", you may decide to do the task, thinking that you will shrink some other tasks because the management of time is flexible. Your perception of time is fuzzy in this case. If your beloved friend will come to visit you tomorrow, you think that waiting a full day is very long; but if you have a difficult project to complete by tomorrow, you will feel that a full day is very short. Cayol [3] explains that the perception of time depends on the culture and she shows that Chinese people see the time as continuous, and represent it as a wave, while westerners perceive time as a sequence of moments and events.

It is necessary to distinguish several aspects of the imprecision of time. The first one is due to the tools used to measure time. If you use a sundial, a mechanical watch or an electronic clock to know the duration of an event, you can obtain different numbers and the precision of this duration will not be the same. The second aspect of the imprecision of time depends on the linguistic expressions used to describe time and duration. "*He is 24 years old*" clearly corresponds to an approximation of the age of this person, not to a precise period of time. The third aspect deals with the management of time. "*I need two days to do it*" does not mean 48 h and 0 min in common language, it must be interpreted in a more imprecise way as an estimation of the duration necessary to complete the task, both imprecise and subject to an uncertainty. The fourth aspect of the imprecision of time is due to the lack of knowledge or the uncertainty on a date or a length of time, as in the example of the origin of the universe or the date of an expected birth.

In this paper, we present various solutions to use fuzzy set-based knowledge representation to cope with the imprecision and uncertainty of "local" time, the time underlying our lives, our projects, our tasks.

2 Temporal Reasoning

Time is among the main concepts represented by fuzzy sets. Temporal reasoning was one of the first directions where a fuzzy representation of time was taken into account, in order to process flexible scheduling, planning and process management.

Uncertainty and imprecision in temporal knowledge has been represented by possibility distribution [4] or fuzzy sets [5]. Extensions of Allen's temporal interval algebra [6] enable the user to introduce more flexibility [7] in the management of dates and length of time. A fuzzy set-based representation of time has been applied to the management of flexible constraints and the construction of fuzzy temporal constraint networks [8], with applications in job shop scheduling [9] or medical environments [10]. Possibility theory has been used to incorporate preferences and to manage uncertainties on the processing time in scheduling [11].

Fuzzy temporal reasoning expert systems have also been used for the development of intelligent monitoring systems [12, 13]. A specific form of propositions dedicated to temporal information has been proposed [14] and managed by means of a grammar enabling experts to express their knowledge in a natural way (e.g. *"Temperature has been very low for a few seconds"*), with the help of a time ontology.

3 Linguistic Summarization of Data

Time series are very common in databases and big data repositories. In data science, forecasting from big data has given rise to various kinds of methods, mainly based on machine learning and statistics. Nevertheless, the interest of a soft analysis of data appears in the interpretability of time series and in the research of a natural language-like description of specific aspects of the considered time series, for instance periodicity, trends or period of validity of a characteristic behaviour. Linguistic summaries of numerical data have been introduced in a more general framework [15, 16], but they have often been used for the analysis of data varying over time. The most classic form of description is based on protoforms such as "Q x are A" or Q B x are A", where Q is a quantifier such as "most" or "few", x the elements to describe, A and B fuzzy modalities respectively called summarizer and qualifier. In the case of temporal data [17], quantifiers are replaced by temporal adverbs, for instance "often", roughly associated with the quantifier "most", "seldom" with the quantifier "few". Protoforms can take specific forms based on temporal attributes such as the trend or the duration. A linguistic summary can also use a temporal indication such as "regularly" or "from time to time".

The evaluation of protoforms takes into account many criteria generally considered to evaluate fuzzy modelling. At the level of the protoforms, their accuracy, compactness, completeness, consistency, transparency can be considered. At the level of the fuzzy modalities used in the protoforms, coverage, normality, distinguish ability are relevant [18]. Since the summarization of temporal data can take various forms going beyond classic protoforms, the interpretability of linguistic summaries can be more sophisticated and depends on the form of linguistic summary [17]. Genetic algorithms prove to be efficient [19] to perform a multi-criterion optimisation coping with several criteria involved in linguistic summarization evaluation, namely accuracy, brevity and coverage.

Among the various works on the linguistic summarization of time series, we must highlight the fact that trend attributes have been taken into account with descriptions such as *"Most slowly decreasing trends are of a very low variability"* [20], giving rise to many developments [21, 21] and applications in finance engineering, exploring various aggregation methods such as Sugeno integral [22], Choquet integral [23] or OWA operators [24].

Comparing time series leads to very different forms of descriptions. The similarity between time series can help to identify local changes [25] (e.g. *"Most days of year 2001, both series exhibit a local change with the same sign"*). The comparison of time series can also point out a temporal contextualisation in summaries of the form "Qy's are P Q_t times", where Q_t is a time quantifier, for instance *"Many patients have high blood pressure most of the time"* [26]. It can highlight differential behaviours such as *"few male patients have a low value of heart rate half of the time, while female patients do not"* [26].

The periodicity of time series is one of their main features. It corresponds to the identification of regularly spaced high and low value groups of approximately constant size in time series. The definition itself being fuzzy, using fuzzy modelling looks appropriate, even though there are many classic methods to deal with periodicity. An automatic adaptive method based on mathematical morphology has been proposed to identify high value and low value groups, requiring no parameters [27]. It is completed by the evaluation of a periodicity degree and an approximate period in the alternation of high and low value groups in the method called "Detection of Periodic Events" [28]. The capacity of detecting locally periodic events, the variety of linguistic rendering of the information extracted from the considered time series and the scalability of the method are among its most interesting features [29]. Examples of summaries are *"Approximately every day, the amount of CO_2 is high"* or *"Approximately from the second third to the end, the series is highly periodic (0.83) with a period of approximately 1 day"*.

4 Forecasting, Prediction and Scoring

Forecasting is obviously taking time into account and many approaches can be considered in a fuzzy modelling framework. Fuzzy time series forecasting has long been studied [30] and various aspects of time can be taken into account, for instance seasonality [31, 32].

Other methods use scenarios and templates to describe generic situations to which observations are compared by means of similarities in order to make predictions. It is the case for instance of a dynamic early warning system for ethno-political risks

[33] in which various attributes involved in the description of the world, including imprecise temporal constraints such as dates, duration or frequency, are represented by fuzzy sets in order to assess a degree of crisis risk to a given part of the world.

Another environment in which time is strongly involved is the evaluation of digital information quality. In addition to the trustworthiness attached to a source of information, such as a local website or an international news channel, criteria attached to the informational context must be taken into account in the assessment of the veracity of a piece of information on a given event [34]. Among such criteria, the compatibility between pieces of information provided by several sources must be evaluated in considering an acceptable range of imprecision, in particular regarding dates and duration. Moreover, it is important to consider the evolution of a situation over time. Differences between descriptions of a given attribute provided by several sources may be due to the moments of the descriptions and a fuzzy knowledge representation can help to manage such differences. Finally, the evolution of the situation can make some pieces of information obsolete and a temporal weighting [34] can be associated with each piece of information, decreasing over time until the piece of information is no more taken into account, becoming too old with respect to more recent ones related to the same event.

5 Spatio-Temporal Modelling

Time is often related to space, from the visualisation of geological periods by geological layers, to the association between location and time in the analysis of movements. Such relations are often represented in a fuzzy framework to take into account their imprecision. Examples of applications of fuzzy spatio-temporal relations can be found in human facial expression recognition [35], in geographical information systems [36], in surveillance systems to provide a qualitative interpretation of the behaviour of a person or a vehicle [37], or a military activity in a given environment [38], in hand-gesture recognition [39].

Fuzzy ontologies have been exhibited to manage spatio-temporal occurrences, for instance specific ontologies for the identification of human behaviour [40] or the fuzzy ontology reasoner *fuzzyDL* [41] on a more general level.

6 Conclusion

The purpose of this paper is to show that, even though the fuzzy nature of time can be discussed, its management benefits from a fuzzy modelling under various forms. Fuzzy temporal reasoning attempts to manage time with a flexibility imitating the natural ability of human beings to manage it. Linguistic summarization of data transforms complex time series into natural language-like descriptions easily understandable by a user. Forecasting, prediction and scoring are among the tasks where a crisp management of time may not be meaningful because of imprecision due to measurement and approximations expressed by observers. Moreover, time underlies data collection and the evolution of situations. It must be taken into account even when it is not one of the features considered to describe a situation or a state of the world. Other such tasks could have been considered, for instance natural language understanding, question answering systems, video summarization, to cite but a few. Fuzzy spatio-temporal modelling is necessary in all tasks where time and space are intertwined in a complex system, in particular in those analysing movements, dynamic flows of information. We could have mentioned many other domains where time is significant and fuzzy methods bring solutions, such as fuzzy controllers or evolving fuzzy systems, although time is generally not considered as fuzzy. This paper does not pretend to give a general overview of all fuzzy approaches of systems based on time, which covers most of the real-world problems, but to show that time benefits of fuzzy set-based representations even though time seems precisely comprehended.

References

1. https://physics.nist.gov/cuu/Units/second.html
2. http://www.hawking.org.uk/the-beginning-of-time.html
3. C. Cayol, Pourquoi les Chinois ont-ils le temps? Tallandier (2017)
4. D. Dubois, H. Prade, Processing fuzzy temporal knowledge. IEEE Trans. Syst. Man Cybern. **19**(4), 729–743 (1989)
5. S. Badaloni, M. Giacomin, The algebra *IAfuz*: a framework for qualitative fuzzy temporal reasoning. Artif. Intell. **170**(10), 872–908 (2006)
6. J. Allen, Maintaining knowledge about temporal intervals. Commun. ACM **26**(11), 832–843 (1983)
7. S. Schockaert, M. De Cock, Temporal reasoning about fuzzy intervals. Artif. Intell. **172**(8–9), 1158–1193 (2008)
8. L. Vila, L. Godo, On fuzzy temporal constraint networks. Mathware Soft Comput. **1–3**, 315–334 (1994)
9. D. Dubois, H. Fargier, H. Prade, Fuzzy constraints in job-shop scheduling. J Intell. Manuf. **6**, 215–235 (1995)
10. M.A. Cardenas Viedma, R. Main Morales, I. Navarrete Sanchez, Fuzzy temporal constraint logic: a valid resolution principle. Fuzzy Sets Syst. **117**(2), 231–250 (2001)
11. D. Dubois, H. Fargier, P. Fortemps, Fuzzy scheduling: modelling flexible constraints vs. coping with incomplete knowledge. Eur. J. Oper. Res. 147(2), 231–252 (2003)
12. S. Barro, R. Marín, J. Mira, A. Patón, A model and a language for the fuzzy representation and handling of time. Fuzzy Sets Syst. **61**(2), 153–175 (1994)
13. R. Marín, S. Barro, F. Palacios, R. Ruiz, F. Martin, An approach to fuzzy temporal reasoning in medicine. Mathware Soft Comput. **3**, 265–276 (1994)
14. P. Cariñena, A. Bugarín, M. Mucientes, S. Barro, A language for expressing fuzzy temporal rules. Mathware Soft Comput. **VII**(2–3), 22 (2000)
15. R. Yager, A new approach to the summarization of data. Inf. Sci. **28**(1), 69–86 (1982)
16. J. Kacprzyk, R. Yager, "Softer" optimization and control models via fuzzy linguistic quantifiers. Inf. Sci. **34**(2), 157–178 (1984)
17. M.-J. Lesot, G. Moyse, B. Bouchon-Meunier, Interpretability of fuzzy linguistic summaries. Fuzzy Sets Syst. **292**, 307–317 (2016)

18. J. Casillas, O. Cordón, F. Herrera, L. Magdalena, in *Interpretability Improvements to Find the Balance Interpretability-Accuracy in Fuzzy Modeling: An Overview*, ed. by J. Casillas, O. Cordón, F. Herrera, L. Magdalena. Interpretability Issues in Fuzzy Modeling. Studies in Fuzziness and Soft Computing, vol. 128 (Springer, Berlin, 2003), pp. 3–22
19. R. Castillo-Ortega, N. Marín, D. Sánchez, A. Tettamanzi, in *Linguistic Summarization of Time Series Data using Genetic Algorithms, EUSFLAT 2011.145* (2011)
20. J. Kacprzyk, A. Wilbik, S. Zadrozny, Linguistic summarization of time series using a fuzzy quantifier driven aggregation. Fuzzy Sets Syst. **159**(12), 1485–1499 (2008)
21. J. Kacprzyk, A. Wilbik, S. Zadrożny, Linguistic summarization of trends: a fuzzy logic based approach, in *Proceedings of the 11th International Conference Information Processing and Management of Uncertainty in Knowledge-based Systems* (IPMU 2006, Paris, France, 2006), pp. 2166–2172
22. J. Kacprzyk, A. Wilbik, S. Zadrożny, Linguistic summaries of time series via a quantifier based aggregation using the Sugeno integral, in *Proceedings of 2006 IEEE World Congress on Computational Intelligence* (WCCI 2006, Vancouver, BC, Canada, 2006), pp. 3610–3616
23. J. Kacprzyk, A. Wilbik, S. Zadrożny, Linguistic summarization of time series by using the Choquet integral, in *IFSA 2007*, vol. 4529, LNCS (LNAI), ed. by P. Melin, O. Castillo, L.T. Aguilar, J. Kacprzyk, W. Pedrycz (Springer, Heidelberg, 2007), pp. 284–294
24. J. Kacprzyk, A. Wilbik, S. Zadrożny, Linguistic summaries of time series via an OWA operator based aggregation of partial trends, in *Proceedings of the IEEE International Conference on Fuzzy Systems* (FUZZ-IEEE 2007, London, UK; IEEE Press, Los Alamitos, 2007), pp. 467–472
25. R. Castillo-Ortega, N. Mann, D. Sánchez, Linguistic local change comparison of time series, in *2011 IEEE International Conference on Fuzzy Systems (FUZZ-IEEE 2011)*, (Taipei, 2011), pp. 2909–2915
26. R. Almeida, M.-J. Lesot, B. Bouchon-Meunier, U. Kaymak, G. Moyse, Linguistic summaries of categorical time series for septic shock patient data, in *Fuzz-Ieee 2013—IEEE International Conference on Fuzzy Systems*, Hyderabad, India (2013), pp. 1–8
27. G. Moyse, M.-J. Lesot, B. Bouchon-Meunier, Linguistic summaries for periodicity detection based on mathematical morphology, in *Proceedings of IEEE SSCI FOCI'13* (2013), pp. 106–113
28. G. Moyse, M.-J. Lesot, Linguistic summaries of locally periodic time series. Fuzzy Sets Syst. **285**, 94–117 (2016)
29. G. Moyse, M.J. Lesot, Fast and incremental erosion score computation, in *IPMU 2014 - International Conference on Information Processing and Management of Uncertainty in Knowledge-Based Systems, Communications in Computer and Information Science, Montpellier*, vol. 442, France (Springer, Berlin, 2014), pp. 376–385
30. Q. Song, B. Chissom, Fuzzy time series and its model. Fuzzy Sets Syst. **54**, 269–277 (1993)
31. E. Bulut, Modeling seasonality using the fuzzy integrated logical forecasting (FILF) approach. Expert Syst. Appl. **41**(4), 1806–1812 (2014)
32. V. Novák, M. Štěpnička, A. Dvořák, I. Perfilieva, V. Pavliska, L. Vavříčková, Analysis of seasonal time series using fuzzy approach. Int. J. Gen Syst. **39**(3), 305–328 (2010)
33. T. Delavallade, L. Mouillet, B. Bouchon-Meunier, E. Collain, Monitoring event flows and modelling scenarios for crisis prediction, application to ethnic conflicts forecasting. Int. J. Uncertainty Fuzziness Knowl. Based Syst. **15**(S1), 83–110 (2007)
34. M.-J. Lesot, T. Delavallade, F. Pichon, H. Akdag, B. Bouchon-Meunier, P. Capet, Proposition of a semi-automatic possibility information scoring process, in *Proceedings of the 7th Conference of the European Society for Fuzzy Logic and Technology (EUSFLAT-2011) and LFA-2011*, Aix-les-Bains, France (Atlantis Press, 2011), pp. 949–956
35. T. Xiang, M.K.H. Leung, S.Y. Cho, Expression recognition using fuzzy spatio-temporal modeling. Pattern Recogn. **41**(1), 204–216 (2008)
36. A. Sözer, A. Yazıcı, H. Oğuztüzün, O. Taş, Modeling and querying fuzzy spatiotemporal databases. Inf. Sci. **178**(19), 3665–3682 (2008)
37. J.M. Le Yaouanc, J.P. Poli, in *A Fuzzy Spatio-temporal-Based Approach for Activity Recognition*, ed. by S. Castano, P. Vassiliadis, L.V. Lakshmanan, M.L. Lee. Advances in Conceptual Modeling. ER 2012. Lecture Notes in Computer Science, vol. 7518 (Springer, Berlin, 2012)

38. V. Eude, Modélisation spatio-temporelle floue pour la reconnaissance d'activités militaires, Doctoral dissertation. Université Paris **6**, 1998 (1998)
39. M.-C. Su, A fuzzy rule-based approach to spatio-temporal hand gesture recognition. IEEE Trans. Syst. Man, Cybern. Part C (Appl. Rev.) **30**(2), 276–281 (2000)
40. N. Díaz Rodríguez, M. Cuéllar. J. Lilius, M. Delgado Calvo-Flores, A fuzzy ontology for semantic modelling and recognition of human behaviour. Knowl. Based Syst. (2014)
41. F. Bobillo, U. Straccia, The fuzzy ontology reasoner *fuzzyDL*. Knowl. Based Syst. **95**, 12–34 (2016)

Hung Jury: The Verdict on Uncertainty

William M. Briggs

Abstract Classical probability and the statistical methods built around it, like hypothesis testing, have been shown to have many glaring weaknesses, as the work of Hung Nguyen has shown with clarity and vigor. It is time for a major renovation in probability. The need for new methods is pressing. Older ways of thinking about probability and decision are inadequate, as two examples will show, one from jury trials, and one about hypothesis testing and the so-called problem of old evidence. In particular, hypothesis testing needs to be abandoned forthwith. The Hung jury is in, and the verdict about p-values is Guilty. Time for them to go.

Keywords Fuzziness · Hypothesis testing · P-values · Uncertainty

1 A Bang up Time

Seven months before Lee Harvey Oswald became famous for his encounter with President Kennedy, it is claimed he popped off his Mannlicher-Carcano rifle at the head of one Major General Edwin Walker, at Walker's residence. Oswald's political ties might have been the motive. According to *Smithsonian Magazine* [1], "Walker was a stark anti-communist voice and an increasingly strident critic of the Kennedy's, whose strong political stances had him pushed out of the army in 1961."

This incident was cited in an early work of Hung's, with Irwin Goodman, *Uncertainty Models for Knowledge-Based Systems; A Unified Approach to the Measurement of Uncertainty*, [2]. The example given in this book is just as relevant today as it was then to the understanding of uncertainty.

> In deciding the culpability of Oswald in the assassination attempt upon General Walker, an expert ballistics analysis group indicated "could have come, and even perhaps a little stronger, to say that it probably came from this ... (gun)", while the FBI investigating team, as a matter of policy, avoiding the category of "probable" identification, refused to come

W. M. Briggs (✉)
New York, USA
e-mail: matt@wmbriggs.com

© The Editor(s) (if applicable) and The Author(s), under exclusive license
to Springer Nature Switzerland AG 2021
V. Kreinovich (ed.), *Statistical and Fuzzy Approaches to Data Processing, with Applications to Econometrics and Other Areas*, Studies in Computational Intelligence 892,
https://doi.org/10.1007/978-3-030-45619-1_5

to a conclusion [256]. Other corroborative evidence included a written note, also requiring an expert verification of authenticity, and verbal testimony of witnesses. Based upon this combination of evidence, the Warren Commission concluded that the suspect was guilty.

To conclude a suspect's guilt is to make a decision. Decisions are based on probabilities. And probabilities are calculated with respect to the evidence, and only the evidence, deemed probative to the decision. Picking which evidence is considered probative is itself often a matter of a decision, one perhaps external to the situation, as when a judge in a trial deems certain evidence admissible or inadmissible.

It should be clear that each of these steps is logically independent of each other, even if there are practical overlaps, as with a judge ruling a piece of relevant evidence inadmissible because of a technicality. It is also obvious that the standard classical methods used to form probabilities and make decisions are inadequate to this sequence. Yet these kinds of situations and decisions are extremely common and form the bulk of reasoning people use to go about their daily business. Everything from deciding whether to invest—in *anything*, from a stock to a new umbrella—to making inferences about people's behavior based on common interactions, to guessing which team will win to jurors deciding questions of guilt.

For instance, there is no way to shoehorn p-values, the classical way of simultaneously forming a probability and making a one-size-fits-all decision, into "economic" decisions of the kind found in assessing guilt or innocence, [3]. P-values first assess the probability of an event not of interest, then conflates that probability with an event which is of interest, then they make a decision designed to fit all situations, regardless of the consequences. This will be made clearer in the examples below. It does not make sense to use p-values when other measures designed to do the exact job asked of them are available and superior in every way. Hung has been one of the major forces pushing p-values into failed bin of history, e.g. [4–7].

P-values rely on standard frequentist probability. It's becoming more obvious ordinary probability in frequency theory is inadequate for many, or even most, real-life decisions, especially economic decisions based on the outmoded idea of "rational actors". In order to use frequentist theory, an "event" has to be embedded, or embeddable, in a unique infinite sequence. Probability in frequentist theory is defined as limits of subsequences in infinite sequences, cf. [8]. No infinite sequences, no probability. In what sequence do we embed the General Walker shooting to form a probability of the proposition or event "Oswald took the shot"? All men who took shots at generals? All white men? All communist men? All men who took shots at officers of any rank? All men who took shots at other men of any kind? At women too? All those who used rifles and not guns? Bows and arrows, too? Only in America? Any country? Only at night? Only in Spring? Since a certain date?

To make frequentist probability work an *infinite* sequence is required; a merely long one won't do. In some physical cases, it might make sense to speak of "very long" sequences, but for many events important to people, it does not. Unique or finite events are ruled out by fiat in frequentist theory, e.g. [9]. And even when events are tacitly embedded in sequences, where little thought is given to the precise character of that sequence, frequentist probability can fail. The well known example of context effects produced by question order in surveys reveals commutativity estimates to fail, e.g. [10].

Hung has been at the forefront of quantum probability as a replacement to ordinary frequentist probability [11–13], especially when applied to human events such as economic actions. This isn't the place to review quantum probability, but I do hope to show through two small examples the inadequacy of classical probability to certain human events. And no event is more human than a trial by jury. Forming probabilities of guilt or innocence in individual trials, and then making decisions whether to judge guilt or innocence, are acts entirely unfit to analysis by ordinary statistical methods. Especially in the face of constantly shifting evidence, unquantifiable complexities, and ambiguity of language, where "fuzzy" notions of terms are had by jury members, another area in which Hung has made fundamental contributions, e.g. [14, 15].

2 New Evidence

Consider an example, similar to the Oswald scenario, provided by Laudan [16, 17], a philosopher who writes on jury trials. He investigates the topic of the traditional Western instructions to jurors that the jurors must start with the belief in the defendant's innocence, and what this means to probability, and why ordinary probability is not up to the task of modeling these situations.

Judging a man guilty or innocent, or at least not guilty, is a decision, an act. It is not probability. Like all decisions it uses probability. The probability formed depends on the evidence assumed or believed by each juror first individually, and finally corporately. Probability is the deduction, not always quantified, from the set of assumed evidence of the proposition of interest. In this case the proposition is "He's guilty."

When jurors are empaneled they enter with minds full of chaos. Some might have already formed high probabilities of guilt of the defendant ("Just look at him!"); some will have formed low ("I like his eyes"). All will have different assumed background evidence, much of it loose and unformed. But it is still evidence probative to the question of Guilt. Yet most, we imagine, will accept the proposition given by a judge that "There's more evidence about guilt that you have not yet heard." Adding that to what's in the jurors' minds, perhaps after subtracting some wayward or irrelevant beliefs based on other judge's orders ("You are to ignore the venue"), and some jurors might form a low initial probability of Guilt.

Now no juror at this point is ever asked to form the decision from his probability to Guilty or Not Guilty. Each could, though. Some do. Many jurors and also citizens do when reading of trials in the news, for instance. There is nothing magical that turns the evidence at the final official decision into the "real probability". Decisions could be and are made at any time. It is only that the law states only one decision counts, the one directed by the judge at the trial's end.

What's going on in a juror's mind (I speak from experience) is nearly constantly shifting. One moment a juror believes or accepts this set of evidence, the next moment maybe something entirely different. Jurors are almost always ready to judge based on the probability they've formed at any instance. "He was near the school? He's Guilty!" Hidden is the step which moves from probability to decision; but it's still and must be there. Then they hear some new evidence and they alter the probability and the decision to "Not Guilty." The judge may tell jurors to ignore a piece of evidence, and maybe jurors can or maybe they can't. (Hence the frequent "tricks" used by attorneys to plant evidence in jurors' minds ruled inadmissible.) Some jurors see a certain mannerism in the defendant, or even the defendant's lawyer, and interpret it in a certain way, some didn't see. And so on.

At trial's end, jurors retire to their room with what they started with: minds full of augmented chaos—a directed chaos now. The direction is honed by the discussion jurors have with each other. They will try to agree on two things: a set of evidence, which necessarily leads to a deduction of a *non-quantified* probability of "Guilty". This won't be precisely identical for each juror, because the set of evidence considered can never be *precisely* identical, but the agreed-to evidence will be shared, and the probability is calculated with respect to that. Even if individuals jurors differ from the corporate assessment. After the probability is formed, then comes the decision based on the probability. Decisions are above probability. They account for thinking about being right and wrong, and what consequences flow from that. Each juror might come to a high probability of Guilty, but they might decide Not Guilty because they think the law is stupid, or "too harsh", or in other ways deplorable. The opposite may also happen.

That's the scheme. This still doesn't account for the judge's initial directive of "presuming innocence". Jurors hear "You must presume the defendant innocent." That can be taken as a judgement, i.e. a *decision* of innocence, or a command to clear the mind of evidence probative to the question of guilt. Or both. If it's a decision, it is nothing but a formality. Jurors don't get a vote at the beginning of a trial anyway, so hearing they would have to vote Not Guilty at the commencement of the trial, were they were allowed to vote, isn't much beyond legal theater. If it is a decision (by the judge), then conditional on that decision, every juror would and must also judge the probability of Guilt to be 0. Therefore, the judge's command is properly taken as guide for juror's to ignore all non-official evidence.

Again, if it's a command by the judge to clear the mind, or a command to at least implant the evidence "I don't know all the evidence, but know more is on its way", and to the extent each juror obeys this command, it is treated as a piece of evidence, and therefore forms part of each juror's total evidence, which itself implies a (non-quantified) probability for each juror.

This means the command is not a "Bayesian prior" per se. A "prior" is a probability, and probability is the deduction *from* a set of evidence. That the judge's command is used in forming a probability (of course very informally), does make it prior evidence, though. Prior to the trial itself. Thus, priors, which will certainly be formed in the minds of each juror, or formed with the set of evidence still allowed by the judge, or by evidence jurors find pleasing.

Probabilities are eventually changed, or "updated". But this does not necessarily mean in a Bayesian sense. Bayes is not necessary; Bayes theorem, that is. The theorem is only a helpful way to chop evidence into computable bits. What's always wanted in any and all situations is the probability represented by this schematic equation:

$$\Pr(Y|\text{All probative evidence}), \tag{1}$$

where Y represents the proposition of interest; here Y = "Guilty". All Bayes does is help to partition the "All probative evidence" into smaller chunks so that numerical estimates can be reached. Numerical probabilities *won't* be had in jury trials, however. And certainly almost no juror will know how to use a complicated formula to form probabilities. Quantum probability, for instance, might be used by researchers after the fact, in modeling juror behavior, but what's going on inside the minds of jurors is anything but math.

The reader can well imagine what would happen if the criminal justice system adopted a set value, such as 0.95, above which Guilt must be decided. Some judges understanding the dire consequences which could result from this hyper-numeracy have banned the use of formal mathematical probability arguments, such as Bayes's theorem, [18].

Laudan says the judge's initial command is "an instruction about [the jurors'] *probative* attitudes". I agree with that, in the sense just stated. But Laudan amplifies: "asking a juror to begin a trial believing that defendant did not commit a crime requires a doxastic act that is probably outside the jurors' control. It would involve asking jurors to strongly believe an empirical assertion for which they have no evidence whatsoever."

That jurors have "no evidence whatsoever" is false, and not even close to true. For instance, I like many jurors walked into my last trial with the thought, "The guy probably did it because he was arrested and is on trial." That is positive evidence for Guilty. I had lots of other thought-evidence, as did each other juror. Surely some jurors came in thinking Not Guilty for any number of other reasons, which is to say other evidence. The name of the crime itself, taken in its local context, is always taken as evidence by jurors. Each juror could commit, as I said, his "doxastic act" (his decision, which is not his probability), at any time. Only his decision doesn't count until the end.

Laudan further says

> asking jurors to believe that defendant did not commit the crime seems a rather strange and gratuitous request to make since at *no* point in the trial will jurors be asked to make a judgment whether defendant is materially innocent. The key decision they must make at the end of the trial does not require a determination of factual innocence. On the contrary, jurors must make a probative judgment: has it been proved beyond a reasonable doubt that defendant committed the crime? If they believe that the proof standard has been satisfied, they issue a verdict of guilty. If not, they acquit him. It is crucial to grasp that an acquittal entails *nothing* about whether defendant committed the crime, [*sic*]

We have already seen how each juror forms his probability and then decision based on the evidence; that's Laudan's "probative judgement". That evidence could very well start with the evidence provided by the judge's command; or, rather, the evidence left in each juror's mind after clearing away the debris as ordered by the judge. Thus Laudan's "at *no* point" also fails. Many jurors, through the fuzziness of language (see [19]), take the vote of Not Guilty to mean *exactly* "He didn't do it!"—by which they mean they believe the defendant is innocent. Anybody who has served on a jury can verify this. Some jurors might say, of course, they're not sure, not convinced of the defendant's innocence, even though they vote that way. To insist that "an acquittal entails *nothing* about whether defendant committed the crime" is just false—except in a narrow, legal sense. It is a mistake to think every decision every person makes is based on extreme probabilities (i.e. 0 or 1).

Laudan says "Legal jurisprudence itself makes clear that the presumption of innocence must be glossed in probatory terms." That's true, and I agree the judge's statement is often taken as theater, part of the ritual of the trial. But it can, and in the manner I showed, be taken as evidence, too.

It seems Laudan is not a Bayesian (and neither am I):

> Bayesians will of course be understandably appalled at the suggestion here that, as the jury comes to see and consider more and more evidence, they must continue assuming that defendant did not commit the crime until they make a quantum leap and suddenly decide that his guilt has been proven to a very high standard. This instruction makes sense if and only if we suppose that the court is not referring to belief in the likelihood of material innocence (which will presumably gradually decline with the accumulation of more and more inculpatory evidence) but rather to a belief that guilt has been proved.

> As I see it, the presumption of innocence is nothing more than an instruction to jurors to avoid factoring into their calculations the fact that he is on trial because some people in the legal system believe him to be guilty. Such an instruction may be reasonable or not (after all, roughly 80% of those who go to trial are convicted and, given what we know about false conviction rates, that clearly means that the majority of defendants are guilty). But I'm quite prepared to have jurors urged to ignore what they know about conviction rates at trial and simply go into a trial acknowledging that, to date, they have seen no proof of defendant's culpability.

I can't say what Bayesians would be appalled by, though the ones I have known have strong stomachs. That Bayesians see an accumulation of evidence leading to a point seems to me to be exactly what Bayesians do think, though. How to think of the initial instruction (command), we have already seen.

I agree that the judge's command is used "to avoid factoring into their calculations the fact that he is on trial because some people in the legal system believe him to be guilty." That belief is evidence, though, which he just said jurors didn't have. Increasing the probability of Guilty *because* the defendant is on trial is what many jurors do. Even Laudan does that. That's why he quotes that "80%". The judge's command (sometimes) removes this evidence, sometimes not. In his favor, Laudan may be using *evidence* as synonymous with *true statements of reality*. I do not and instead call it the premises the jury believes true. After all, some lawyers and witnesses have been known to lie about evidence.

Laudan reasons in a frequentist fashion, but we have seen how that theory fails here. Jury trials are thus perfect at illuminating the weakness of frequentism as a theory or definition of probability people actually use in real-life decisions. Again, in frequentist theory, probabilities are defined by infinite sequences of positive (guilty) measurements embedded in infinite sequences of positive and negative (guilty and not guilty) measurements.

No real-life trial is part of an exact unique no-dispute no-possibility-of-other¡ infinite sequence, just the Walker shooting was not. Something more complex is happening in the minds of jurors as they form probabilities then just tallying whether this or that piece of evidence adds to the tally of an infinite sequence.

3 Old Evidence

When jurors hear a piece of evidence, it is new evidence. However, they come stocked (in their minds) with what we can call old evidence. We have seen mixing the two is no difficulty. However, some say there is a definite problem of how to understand old evidence and how it fits into probability, specifically probability when using Bayes's theorem. We shall see here that there is no problem, and that probability always works.

Howson and Urbach [20] is an influential book showing many errors of frequentism, though it introduced a few new ones due to emphasis on subjectivity; i.e. the theory that probability is always subjective. If probability were subjective, then probability would depend on how many scoops of ice cream the statistician had before modeling. There is also under the heading of subjectivity the so-called problem of old evidence, [21].

The so-called problem is this, quoting from Howson:

> The 'old evidence problem' is reckoned to be a problem for Bayesian analyses of confirmation in which evidence E confirms hypothesis H just in case $Pr(H|E) > Pr(H)$. It is reckoned to be a problem because in such classic examples as the rate of advance of Mercury's perihelion (M) supposedly confirming general relativity (GR), the evidence had been known before the theory was proposed; thus, before GR was developed $Pr(M)$ was and remained equal to 1, and Bayes's Theorem tells us that therefore $Pr(GR|M) = Pr(GR)$. The failure is all the more embarrassing since M was not used by Einstein in constructing his theory...

The biggest error, found everywhere in uses of classical probability, is to only partially write down the evidence one has for a proposition, and then to allow that information "float", so that one falls prey to an equivocation fallacy. It is seen in this description of the so-called problem. How will become clear below.

A step in classical hypothesis testing is to choose a statistic, here following Kadane [22] $d(X)$, the distribution of which is known when a certain hypothesis H nobody believes is true is true, i.e. when the "null" is true. The p-value is the probability of more extreme values of $d(X)$ given this belief. The philosopher of statistics Mayo [23] quotes Kadane as saying the probability statement: $Pr(d(X) >= 1.96) = .025$ "is a statement about $d(X)$ before it is observed. After it is observed, the event $\{d(X) >= 1.96\}$ either happened or did not happen and hence has probability either one or zero (2011, p. 439)."

Mayo following Glymour [24] then argues that if

the probability of the data x is 1, then $\Pr(x|H)$ also is 1, but then $\Pr(H|x) = \Pr(H)\,\Pr(x|H)/\Pr(x) = \Pr(H)$, so there is no boost in probability for a hypothesis or model arrived at after x. So does that mean known data doesn't supply evidence for H? (Known data are sometimes said to violate *temporal novelty*: data are temporally novel only if the hypothesis or claim of interest came first.) If it's got probability 1, this seems to be blocked. That's the old evidence problem. Subjective Bayesianism is faced with the old evidence problem if known evidence has probability 1, or so the argument goes.

There are number of difficulties with this reasoning. To write "$\Pr(d(X) > 1.96)$" is strictly to make a mistake. The proposition "$d(X) > 1.96$" has no probability. Nothing *has* a probability. Just like all logical argument require premises, so do all probabilities. They are here missing, and they are later supplied in different ways, which is when equivocation occurs and the "problem" enters.

In other words, we need a right hand side. We might write

$$\Pr(d(X) > 1.96|H), \tag{2}$$

where H is some compound, complex proposition that supplies information about the observable $d(X)$, and what the (here anyway) *ad hoc* probability model for $d(X)$ is. If this model allows quantification, we can calculate a value for (2). Unless that model insists "$d(X) > 1.96$" is impossible or certain, the probability will be non-extreme (i.e. not 0 or 1).

Suppose we actually observe some $d(X_o)$ (o-for-observed). We can calculate

$$\Pr(d(X) > d(X_o)|H), \tag{3}$$

and unless $d(X_o)$ is impossible or certain (given H), then again we'll calculate some non-extreme number. Equation (3) is almost identical with (2) but with a possibly different number than 1.96 for $d(X_o)$. The following equation is *not* the same:

$$\Pr(1.96 >= 1.96|H), \tag{4}$$

which indeed has a probability of 1. Of course it does! "I observed what I observed" is a tautology where knowledge of H is irrelevant. The problem comes in *where* to put the actual observation, of the right or left hand side.

Take the standard evidence of a coin flip, the proposition C = "Two-sided object which when flipped must show one of h or t", then $\Pr(h|C) = 1/2$. One would not say because one just observed a tail on an actual flip that, suddenly, $\Pr(h|C) = 0$. $\Pr(h|C) = 1/2$ because that 1/2 is deduced from C about h. Recall h is the proposition "A head will be observed".

However, and this is the key, $\Pr(\text{I saw an } h|\text{I saw an } h \,\&\, C) = 1$, and $\Pr(\text{A new } h|\text{I saw an } h \,\&\, C) = 1/2$. It is not different from 1/2 because C says *nothing* about how to add evidence of new flips. In other words, $\Pr(h|C)$ stays 1/2 forever, regardless what data is seen. There is nothing about data among the conditions. The same is true for any proposition, such as knowing about the theory of general relativity above, or in mathematical theorems, as in [25]. It may be true that at some later date new evidence for some proposition is learned, but this is *no* way changes the probability of the proposition given the old evidence, and *only* old evidence. The probability of proposition can indeed change given the old plus the new evidence, but this probability is in no way the same as the probability of the proposition given only the old evidence. Thus the so-called problem of old evidence is only a problem because of sloppy or careless notation. Probability was never in any danger.

Suppose, for ease, $d()$ is "multiply by 1" and H says X follows a standard normal. Then

$$\Pr(X > 1.96|H) = 0.025, \tag{5}$$

If an X of (say) 0.37 is observed, then what does (5) equal? The same. But this is not (5):

$$\Pr(0.37 > 1.96|H) = 0, \tag{6}$$

but because of the assumption H includes, as it always does, tacit and implicit knowledge of math and grammar.

Or we might try this:

$$\Pr(X > 1.96|\text{I saw an old } X = 0.37 \,\&\, H) = 0.025, \tag{7}$$

The answer is also the same because H like C says nothing about how to take old X and modify the model of X.

Now there are problems in this equation, too:

$$\Pr(H|x) = \frac{\Pr(H)\Pr(x|H)}{\Pr(x)} = \Pr(H), \tag{8}$$

There is no such thing as "$\Pr(x)$" nor does "$\Pr(H)$" exist, and we already seen it is false that "$\Pr(x|H) = 1$". This is because nothing *has* a probability. Probability does not exist. Probability, like logic, is a measure of a proposition of interest with respect to premises. If there are no premises, there is no logic and no probability. Thus we can never write, for any H, $\Pr(H)$.

Better notation is:

$$\Pr(H|xME) = \Pr(x|HME)\Pr(H|ME)/\Pr(x|ME), \tag{9}$$

where M is a proposition specifying information about the *ad hoc* parameterized probability model, H is usually a proposition saying something about one or more of the parameters of M, but it could also be a statement about the observable itself, and x is a proposition about some observable number. And E is a compound proposition that includes assumptions about all the obvious things.

There is no sense that $\Pr(x|HME)$ nor $\Pr(x|ME)$ equals 1 (unless we can deduce that via H or ME) before or after any observation. To say so is to swap in an incorrect probability formulation, like in (6) above.

There is therefore no old evidence problem. There are many self-created problems, though, due to incorrect bookkeeping and faulty notation, which leads to equivocation fallacies. This solution to the so-called old evidence problem is thus yet another argument against hypothesis testing.

What we always want, is what we wanted above in (1); i.e. Pr(Y|All probative evidence). And where Y is the relevant proposition of actual interest. Such as "Guilty" or "Buy now" and so on and so forth.

4 The Future

It is a very interesting time in probability and statistics. We are at a point similar to the 1980s when Bayesian statistics was being rediscovered, as it were. Yet we have roughly a century of methods developed for use in classical hypothesis. These methods are relied on by scientists, economists, governments, and regulatory agencies everywhere. They do not know of anything else. Hypothesis testing in particular is given far too much authority. The classical methods in use all contain fatal flaws, especially in the understanding of what hypothesis testing and probability are; see [26].

We therefore need a comprehensive new program to replace all these older, failing methods, with new ones which respect the way people actually act and make decisions. Work being led by our celebrant will, it is hoped, change the entire practice in the field within the next decade.

References

1. J.F.K. Before, Lee Harvey Oswald tried to kill an army Major General. www.smithsonianmag. com/smart-news/before-jfk-lee-harvey-oswald-tried-to-kill-an-army-major-general-609517/. *Smithsonian Magazine*. Accessed 15 Aug 2019
2. I.R. Goodman, H.T. Nguyen, *Uncertainty Models for Knowledge-Based Systems; A Unified Approach to the Measurement of Uncertainty* (Elsevier Science Inc., New York, NY, USA, 1985)
3. W.M. Briggs, in *Beyond Traditional Probabilistic Methods in Economics*, ed. by V. Kreinovich, N. Thach, N. Trung, D. Thanh (Springer, New York, 2019), pp. 22–44
4. H.T. Nguyen, in *Integrated Uncertainty in Knowledge Modelling and Decision Making* (Springer, 2016), pp. 3–15
5. H.T. Nguyen, A. Walker, in *Advances in the Dempster-Shafer Theory of Evidence* (Wiley, 1994) (forthcoming)
6. D. Trafimow, V. Amrhein, C.N. Areshenkoff, C.J. Barrera-Causil, E.J. Beh, Y.K. Bilgiç, R. Bono, M.T. Bradley, W.M. Briggs, H.A. Cepeda-Freyre, S.E. Chaigneau, D.R. Ciocca, J.C. Correa, D. Cousineau, M.R. de Boer, S.S. Dhar, I. Dolgov, J. Gómez-Benito, M. Grendar, J.W. Grice, M.E. Guerrero-Gimenez, A. Gutiérrez, T.B. Huedo-Medina, K. Jaffe, A. Janyan, A. Karimnezhad, F. Korner-Nievergelt, K. Kosugi, M. Lachmair, R.D. Ledesma, R. Limongi, M.T. Liuzza, R. Lombardo, M.J. Marks, G. Meinlschmidt, L. Nalborczyk, H.T. Nguyen, R. Ospina, J.D. Perezgonzalez, R. Pfister, J.J. Rahona, D.A. Rodríguez-Medina, X. Romão, S. Ruiz-

Fernández, I. Suarez, M. Tegethoff, M. Tejo, R. van de Schoot, I.I. Vankov, S. Velasco-Forero, T. Wang, Y. Yamada, F.C.M. Zoppino, F. Marmolejo-Ramos, Front. Psychol. **9**, 699 (2018). https://doi.org/10.3389/fpsyg.2018.00699. https://www.frontiersin.org/article/10.3389/fpsyg.2018.00699

7. W.M. Briggs, H.T. Nguyen, Asian J. Bus. Econ. **1** (2019) (accepted)
8. P. Billingsley, *Probability and Measure* (Wiley, New York, 1995)
9. T.L. Fine, *Theories of Probability: An Examination of Foundations* (Academic Press, New York, 1973)
10. Z. Wang, T. Solloway, R.M. Shiffrin, J.R. Busemeyer, PNAS **111**, 9431 (2014)
11. H.T. Nguyen, S. Sriboonchitta, N.N. Thac, in *Structural Changes and Their Econometric Modeling* (Springer, Berlin, 2019) (forthcoming)
12. V. Kreinovich, H.T. Nguyen, S. Sriboonchitta, in *Econometrics for Financial Applications* (Springer, Berlin, 2017), pp. 146–151
13. H. Nguyen, Short Course at BUH. New Mexico State University (2018)
14. H.T. Nguyen, B. Wu, *Fundamentals of Statistics with Fuzzy Data*, vol. 198 (Springer, Berlin, 2006)
15. H.T. Nguyen, C.L. Walker, E.A. Walker, *A First Course in Fuzzy Logic* (CRC Press, 2018)
16. L. Laudan, Why presuming innocence is not a Bayesian prior. errorstatistics.com/2013/07/20/guest-post-larry-laudan-why-presuming-innocence-is-not-a-bayesian-prior/. *Error Statistics Philosophy*. Accessed 15 Aug 2019
17. L. Laudan, *Truth, Error and Criminal Law: An Essay in Legal Epistemology* (Cambridge University Press, Cambridge, 2006)
18. Nulty v Milton Keynes Borough Council (2013) EWCA Civ 15; Britain Court of Appeal
19. H.T. Nguyen, J. Math. Anal. Appl. **64**, 369 (1978)
20. C. Howson, P. Urbach, *Scientific Reasoning: The Bayesian Approach*, 2nd edn. (Open Court, Chicago, 1993)
21. C. Howson, Philos. Sci. **84**, 659 (2017)
22. J.B. Kadane, *Principles of Uncertainty* (Chapman & Hall, London, 2011)
23. D. Mayo, The conversion of subjective Bayesian, Colin Howson, & the problem of old evidence. errorstatistics.com/2017/11/27/the-conversion-of-subjective-bayesian-colin-howson-the-problem-of-old-evidence/ (2017)
24. C. Glymour, *Theory and Evidence* (Princeton University Press, Princeton, 1980)
25. J. Franklin, Math. Intell. **38**, 14 (2016)
26. W.M. Briggs, Asian J. Bus. Econ. **1**, 37 (2019)

Applications of Type-2 Fuzzy Sets with Set Approximation Approaches: A Summary

Chia-Wen Chang and Chin-Wang Tao

Abstract Although fuzzy control technique has been widely developed in main scientific applications and engineering system, a challenging research problem how to design the appropriate parameters of fuzzy controller in the different applications still has drawn attention of researchers in various fields. Two approximating-based methods included in this paper are summarized from our previous work. One is that the parameters of Gaussian membership function can be approximately calculated when the range and mean of the input data are given. Another is that a best crisp approximation of fuzzy set can be calculated such that an interval type-2 fuzzy set can be reduced to a type-1 fuzzy set.

1 Introduction

It is known that the exact mathematical model of the controlled plant is not always available in the real world. That is, the mathematical plant model includes uncertainties. The structured (parametric) uncertainties may arise from the major variations in the parameters due to mass series production, to different operating environment, and to aging problem. The unstructured uncertainties appear when a parameterized mathematical model fails to specify the system with dynamic models. Many mechanisms (e.g., intervals, linguistic information, etc.) can be used as the representations of uncertainties to describe the difference between the models and real plants.

It is difficult to combine the linguistic and numerical information [1]. In order to include the linguistic information in the numerical system, a proper interpretation of the linguistic terms is necessary. One of the popular way to interpret the

C.-W. Chang
Ming Chuan University, Taipei, Taiwan
e-mail: cwchangg@mail.mcu.edu.tw

C.-W. Tao (✉)
National Ilan University, Yilan City, Taiwan
e-mail: cwtao@niu.edu.tw

V. Kreinovich (ed.), *Statistical and Fuzzy Approaches to Data Processing, with Applications to Econometrics and Other Areas*, Studies in Computational Intelligence 892,
https://doi.org/10.1007/978-3-030-45619-1_6

linguistic terms is to represent the linguistic terms as fuzzy sets [2]. For each fuzzy set, a membership function is defined to assign a value (from [0 1]) to every element in the input universe of discourse [3]. The fuzzy representation of uncertainties not only indicates the interval of the variations (by the support of the fuzzy set), but also describe the possibility of each different value in the variation interval (by the membership function). With this concept, Nguyen and Kreinovich design an algorithm to calculate the degree of belief that the chosen control strategy can stabilize the linear system with fuzzy representation of uncertainties [4]. Even the control strategies can be evaluated by the corresponding degree of belief that the control system is stable, it is computational time consuming to develop the satisfactory controller with this approach. Recently, to minimize the degree of the difference between a fuzzy set and a approximated crisp set, Nguyen derives a sufficient condition for a crisp set to be the best approximation of a fuzzy set [5]. With the fuzzy sets appropriately approximated by crisp sets, many techniques mentioned above can be applied to design a robust controller since the crisp sets are bounded. Since the concept "every element in the defined universe of discourse is possible for the uncertain parameter" is used to be implicitly expressed in the expert's linguistic information, the support of the fuzzy set may come out to be a very large or even an infinite set. However, the experts might not really mean that every possible element for the parameter will occur when the system is in the normal or expected situation. The best crisp approximation of a fuzzy set theoretically indicates an interval with high possibility (membership values) elements which may actually represents the region of the regular occurrence of parameter. Because the approximated crisp sets contain only the values of variations with high possibility of occurrence, the controller based on the approximated crisp sets will not lead to a conservative design. Thus, the control system is expected to have satisfactory performance and reasonable robustness to the variations of the system plant. Even so, the sufficient condition in [5] cannot specify an unique crisp set. That is, there are more than one crisp set which can satisfies the sufficient condition. This phenomenon leads to the confusion for the design of a robust controller.

Fuzzy logic technique has been successfully applied on the control system without exact mathematical plant model. The fuzzy control systems are shown to have satisfactory system performance. With the uncertainties described by linguistic information and represented as fuzzy sets, the uncertain system is said to be an SFRU. An approach to implement Nguyen's idea is developed to best approximate the fuzzy sets as intervals [6]. Then a system with fuzzy representation of uncertainties is approximated by the model with intervals. The best approximation intervals of fuzzy sets are shown to be more related to the possibility distribution of the elements in the universe of discourse of fuzzy sets than the type of membership functions used for fuzzy sets.

Fuzzy control technique has been widely developed and utilized in many scientific applications and engineering systems over the past several decades, especially in control systems with unexpected complex dynamics and external disturbances [7–9]. However, the crisp membership grades might make FC^{TI}s have problems in fully handling and accommodating the linguistic and numerical uncertainties associated with the changing and dynamic unstructured environments. Moreover, FC^{TI}s cannot directly deal with uncertainties in the parameters of the antecedent and the consequent membership functions (MFs) due to the noisy measurements and the environmental variances [10, 11].

The concept of type-2 fuzzy sets (FS^{T2}s) was introduced by Zadeh [12] as an extension of the ordinary type-1 fuzzy set (FS^{T1}). Unlike the FS^{T1}s, FS^{T2}s adopt type-2 membership functions (MFs) to indicate the uncertainties in MFs. Type-2 fuzzy techniques with type-2 antecedent and consequent MFs have attracted a considerable interest. The only difference in the structure of a FC^{T1} and a type-2 fuzzy controller (FC^{T2}) is that the output processor in a FC^{T2} consists of a type reducer and a defuzzifier, but the output processor in a FC^{T1} is just a defuzzifier [13, 14]. Since FC^{T2}s require type reducers for type reduction, FC^{T2}s are more complex and more computationally expensive than FC^{T1}s. To alleviate the heavy computational load, the interval-valued fuzzy sets are introduced [15]. Moreover, the Extended-Karnik-Mendel (EKM) approach is presented in [16] to reduce the complexity in the type reduction procedure for interval-valued fuzzy sets. With the interval-valued fuzzy sets, the interval-valued FCs (FC^{IV}s) become realizable in numerous applications. There are several different approaches have been presented in the literature to obtain crisp outputs for FC^{IV}s. Although FC^{IV}s have been successfully applied in control systems as well as in many other applications due to their ability to model uncertainties [17, 18], FC^{IV}s are still noted to have heavy computational loads [19]. The computational-intensive characteristic of FC^{IV}s limits their practical application [20]. Unlike the approaches in [21–23] and [17], based on the combination of the multi-information represented using interval-valued MFs, a type-reduction with the crisp interval operations is proposed in this paper to avoid the complex computations in FC^{IV}. With the integrated information, an interval-valued fuzzy set is reduced to be a simple fuzzy set.

2 Approximation of Type-1 Fuzzy Sets

2.1 Approximation of Fuzzy Sets [24]

Let M be a fuzzy set with a general asymmetric Gaussian membership function $\mu_M(m, \sigma_1, \sigma_2, x)$,

$$\mu_M(m, \sigma_1, \sigma_2, x) = \begin{cases} \mu_M(x) = e^{\frac{-(x-m)^2}{\sigma_1^2}}, & \text{if } x \leq m \\ \mu_M(x) = e^{\frac{-(x-m)^2}{\sigma_2^2}}, & \text{if } x > m \end{cases} \tag{1}$$

where m is the mean, $\sigma_i^2, i = 1, 2$, are the variance and x is the input of the Gaussian function. Define a very small positive constant value ϵ which is nearly zero. For the given m, σ_1, and σ_2, $\mu_M(m, \sigma_1, \sigma_2, x) = \epsilon$ can be obtained when x is chosen as

$$x_1 = m - \sigma_1\sqrt{|ln\epsilon|} \text{ and } x_2 = m + \sigma_2\sqrt{|ln\epsilon|},$$

with the corresponding σ_1 and σ_2, respectively. It means that the main support of Gaussian membership function, μ_M, is $[x_1 \ x_2]$. Based on above discussion, it clear that the variance, σ_1 and σ_2, can be approximately calculated as follows

$$\sigma_1^2 = \frac{(m - x_1)^2}{|ln\epsilon|} \text{ and } \sigma_2^2 = \frac{(m - x_2)^2}{|ln\epsilon|}, \tag{2}$$

when the range of the input $x \in [x_1 \ x_2]$, m, and ϵ are given.

2.2   Interval Approximation of Fuzzy Sets [25]

Since the intervals are crisp sets, the main idea to approximate a fuzzy set M by an interval D is to find a crisp set D which is the best crisp approximation of the fuzzy set M [6]. Unlike the Hamming distance $H(M, D)$ between a fuzzy set M and an interval D in [26], the absolute distance $abs(M, D)$,

$$abs(M, D) = |\int_{-\infty}^{\infty} \mu_D(x) - \mu_M(m, \sigma, x)dx|$$

with

$$\mu_D(x) = \begin{cases} 1, & \text{for } x \in D \\ 0, & \text{otherwise} \end{cases}$$

is utilized in this paper. To best approximate a fuzzy set M by an interval D is then to minimize the absolute distance $abs(M, D)$ [6]. It can be seen that when $abs(M, D)$ is minimized, the fuzzy set M and its best approximated interval D have the same cardinality. Based on this idea, the sufficient condition for the best crisp interval D is

$$\int_{D^c} \mu_M(m, \sigma, x)dx = \int_D (1 - \mu_M(m, \sigma, x))dx \tag{3}$$

where D^c is the complement set of D and $\mu_M(m, \sigma, x)$ is the MF of the fuzzy set M. Let the interval D in (3) be specified as $D = [d_1 \ d_2]$ and the interval $A = [a_1 \ a_2]$ be a region such that

$$\mu_M(m, \sigma, x) = 1 \ \forall x \in [a_1 \ a_2]$$

with the center $a_c = (a_1 + a_2)/2$. Based on the common sense that the best crisp approximation interval D covers the region with high membership values, it is reasonable to assume $A \subset D$. Then it is straightforward to see that

$$\int\limits_{D^c} \mu_M(m, \sigma, x)dx = \int\limits_{-\infty}^{d_1} \mu_M(m, \sigma, x)dx + \int\limits_{d_2}^{\infty} \mu_M(m, \sigma, x)dx$$

and

$$\int\limits_{D} (1 - \mu_M(m, \sigma, x))dx = \int\limits_{d_1}^{a_c} (1 - \mu_M(m, \sigma, x))dx + \int\limits_{a_c}^{d_2} (1 - \mu_M(m, \sigma, x))dx$$

Equation (3) can be satisfied if

$$\int\limits_{-\infty}^{d_1} \mu_M(m, \sigma, x)dx = \int\limits_{d_1}^{a_c} (1 - \mu_M(m, \sigma, x))dx \tag{4}$$

and

$$\int\limits_{d_2}^{\infty} \mu_M(m, \sigma, x)dx = \int\limits_{a_c}^{d_2} (1 - \mu_M(m, \sigma, x))dx. \tag{5}$$

Therefore, from (4) and (5), d_1 and d_2 can be determined as

$$d_1 = a_c - \int\limits_{-\infty}^{a_c} \mu_M(m, \sigma, x)dx \tag{6}$$

and

$$d_2 = a_c + \int\limits_{a_c}^{\infty} \mu_M(m, \sigma, x)dx, \tag{7}$$

if $\int\limits_{-\infty}^{a_c} \mu_M(m, \sigma, x)dx < \infty$ and $\int\limits_{a_c}^{\infty} \mu_M(m, \sigma, x)dx < \infty$.

Example 1 Let the best interval approximation of M in (1) is defined as $D = [d_1 \, d_2]$. Since $a_c = m$, and

$$\int\limits_{-\infty}^{m} \mu_M(m, \sigma_1, \sigma_2, x)dx = \frac{\sigma_1}{2}\sqrt{\pi} \tag{8}$$

$$\int\limits_{m}^{\infty} \mu_M(m, \sigma_1, \sigma_2, x)dx = \frac{\sigma_2}{2}\sqrt{\pi} \tag{9}$$

d_1 and d_2 are derived as

$$d_1 = m - \frac{\sigma_1}{2}\sqrt{\pi} \tag{10}$$

$$d_2 = m + \frac{\sigma_2}{2}\sqrt{\pi} \tag{11}$$

according to (6) and (7). Thus, the best interval approximation of the fuzzy set M is

$$D = \left[m - \frac{\sigma_1}{2}\sqrt{\pi} \quad m + \frac{\sigma_2}{2}\sqrt{\pi}\right]. \tag{12}$$

Considering a symmetric Gaussian membership function ($\sigma = \sigma_1 = \sigma$ in (1)), Eq. (12) can be obtained as follows

$$D = \left[m - \frac{\sigma}{2}\sqrt{\pi} \quad m + \frac{\sigma}{2}\sqrt{\pi}\right].$$

Example 2 Let M be a fuzzy set with the membership function $\mu_F(x) = exp(-x^2)$, and the best crisp approximation of M be defined as $D = [d_1 \ d_2]$. From (3), we have

$$\int_{-\infty}^{d_1} \mu_M(x)dx + \int_{d_2}^{\infty} \mu_M(x)dx = \int_{d_1}^{d_2} (1 - \mu_M(x))dx$$

By knowing that

$$\int_{-\infty}^{0} \mu_M(x)dx = \int_{0}^{\infty} \mu_M(x)dx = 0.5\sqrt{\pi},$$

we can find the best crisp approximation of the fuzzy set M as

$$D = [-0.5\sqrt{\pi} \ \ 0.5\sqrt{\pi}].$$

3 Approximation of Type-2 Fuzzy Sets

3.1 Type-2 Fuzzy Sets

It is known that type-2 fuzzy controllers actually combine the type-2 information to make decisions and type-2 fuzzy controllers make reasonable decisions with type-2 fuzzy sets in fuzzy rules. However, the computational complexity of a type-2 fuzzy controller is very heavy [20]. In stead of using Enhanced Karnik-Mendel (EKM) [16] to combine the interval type-2 information for an interval type-2 fuzzy controller, interval type-2 information is combined with fuzzy set operations in this paper. With the interval type-2 information combined, an interval type-2 fuzzy set is reduced to be a type-1 fuzzy set.

As indicated in [17], the primary MF $\mu_{\tilde{A}}(x)$ of a type-2 fuzzy set $\tilde{A}$ can be considered as a collection of an infinite number of embedded type-1 MFs $\mu_{A_i}(x)$. If a type-2 fuzzy set $\tilde{A}$ (ex. "small") is required to be simplified by a type-1 fuzzy set A (as the same "small"), each embedded type-1 MF can be used as the MF of A with a suitable factor at input x. The suitable factor is so-called the secondary membership value. Therefore, unlike a type-1 fuzzy set adopting one fixed MF, a type-2 fuzzy set provides multi-information in the form of multi-possible MFs.

An interval type-2 fuzzy set (FS^{IT2}) $\tilde{A}$ is characterized by it MF $\mu_{\tilde{A}}(x, u)$ as

$$\tilde{A} = \int_{x \in D_{\tilde{A}}} \int_{u = \in J_x \subseteq [0,1]} \mu_{\tilde{A}}(x, u)/(x, u) \tag{13}$$

where $x \in D_{\tilde{A}}$ is the primary variable in domain $D_{\tilde{A}}$, $u \in J_x$ is the secondary variable, J_x is called the primary membership of x, and the amplitude of $\mu_{\tilde{A}}$ is the secondary grades of $\tilde{A}$. In FS^{IT2}s, the secondary grades of $\tilde{A}$ are all equal to one $\forall x \in D_{\tilde{A}}$ and $\forall u \in J_x \subseteq [0, 1]$. The uncertainty about the MF of $\tilde{A}$ can be described by the union of all the primary memberships which is call the footprint of uncertainty (FOU) of $\tilde{A}$:

$$FOU(\tilde{A}) = \bigcup_{\forall x \in D_{\tilde{A}}} J_x = \{(x, u) : u \in [\underline{\mu}_{\tilde{A}}(x), \bar{\mu}_{\tilde{A}}(x)]\}$$

where $\underline{\mu}_{\tilde{A}}(x)$ is the lower membership function (LMF) and $\bar{\mu}_{\tilde{A}}(x)$ is the upper membership function (UMF). An example of an FS^{IT2} is illustrated in Fig. 1. Observe that an FS^{IT2} is bounded by two FS^{T1}, $\bar{A}$ and $\underline{A}$. The area between $\bar{A}$ and $\underline{A}$ is the footprint of uncertainty.

Fig. 1 The MF of a FS^{IT2} $\tilde{A}$

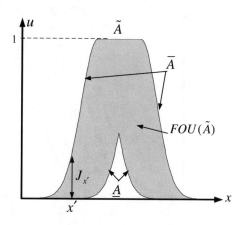

3.2 Approximation of Type-2 Fuzzy Sets [24]

For the design of an effective controller, the interval type-2 fuzzy set $\tilde{M}$ discussed in this paper is assumed to have symmetric Gaussian primary MF $\mu_{\tilde{M}}(m, \sigma_1 = \sigma_2 = \sigma_1, x)$ with uncertain mean $m \in [m_{min}\ m_{max}]$. Also, the embedded type-1 MFs $\mu_{M_i}(m_i, \sigma, x), i \geq 1$, in the primary MF $\mu_{\tilde{M}}$ are assumed to be symmetric Gaussian function with the variance σ^2 and the mean $m_{min} \leq m_i \leq m_{max}$, shown as in Fig. 2. Let type-1 fuzzy sets with embedded type-1 MFs be called embedded type-1 fuzzy sets. The union of all type-1 fuzzy sets M_i with MFs μ_{Mi} is defined as a fuzzy set $\bar{M}$ in Fig. 2. It is easy to see that the MF $\mu_{\bar{M}}$ of $\bar{M}$ is the upper MF of $\mu_{\tilde{M}}$. Similarly, the intersection of all embedded type-1 fuzzy sets is a type-1 fuzzy set $\underline{M}$ with the MF $\mu_{\underline{M}}$ which is the lower MF of $\mu_{\tilde{M}}$.

The main support of $\mu_{\bar{M}}$ can be roughly calculated as $[\bar{x}_m\ \bar{x}_M]$ to satisfy $\mu_{\bar{M}}(\bar{x}_m) < \epsilon$ and $\mu_{\bar{M}}(\bar{x}_M) < \epsilon$, where $\epsilon > 0$ is a positive constant nearly zero. Since the membership functions considered here are mainly in the Gaussian shape, Gaussian approximations $\mu_{\bar{M}}^g$ and $\mu_{\underline{M}}^g$ (in Fig. 3) of $\mu_{\bar{M}}$ and $\mu_{\underline{M}}$ are provided as the MFs for the union $\bar{M}$ and the intersection $\underline{M}$, respectively. With the main support of $\mu_{\bar{M}}$ obtained, the variance $\bar{\sigma}^2$ of the type-1 Gaussian membership function $\mu_{\bar{M}}^g$ can be calculated by (2). Similarly, the main support of $\mu_{\underline{M}}$ can be limited into $[\underline{x}_m\ \underline{x}_M]$ to satisfied $\mu_{\underline{M}}(\underline{x}_m) < \epsilon$ and $\mu_{\underline{M}}(\underline{x}_M) < \epsilon$. To approximate the membership function $\mu_{\underline{M}}(x)$ of the intersection $\underline{M}$ into the type-1 Gaussian membership function $\mu_{\underline{M}}^g$, the variance, $\underline{\sigma}^2$, can again be derived by (2).

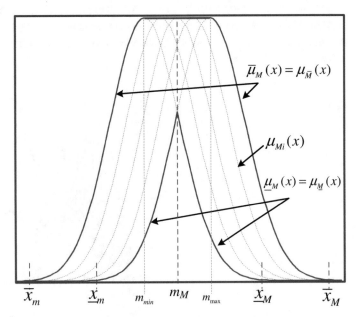

Fig. 2 Gaussian primary MF $\mu_{\tilde{M}}$ for $\tilde{M}$

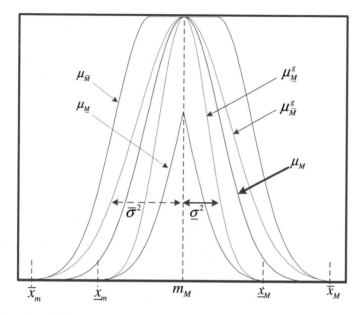

Fig. 3 Gaussian MF for M

The idea to reduce an interval type-2 fuzzy set $\tilde{M}$ to a type-1 fuzzy set M is to find the fuzzy set M with a symmetric Gaussian MF $\mu_M(m_M, \sigma_M, x)$ from $\mu_{\tilde{M}}$, which represent the collection of the type-1 fuzzy information. To retrieve $\mu_M(m_M, \sigma_M, x)$ from the fuzzy sets $\overline{M}$ and $\underline{M}$, the mean of $\mu_M(m_M, \sigma_M, x)$ is reasonably selected as

$$m_M = \frac{m_{max} + m_{min}}{2}.$$

The approach to determine the σ_M of $\mu_M(m_M, \sigma_M, x)$ is more complicated. Based on the MFs, $\mu_{\overline{M}}^g$ and $\mu_{\underline{M}}^g$ of $\overline{M}$ and $\underline{M}$, σ_M is defined as the root mean square of $\overline{\sigma}$ and $\underline{\sigma}$,

$$\sigma_M = \sqrt{\frac{\overline{\sigma}^2 + \underline{\sigma}^2}{2}} \tag{14}$$

where $\overline{\sigma}$ and $\underline{\sigma}$ are obtained in the following,

$$\overline{\sigma} = \sqrt{\frac{m_M - \overline{x}_m}{|ln\epsilon|}}, \text{ and } \underline{\sigma} = \sqrt{\frac{m_M - \underline{x}_m}{|ln\epsilon|}}$$

3.3 Interval Approximation of Type-2 Fuzzy Sets [25]

Since the interval-valued membership degrees of an interval-valued fuzzy set $\tilde{M}$ are equivalent to one, intervals D_i, $i \to \infty$, are equivalently suitable. Therefore, it is reasonable to use the union operation to combine the interval-valued information. Let the union of the infinite number of intervals be D_U. That is,

$$D_U = \bigcup_{i=1}^{\infty} D_i = [d_{UL} \quad d_{UR}]$$

Let the crisp approximation intervals for fuzzy sets with $m_L = m_{min}$ and $m_R = m_{max}$ be $D_L = [d_{L1} \quad d_{L2}]$ and $D_R = [d_{R1} \quad d_{R2}]$, respectively. Then the interval-valued information for $\tilde{M}$ with the primary MF in Fig. 4 can be combined using union operation as the interval $D_U = [d_{UL} \quad d_{UR}] = [d_{L1} \quad d_{R2}]$. On the other hand, the intersection of the infinite number of intervals,

$$D_U = \bigcap_{i=1}^{\infty} D_i = [d_{IL} \quad d_{IR}],$$

can be used to indicate the more important and essential information in the primary MF of an interval-valued fuzzy set $\tilde{M}$. For $\tilde{M}$ with the primary MF in Fig. 4, it can be seen that $D_I = [d_{IL} \quad d_{IR}] = [d_{R1} \quad d_{L2}]$.

Note that the intersection D_I would be an empty set when $\Delta m = |m_{max} - m_{min}|$ is large. As an example, $D_I = \phi$ for $\tilde{M}$ with the primary MF in Fig. 5a.

The idea to reduce an interval-valued fuzzy set $\tilde{M}$ to a type-1 fuzzy set M is to find the fuzzy set M from the interval which represent the combination of the

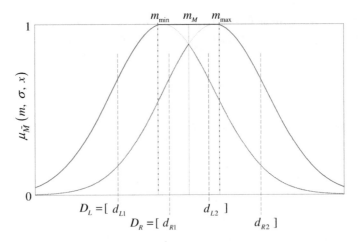

Fig. 4 D_L and D_R of fuzzy sets with means m_{min} and m_{max} in MFs

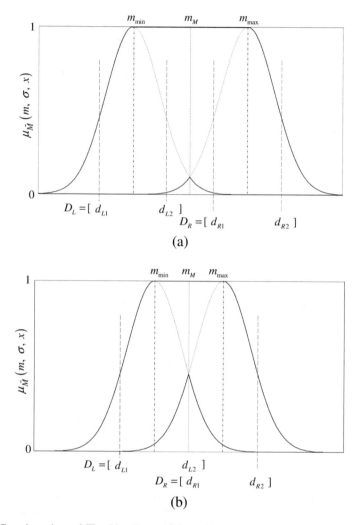

Fig. 5 Gaussian primary MFs with **a** $D_I = \phi$ **b** $D_I = [m_M]$

interval-valued information. For simplicity, the fuzzy set M is assumed to have a symmetric MF for the following derivation. It can be seen that the provided derivation is applicable to the fuzzy set M with asymmetric MF. To retrieve a Gaussian MF $\mu_M(m_M, \sigma_m, x)$ from the intervals D_U and D_I, m_M of $\mu_M(m_M, \sigma_m, x)$ is reasonably selected as

$$m_M = \frac{m_{max} + m_{min}}{2} \qquad (15)$$

Then σ_M has to be found to determine $\mu_M(m_M, \sigma_m, x)$. Although the union D_U characterizes the combination of the interval-valued information represented by equivalent suitable intervals, the intersection D_I is also necessary to be considered for the derivation of σ_M to emphasize the more important part of the interval-valued information. Therefore, when $D_I \neq \phi$, based on the intervals D_U and D_I, σ_U and σ_I are obtained with (10), (11), and the mean m_M in (15),

$$\sigma_U = \frac{2}{\sqrt{\pi}}(m_M - d_{UL}) = \frac{2}{\sqrt{\pi}}(d_{UR} - m_M) \tag{16}$$

$$\sigma_I = \frac{2}{\sqrt{\pi}}(m_M - d_{IL}) = \frac{2}{\sqrt{\pi}}(d_{IR} - m_M) \tag{17}$$

Then σ_M is defined as the root mean square of σ_U and σ_I,

$$\sigma_M = \sqrt{\frac{\sigma_U^2 + \sigma_I^2}{2}} \tag{18}$$

If $\Delta m = |m_{max} - m_{min}|$ is larger, the length of the intersection D_I, $|d_{IR} - d_{IL}|$, would become smaller. When Δm equals to the critical value Δm_c,

$$\Delta m_c = (m_{max} - d_{R1}) + (d_{L2} - m_{min}),$$

the intersection D_I contains only one points, i.e.,

$$D_I = [d_{IR} = d_{IL} = m_M] \text{ (as in Fig. 5b)}$$

and

$$\sigma_M = \sqrt{\frac{\sigma_U^2}{2}} = \sigma_{Mm},$$

because $\sigma_I = 0$. The σ_{Mm} is defined to be the marginal value of σ_M. It is easy to find that $D_I = \phi$ if $\Delta m > \Delta m_c$ (see Fig. 5a). When $D_I \neq \phi$, (18) can no longer be used for σ_M. Moreover, no intersection indicates some information contradiction occurs between intervals even though all the intervals are equivalently suitable. To avoid the inconsistent information being taken into account, the σ_M is defined to be its marginal value when $D_I = \phi$. To indicate the result of the proposed type reduction approach, Fig. 6 presents the MF μ_M of the type-reduced fuzzy set M for the interval-valued fuzzy set $\tilde{M}$ with different symmetrical Gaussian primary MF $\mu_{\tilde{M}}$. From Fig. 6, it is found that μ_M is inside the FOU of the primary MF $\mu_{\tilde{M}}$. Therefore, the fuzzy set M with μ_M can be considered as one suitable type-reduced set for the interval-valued fuzzy set $\tilde{M}$.

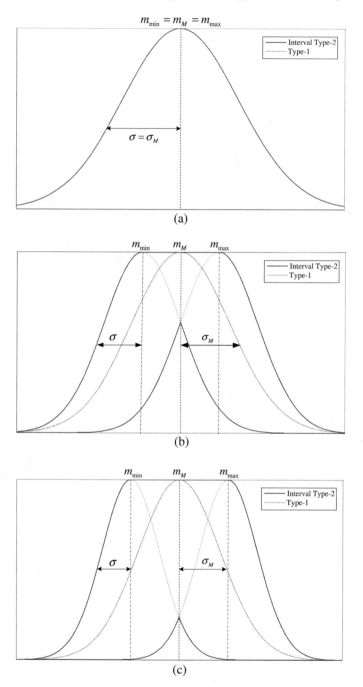

Fig. 6 The Gaussian MFs of $\tilde{M}$ and M for **a** $\Delta m = 0$, **b** $\Delta m < \Delta m_c$, **c** $\Delta m > \Delta m_c$

4 Applications

4.1 Ball and Beam Control System

The configuration of the end-point driven ball and beam system (BBS) is shown in Fig. 7, where o_m is the small gear mounted on a DC motor which provides the necessary torque of interest, o_g represents the large gear that can control the angle of the beam, and o_p is the pivot that connects the beam and the stand [24]. In addition, r is the distance between o_p and the ball, θ is the angle of the large gear, τ_m is a torque of the DC motor, τ_g is a torque of the large gear and τ_b is a torque provided by a DC motor to the beam via gear and linker. This system is typically underactuated and the control objective is to move the ball on the beam to the desired position. Considering o_p as the reference origin, the state vector of the ball and beam system is defined as $X(t) = [x_1(t)\, x_2(t)\, x_3(t)\, x_4(t)]^T$, where $x_1(t) = r(t)$ is the ball position (m), $x_2(t)$ is the velocity of the ball (m/sec), $x_3(t) = \alpha(t)$ is the beam angle (rad), and $x_4(t)$ is the angular velocity of the beam (rad/sec). According to the Euler-Lagrange method, the mathematical model of a ball and beam system can be represented as follows

$$\dot{x}_1 = x_2$$
$$\dot{x}_2 = (x_1 x_4^2 - g \sin x_3) K_1$$
$$\dot{x}_3 = x_4$$
$$\dot{x}_4 = K_4(x_1) \cos x_3 [K_2 x_4 \cos \frac{l x_3}{d} + K_3 \cos \frac{l x_3}{d} u - 0.5 l m_b g - m_B g x_1] - 2 m_B x_1 x_2 x_4 K_4(x_1)$$

$$(19)$$

where $K_1 = (1 + m_B^{-1} J_B r_B^{-2})^{-1}$, $K_2 = K_b K_t l^2 (R_a d^2)^{-1}$, $K_3 = n K_b l (R_a d)^{-1}$, $K_4(x_1) = (J_B + J_b + m_B x_1^2)^{-1}$, u is the input voltage of the DC motor, and the parameters of the system are given in Table 1. The output of hybrid interval type-2 fuzzy sliding controller with simplified type reduction (HFSC^{STR}) is defined to be

$$u_{HFSC^{STR}} = F_{P-FSC} + F_{B-FSC} \qquad (20)$$

where

$$F_{P-FSC} = -(c_2 \beta_2)^{-1} [c_1 (a_{P11}(x_1 - x_{1d}) + a_{P12} x_2) + c_2 (a_{P21}(x_1 - x_{1d}) + a_{P22} x_2)]$$
$$- (c_2 \beta_2)^{-1} K_P \frac{-\sum_{i=1}^{9} |\kappa_P^i| \mathrm{sgn}(S_P) \mu_P^i(S_P)}{\sum_{i=1}^{9} \mu_P^i(S_P)}$$
$$F_{B-FSC} = -(c_4 \beta_4)^{-1} [c_3 (a_{B11} x_3 + a_{B12} x_4) + c_4 (a_{B21} x_3 + a_{B22} x_4)]$$
$$- (c_4 \beta_4)^{-1} K_B \frac{-\sum_{i=1}^{9} |\kappa_B^i| \mathrm{sgn}(S_B) \mu_B^i(S_B)}{\sum_{i=1}^{9} (S_B) \mu_B^i(S_B)}$$

$$(21)$$

Fig. 7 Scheme diagram of the ball and beam system

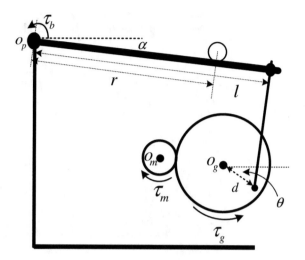

Table 1 Parameters of ball and beam system

Symbol	Definition	Value
m_B	Mass of the ball	0.029 kg
m_b	Mass of the beam	0.334 kg
r_B	Radius of the ball	0.0095 m
l	Beam length	0.4 m
d	Link distance	0.04 m
J_B	Ball inertia	$1.05 \times 10^{-6}\,\text{kgm}^2$
J_b	Beam inertia	$0.0178\,\text{kgm}^2$
K_b	Back-EMF constant	0.1491 Nm/A
K_t	Torque constant	0.1491 Nm/A
R_a	Armature resistance	18.91 Ω
n	Ratio of the gear	4.2
g	Acceleration of gravity	9.8 m/s^2

To show the effectiveness of the proposed algorithm, the performance comparisons between the ball-and-beam system with HFSC^{EKM} which the method of type reduction is the Extended-Karnik-Mendel approach, and the ball-and-beam system with the proposed HFSC^{STR} are provided in this section. To indicate the robustness of HFSC^{STR} and HFSC^{EKM}, an external random noise d ($|d| \leq 1.5$ mm from 0 to 10 s) is added into the ball position state for simulations. The input and output membership functions of HFSC^{STR} controller is shown in Fig. 8, the parameters of BBS are represented in Table 1 and the parameters of HFSC are selected as follows,

$$[c_1 \ c_2 \ c_3 \ c_4] = [10 \ 7 \ 7 \ 1], \ [K_P \ K_B] = [12 \ 24]$$

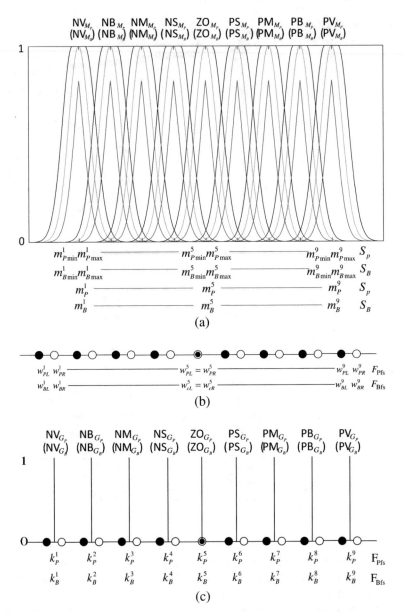

Fig. 8 a Input MFs: solid lines for Pfs^{IT2} and Bfs^{IT2}, dotted lines for Pfs^{STR} and Bfs^{STR}. **b** Precomputed w_L^i and w_R^i, with $1 \leq i \leq 9$, in ascending order. **c** Fuzzy singletons for Pfs^{STR} and Bfs^{STR}

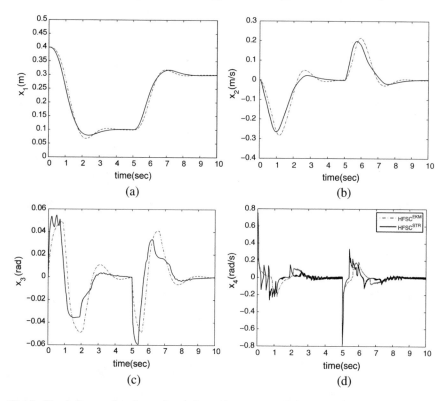

Fig. 9 Simulation results of case 1: **a** ball position, $x_1(t)$, **b** ball velocity, $x_2(t)$, **c** beam angle, $x_3(t)$, **d** beam angular velocity, $x_4(t)$

$$m_{Pmin} = [m^1_{Pmin} \; m^2_{Pmin} \; \cdots \; m^9_{Pmin}] = [-1. \; -0.72 \; -0.43 \; -0.24 \; -0.05 \; 0.14 \; 0.33 \; 0.62 \; 0.9]$$
$$m_{Pmax} = [m^1_{Pmax} \; m^2_{Pmax} \; \cdots \; m^9_{Pmax}] = [-1. \; -0.62 \; -0.33 \; -0.14 \; 0.05 \; 0.24 \; 0.43 \; 0.72 \; 1]$$
$$\sigma_P = [\sigma^1_P \; \sigma^2_P \; \cdots \; \sigma^9_P] = [0.1 \; 0.1 \; 0.1 \; 0.1 \; 0.1 \; 0.1 \; 0.1 \; 0.1 \; 0.1]$$
$$w_{PL} = [w^1_{PL} \; w^2_{PL} \; \cdots \; w^9_{PL}] = [-0.95 \; -0.75 \; -0.45 \; -0.25 \; -0.05 \; 0.25 \; 0.45 \; 0.75]$$
$$w_{PR} = [w^1_{PR} \; w^2_{PR} \; \cdots \; w^9_{PR}] = [-0.75 \; -0.45 \; -0.25 \; 0.05 \; -0.25 \; 0.45 \; 0.75 \; 0.95]$$
$$m_{Bmin} = m_{Pmin}, \; m_{Bmax} = m_{Pmax}$$
$$\sigma_B = \sigma_P, \; w_{BL} = w_{PL}, \; w_{BR} = w_{PR}.$$

Two simulation cases of different condition are addressed later,
case 1: the command of ball position state x_{1d} is set to 0.1 ($t = 0 - 5 \; sec$) and 0.3
($t = 5 - 10 \; sec$), and the commands of other states, x_2, x_3, x_4, are set to 0
case 2: the command of ball position state x_{1d} is set to $x_{1d} = 0.2 - 0.05 \, cos(0.2\pi t)$
($t = 0 - 10 \; sec$) and x_{id}, $i = 2, 3, 4$, are set to 0.

Furthermore, some criteria are selected as comparison metrics, such as the integral
of the absolute value of the error (IAE), the integral of square error (ISE), and the

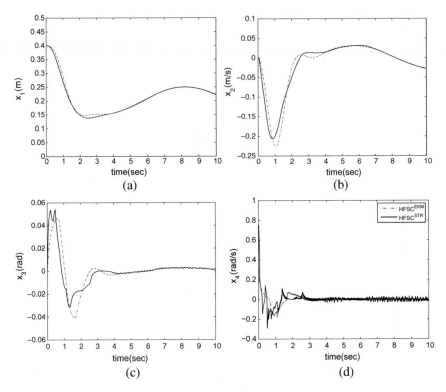

Fig. 10 Simulation results of case 2: **a** ball position, $x_1(t)$, **b** ball velocity, $x_2(t)$, **c** beam angle, $x_3(t)$, **d** beam angular velocity, $x_4(t)$,

integral of the time multiplied by the square value of the error (ITSE). In this paper, the ball-tracking is considered as the main goal of concern. Therefore, the response of ball tracking error is considered to calculate IAE, ISE, and ITSE as follows,

$$IAE = \sum_{t=1}^{N} |x_{1d}(t) - x_1(t)|,$$

$$ISE = \sum_{t=1}^{N} (x_{1d}(t) - x_1(t))^2,$$

$$ITSE = \sum_{t=1}^{N} t(x_{1d}(t) - x_1(t))^2.$$

In each simulation cases, the initial conditions are $X(0) = [0.4\ 0\ 0\ 0]$. The simulation results are illustrated in Figs. 9 and 10. Moreover, the simulation comparisons criteria are calculated and presented in Table 2, where T_{CPU} is the simulation time with MATLAB software.

Table 2 Simulation comparisons

	HFSC^{EKM}		HFSC^{STR}	
	Case 1	Case 2	Case 1	Case 2
IAE	1012	493	991	490
ISE	114	43.1	110	40.7
$ITSE$	4806	1827	4784	1836
T_{CPU}	7.84	7.17	0.47	0.47

The case 1 is investigated to verify the transient response of the BBS with HFSC^{STR}. The simulation results of case 1 are illustrated as Fig. 9, including the response of ball position, ball velocity, beam angle, and beam angular velocity in Fig. 9a–d, respectively. In Fig. 9, the solid lines are the responses of the proposed method HFSC^{STR} and the dash lines are the responses of HFSC^{EKM}. In Fig. 9, it can be seen that the transient convergence of HFSC^{STR} is better than the HFSC^{EKM}. From Fig. 9a, it shows that the responses of ball position are very closely by HFSC^{STR} and HFSC^{EKM}. Moreover, the beam response of using the proposed method is slightly better than using the EKM approach. From Table 2, it can be seen that the criterion values with IAE, ISE, and ITSE are very closely between HFSC^{STR} and HFSC^{EKM}. To compare the computational efficiency between HFSC^{STR} and HFSC^{EKM} in Table 2, the simulation time with the ball-and-beam control system by HFSC^{STR} is 0.47 sec and simulation time of BBS with HFSC^{EKM} need 7.84 sec. It means that the computational cost can be obviously improved by the proposed HFSC^{STR} and the control performance is not be affected. The responses for the BBS with the time-varying tracking trajectory are presented in Fig. 10. Form Fig. 10 and Table 2, it can be seen again that the BBS systems with the proposed HFSC^{STR} provides better performance. Since the proposed type-reduction approach can be completed before its application to the design of the fuzzy sliding controller, the time-consuming difficulty of type reduction can be avoided.

4.2 Double Pendulums and Cart System

The double-pendulum-and-cart system (DPCS) contains double pendulums and a cart [25]. The cart can freely move on a preserved rail limits, and double pendulum can rotate on the vertical plane around an axis that is fixed on the cart. The system dynamics and parameters are presented in [27]. Let the desired values of the state variables in the double-pendulum-and-cart system be assumed as zeros,

$$x_{d1} = x_{d2} = x_{d3} = x_{d4} = x_{d5} = x_{d6} = 0 \tag{22}$$

Then the errors between the current states, $x_{pi}, i = 1, 2, \ldots, 6$, and the desired states, $x_{di}, i = 1, 2, \ldots, 6$, of the double pendulums and the cart are

$$
E_p = \begin{bmatrix} e_{p1} \\ e_{p2} \\ e_{p3} \\ e_{p4} \end{bmatrix} = \begin{bmatrix} x_{p1} - x_{d1} \\ x_{p2} - x_{d2} \\ x_{p3} - x_{d3} \\ x_{p4} - x_{d4} \end{bmatrix} = \begin{bmatrix} x_1 \\ x_2 \\ x_3 \\ x_4 \end{bmatrix}
$$

$$
E_c = \begin{bmatrix} e_{c5} \\ e_{c6} \end{bmatrix} = \begin{bmatrix} x_{c5} - x_{d5} \\ x_{c6} - x_{d6} \end{bmatrix} = \begin{bmatrix} x_5 \\ x_6 \end{bmatrix} \tag{23}
$$

With the change of the state variables and the inclusion of the control action, F_{HFSC^*}, the state equations of the double-pendulum-and-cart system become

$$
\begin{aligned}
\dot{e}_{p1}(t) &= e_{p2} \\
\dot{e}_{p2}(t) &= g_1(E_p) + \beta_{p2} F_{HFSC^*} + d \\
\dot{e}_{p3}(t) &= e_{p4} \\
\dot{e}_{p4}(t) &= g_2(E_p) + \beta_{p4} F_{HFSC^*} + d \\
\dot{e}_{c5}(t) &= e_{p6} \\
\dot{e}_{c6}(t) &= g_3(E_p, E_c) + \beta_{c2} F_{HFSC^*}
\end{aligned} \tag{24}
$$

where $F_{HFSC^*} = F_{p-FSC^*} - F_{c-FSC^*}$ with $F_{p-FSC^*}(F_{c-FSC^*})$ being the control action to balance the double-pendulum (cart) subsystem. As in [27], state equations in (24) for the double-pendulum-and-cart system are reformulated and further processed to be pseudo-decomposed as

$$
\begin{aligned}
\dot{\bar{E}}_p &= \bar{A}_p \bar{E}_p + \Delta \bar{\varphi}_p \bar{E}_p + \bar{\beta}_p F_{p-FSC^*} \\
&= \begin{bmatrix} \bar{A}_{p11} & \bar{A}_{p12} \\ \bar{A}_{p21} & \bar{A}_{p22} \end{bmatrix} \begin{bmatrix} \bar{E}_{p1} \\ \bar{E}_{p2} \end{bmatrix} + \begin{bmatrix} 0 \\ \Delta \bar{\varphi}_{p4} \end{bmatrix} \begin{bmatrix} \bar{E}_{p1} \\ \bar{E}_{p2} \end{bmatrix} + \begin{bmatrix} 0 \\ \bar{\beta}_{p4} \end{bmatrix} F_{p=FSC^*},
\end{aligned} \tag{25}
$$

and

$$
\begin{aligned}
\dot{\bar{E}}_c &= A_c \bar{E}_c + \Delta \varphi_c E_c + \beta_c F_{c-FSC^*} \\
&= \begin{bmatrix} a_{c11} & a_{c12} \\ a_{c21} & a_{c22} \end{bmatrix} E_c + \begin{bmatrix} 0 \\ \varphi_{c2} \end{bmatrix} E_c + \begin{bmatrix} 0 \\ \beta_{c2} \end{bmatrix} F_{c=FSC^*},
\end{aligned} \tag{26}
$$

for the double-pendulum and cart subsystems, respectively. The hybrid interval-valued fuzzy sliding controller with EKM (HFSC^{IV}) to balance a DPCS [27] is first reviewed. Then, the design of the hybrid interval-valued fuzzy sliding controller based on the information combination with crisp-interval-operation type reduction (HFSC^{CIOTR}) is proposed to avoid the computational complexity of the type-reduction process for the interval-valued FCs.

The hybrid interval-valued fuzzy sliding controller with crisp-interval-operation type reduction (HFSC^{CIOTR}) in Fig. 12 is proposed to be the combination of two

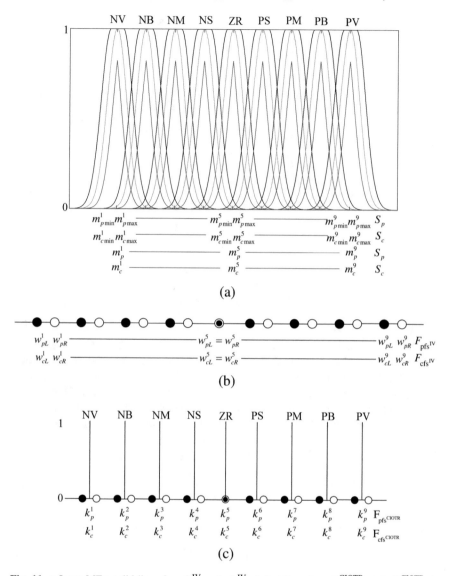

Fig. 11 **a** Input MFs: solid lines for pfs^{IV} and cfs^{IV}, dotted lines for pfs^{CIOTR} and cfs^{CIOTR}. **b** Precomputed w_L^i and w_R^i, with $1 \le i \le 9$, in ascending order. **c** Fuzzy singletons for pfs^{CIOTR} and cfs^{CIOTR}

interval-valued fuzzy sliding subcontrollers with crisp-interval-operation type reduction, p-FSC^{CIOTR} and c-FSC^{CIOTR}, for the double-pendulum and cart subsystems. p-FSC^{CIOTR} and c-FSC^{CIOTR} are designed to have outputs as

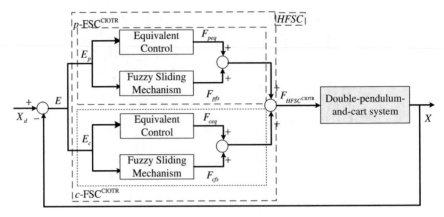

Fig. 12 Block diagram of the DPCS system with HFSC^{CIOTR}

$$F_{p\text{-}FSC^{CIOTR}} = F_{peq} + (\bar{C}_{p2}\bar{\beta}_{p4})^{-1} F_{pfs^{CIOTR}}$$
$$F_{c\text{-}FSC^{CIOTR}} = F_{ceq} + (c_{c6}\beta_{c2})^{-1} F_{cfs^{CIOTR}} \tag{27}$$

where $F_{pfs^{CIOTR}}$ and $F_{cfs^{CIOTR}}$ are the outputs of type-2 fuzzy sliding mechanisms pfs^{CIOTR} and cfs^{CIOTR} with crisp-interval-operation type reduction, respectively. Likewise, the state equations for the double-pendulum-and-cart subsystems with HFSC^{CIOTR} would have the same form as in (25) and (26), with $F_{p\text{-}FSC^{\star}}$ and $F_{c\text{-}FSC^{\star}}$ replaced with $F_{p\text{-}FSC^{CIOTR}}$ and $F_{c\text{-}FSC^{CIOTR}}$. The input MFs for the pfs^{CIOTR}(cfs^{CIOTR}) are provided as regular membership functions (dotted line in Fig. 11a) of type-reduced sets obtained using the crisp-interval-operation type reduction approach for the interval-valued fuzzy sets adopted in pfs^{IV}(cfs^{IV}). The output fuzzy singletons for pfs^{CIOTR} and cfs^{CIOTR} in Fig. 11c are calculated as

$$\tilde{k}_p^i = \frac{w_{pL}^i + w_{pR}^i}{2} \tag{28}$$

and

$$\tilde{k}_c^i = \frac{w_{cL}^i + w_{cR}^i}{2} \tag{29}$$

$i = 1, 2, \ldots, 9$, respectively. The same sliding functions are utilized for p-FSC^{CIOTR} and c-FSC^{CIOTR}. To have $\dot{S}_p = 0$ and $\dot{S}_c = 0$, the equivalent control actions for the double-pendulum and the cart subsystems in the sliding mode are

$$F_{peq} = -(\bar{C}_{p2}\bar{\beta}_{p4})^{-1}[\bar{C}_{p1}(\bar{A}_{p11}\bar{E}_{p1} + \bar{A}_{p12}\bar{E}_{p2}) + \bar{C}_{p2}(\bar{A}_{p21}\bar{E}_{p1} + \bar{A}_{p22}\bar{E}_{p2})]$$
$$F_{ceq} = -(c_{c6}\beta_{c2})^{-1}[c_5(a_{c11}e_{c5} + a_{c12}e_{c6}) + c_6(a_{c21}e_{c5} + a_{c22}e_{c6})]. \tag{30}$$

With the sum-product fuzzy inference and centroid defuzzification, the control actions of the provided fuzzy mechanisms are

$$F_{pfs^{CIOTR}} = -\frac{\sum_{i=1}^{9} k_p^i \mathrm{sign}(S_p) M_p^i(S_p)}{\sum_{i=1}^{9} M_p^i(S_p)}$$

$$F_{cfs^{CIOTR}} = -\frac{\sum_{i=1}^{9} k_c^i \mathrm{sign}(S_c) M_c^i(S_c)}{\sum_{i=1}^{9} M_c^i(S_c)}$$

(31)

where $k_p^i = |\tilde{k}_p^i| > 0$, $k_c^i = |\tilde{k}_c^i| > 0$, and M_p^i, M_c^i are the input MFs in the ith rule for pfs^{CIOTR} and cfs^{CIOTR} respectively, $i = 1, 2, \cdots, 9$. Substituting F_{peq}, F_{ceq}, $F_{pfs^{CIOTR}}$, and $F_{cfs^{CIOTR}}$ into (27), the output of the controllers p-FSC^{CIOTR} and c-FSC^{CIOTR} are

$$F_{p-FSC^{CIOTR}} = F_{peq} + (\bar{C}_{p2}\bar{\beta}_{p4})^{-1} F_{pfs^{CIOTR}}$$
$$= -(\bar{C}_{p2}\bar{\beta}_{p4})^{-1} [\bar{C}_{p1}(\bar{A}_{p11}\bar{E}_{p1} + \bar{A}_{p12}\bar{E}_{p2}) + \bar{C}_{p2}(\bar{A}_{p21}\bar{E}_{p1} + \bar{A}_{p22}\bar{E}_{p2})]$$
$$- \frac{\sum_{i=1}^{9} k_p^i \mathrm{sign}(S_p) M_p^i(S_p)}{\sum_{i=1}^{9} M_p^i(S_p)}$$

$$F_{c-FSC^{CIOTR}} = F_{ceq} + (c_{c6}\beta_{c2})^{-1} F_{cfs^{CIOTR}}$$
$$= -(c_6\beta_{c2})^{-1} [c_5(a_{c11}e_{c5} + a_{c12}e_{c6}) + c_6(a_{c21}e_{c5} + a_{c22}e_{c6})]$$
$$- \frac{\sum_{i=1}^{9} k_c^i \mathrm{sign}(S_c) M_c^i(S_c)}{\sum_{i=1}^{9} M_c^i(S_c)}$$

(32)

To balance the double pendulum at the straight upward direction, the control action defined to be $F_{HFSC^{CIOTR}} = F_{p-FSC^{CIOTR}} - F_{c-FSC^{CIOTR}}$ (Fig. 12).

Since the emphasis of this paper is given to the effectiveness of the proposed type reduction approach, hybrid interval-valued fuzzy controller HFSC^{IV} with EKM in [27] is taken as an example. As indicated in [27], the DPCS using HFSC^{IV} with EKM is stable and robust. Therefore, the DPCS using HFSC^{CIOTR} would be also stable if the type reduction approach based on the crisp interval operations is effective. The simulation results in the next section empirically prove that the DPCS with HFSC^{CIOTR} is stable and robust and the proposed type reduction approach is effective.

The main parameters of the DPCS system in [27] are adopted and the limitations for the DPCS system mentioned in [27] are applied in simulations. Also the parameters designed for HFSC^{IV} in [27] are utilized directly. According to the parameters of HFSC^{IV}, the parameters for the proposed HFSC^{CIOTR} are calculated. Moreover, the performance comparisons between HFSC^{IV} and HFSC^{CIOTR} are included.

The interval-valued primary MFs utilized for the subcontrollers p-FSC^{IV} and c-FSC^{IV} in HFSC^{IV} are Gaussian functions with uncertain means $m_p^i \in [m_{pmin}^i \ m_{pmax}^i]$, $m_c^i \in [m_{cmin}^i \ m_{cmax}^i]$ and variances σ_p^i, σ_c^i, $i = 1, 2, \cdots, 9$, respectively. Also, the centroids of the interval-valued consequent sets are denoted as $w_{pL}^i \ w_{pR}^i$, $w_{cL}^i \ w_{cR}^i$, with $i = 1, 2, \cdots, 9$. The parameters in [27] for HFSC^{IV} are,

$$
\begin{cases}
m_{pmin} = [m^1_{pmin} \cdots m^9_{pmin}] = [-0.55 \ -0.3 \ -0.16 \ -0.015 \ -0.002 \ 0.005 \ 0.06 \ 0.1 \ 0.35] \\
m_{pmax} = [m^1_{pmax} \cdots m^9_{pmax}] = [-0.35 \ -0.1 \ -0.06 \ -0.005 \ 0.002 \ 0.015 \ 0.16 \ 0.3 \ 0.55] \\
\sigma_p = [\sigma^1_p \cdots \sigma^9_p] = [0.2 \ 0.1 \ 0.05 \ 0.03 \ 0.003 \ 0.03 \ 0.05 \ 0.1 \ 0.2] \\
w_{pL} = [w^1_{pL} \cdots w^9_{pL}] = [-6.1 \ -3.1 \ -2.4 \ -2.1 \ 0 \ 1.9 \ 2.2 \ 2.9 \ 5.9] \\
w_{pR} = [w^1_{pR} \cdots w^9_{pR}] = [-5.9 \ -2.9 \ -2.2 \ -1.9 \ 0 \ 2.1 \ 2.4 \ 3.1 \ 6.1]
\end{cases}
$$

$$
\begin{cases}
m_{cmin} = [m^1_{cmin} \cdots m^9_{cmin}] = [-10.6 \ -1.3 \ -0.054 \ -0.035 \ -0.005 \ 0.025 \ 0.044 \ 1.1 \ 10.2] \\
m_{cmax} = [m^1_{cax} \cdots m^9_{cmax}] = [-10.2 \ -1.1 \ -0.044 \ -0.025 \ -0.005 \ 0.035 \ 0.054 \ 1.3 \ 10.6] \\
\sigma_c = [\sigma^1_c \cdots \sigma^9_c] = [4 \ 1.7 \ 0.3 \ 0.005 \ 0.01 \ 0.005 \ 0.3 \ 0.17 \ 4] \\
w_{cL} = [w^1_{cL} \cdots w^9_{cL}] = [-26 \ -19.5 \ -7.5 \ -2.1 \ 0 \ 1.9 \ 7.3 \ 19.3 \ 25.2] \\
w_c = [w^1_{cR} \cdots w^9_{cR}] = [-25.2 \ -19.3 \ -7.3 \ -1.9 \ 0 \ 2.1 \ 7.5 \ 19.5 \ 26]
\end{cases}
$$

According to the parameters for HFSC^{IV} in previous subsection, the parameters for the input MFs of HFSC^{CIOTR} are determined (as follow) with the proposed crisp-interval-operations type reduction approach in section II. Also, the parameters for the output singletons in HFSC^{CIOTR} are found with (28) and (29). The parameters can be determined directly before its application to find the control action. Therefore, the time-consuming of the type reduction approach is completely avoided.

$$
\begin{cases}
m_p = [m^1_p \cdots m^9_p] = [-0.45 \ -0.2 \ -0.11 \ -0.01 \ 0 \ 0.01 \ 0.11 \ 0.2 \ 0.45] \\
\sigma_{pM} = [\sigma^1_{pM} \cdots \sigma^9_{pM}] = [0.22 \ 0.16 \ 0.05 \ 0.03 \ 0.003 \ 0.03 \ 0.05 \ 0.12 \ 0.22] \\
k_p = [k^1_p \cdots k^9_p] = [-6 \ -3 \ -2.3 \ -2 \ -2 \ 0 \ 2 \ 2.3 \ 3 \ 6] \\
m_c = [m^1_c \cdots m^9_c] = [-10.4 \ -1.2 \ -0.049 \ -0.03 \ 0 \ 0.03 \ 0.049 \ 1.2 \ 10.4] \\
\sigma_{cM} = [\sigma^1_{cM} \cdots \sigma^9_{cM}] = [3.94 \ 1.74 \ 0.38 \ 0.08 \ 0.01 \ 0.08 \ 0.38 \ 1.74 \ 3.94] \\
k_c = [k^1_c \cdots k^9_c] = [-25.6 \ -19.4 \ -7.4 \ -2 \ 0 \ 2 \ 7.4 \ 19.4 \ 25.6]
\end{cases}
$$

The performance comparisons between the DPCS with HFSC^{IV} (using EKM type-reduction) and the DPCS with the proposed HFSC^{CIOTR} are provided in this subsection. To indicate the robustness of HFSC^{CIOTR} and HFSC^{IV}, an external random noise d ($|d| \leq 0.005$ from 0 to 25 s) is added into two pendulums for simulations. The simulation results are shown in Fig. 13. In Fig. 13, the HFSC^{CIOTR} is indicated to have better performance than HFSC^{IV} in the sense of short convergence time for the DPCS with initial condition $[\pi/6 \ 0 \ \pi/18 \ 0 \ 0 \ 0]$ (note that the convergence time is defined to be the time t_s such that the absolute error of the state is smaller than 0.02 when $t \geq t_s$). Moreover, two criterion indexes [the convergence time and the sum of absolute error (SAE)] are adopted to further evaluate HFSC^{CIOTR} and HFSC^{IV} and the numerical performance comparisons are listed in Table 3. For most of the evaluations in Table 3, it can be seen that the convergence time and SAE for using HFSC^{CIOTR} with noise is smaller than using HFSC^{IV} with noise. Since the proposed

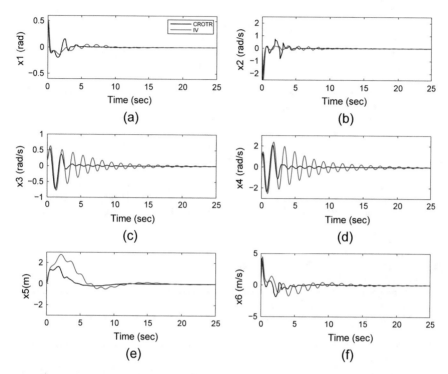

Fig. 13 Simulation results with initial condition $[\pi/6\ 0\ \pi/18\ 0\ 0\ 0]$ and with a random noise of $d = \pm 0.005$

Table 3 Performance comparison of HFSC^{IV} and HFSC^{CIOTR}

	Convergence times			SAE		
	x_1	x_3	x_5	x_1	x_3	x_5
HFSC^{IV}	8.84	20.69	24.18	5.60	28.22	116.56
HFSC^{CIOTR}	5.50	14.00	16.53	6.85	11.29	38.91

type-reduction approach can be completed before its application to the design of the fuzzy sliding controller, the time-consuming difficulty of type reduction can be avoided.

References

1. C.W. Tao, J.S. Taur, Design of fuzzy controllers with adaptive rule insertion. IEEE Trans. Syst. Man Cybern.**SMC-29(3)**, 389–397 (1999)
2. L.A. Zadeh, Fuzzy sets. Inform. Control **8**(3), 338–353 (1965)
3. D. Driankov, H. Helendoorn, M. Reinfrank, *An Introduction to Fuzzy Control* (Springer, New York, 1993)

4. H.-T. Nguyen, V. Kreinovich, How stable is a fuzzy linear system, in *Proceedings od 3rd IEEE Conference on Fuzzy Systems* (1994), pp. 1023–1027
5. H.T. Nguyen, Fuzziness in statistics, in *Presented at the Speech Notes, Taiwan* (2000)
6. C.W. Tao, J.S. Taur, Robust fuzzy control for a plant with fuzzy linear model. IEEE Trans. Fuzzy Syst. **13**(1), 30–41 (2005)
7. C.-C. Lee, Fuzzy logic in control systems: fuzzy logic controller part I. IEEE Trans. Syst. Man Cybern. **20**(2), 404–418 (1990)
8. K.-H. Su, S.-J. Huang, C.-Y. Yang, Development of robotic grasping gripper based on smart fuzzy controller. Int. J. Fuzzy Syst. **17**(4), 595–608 (2015)
9. C.W. Tao, M.L. Chan, T.T. Lee, Adaptive fuzzy sliding mode controller for linear systems with mismatched time-varying uncertainties. IEEE Trans. Syst. Man Cybern. Part B Cybern. **33**(2), 238–294 (2003)
10. K. Hirota, *Industrial Application of Fuzzy Technology* (Springer, New York, 2004)
11. P.-Z. Lin, C.-M. Lin, C.-F. Hsu, T.-T. Lee, Type-2 fuzzy controller design using a sliding-mode approach for application to dc-dc converters. IEE Proc. Electr. Power Appl. **152**(6), 1482–1488 (2005)
12. L.A. Zadeh, The concept of a linguistic variable and its application to approximate reasoning I. Inf. Sci. **8**(3), 199–249 (1975)
13. E. Roventa, T. Spircu, Averaging procedures in defuzzification processes. Fuzzy Sets Syst. **136**, 375–385 (2003)
14. Y. Ogura, S. Li, D. Ralescu, Set defuzzification and choquet integral. Int. J. Uncertain. Fuzziness Knowl. Based Syst. **9**(1), 1–12 (2001)
15. J.M. Mendel, *Uncertain Rule-based Fuzzy Logic Systems: Introduction and New Directions* (Prentice-Hall, Upper Saddle River, 2001)
16. D. Wu, J.M. Mendel, Enhanced-karnik-mendel algorithms. IEEE Trans. Fuzzy Syst. **17**(4), 923–934 (2009)
17. N.N. Karnik, J.M. Mendel, Q. Liang, Type-2 fuzzy logic systems. IEEE Trans. Fuzzy Syst. **7**(6), 643–658 (1999)
18. J.M. Mendel, Advance in type-2 fuzzy sets and systems. Inf. Sci. **177**(1), 84–110 (2007)
19. F.J. Lin, P.H. Chou, Adaptive control of two-axis motion control system using interval type-2 fuzzy neural network. IEEE Trans. Ind. Electron. **56**(1), 178–193 (2009)
20. H. Wu, J.M. Mendel, Uncertainty bounds and their use in the design of interval type-2 fuzzy logic systems. IEEE Trans. Fuzzy Syst. **10**(5), 622–639 (2002)
21. M. Melgarejo, A fast recursive method to compute the generalized centroid of an interval type-2 fuzzy set, in *2007 Annual Meeting of the North American Fuzzy Information Processing Society* (2007), pp. 190–194
22. H. Bernal, K. Duran, M. Melgarejo, A comparative study between two algorithms for computing the generalized centroid of an interval type-2 fuzzy set, in *IEEE International Conference on Fuzzy Systems* (2008), pp. 954–959
23. K. Duran, H. Bernal, M. Melgarejo, Improved iterative algorithm for computing the generalized centroid of an interval type-2 fuzzy set, in *Fuzzy Information Processing Society* (2008), pp. 1–5
24. C.W. Tao, C.-W. Chang, J.S. Taur, A simplify type reduction for interval type-2 fuzzy sliding controllers. Int. J. Fuzzy Syst. **15**(4), 460–470 (2013)
25. C.W. Tao, J.S. Taur, C.-W. Chang, H.-R. Chen, An alternative type reduction approach based on information combination with interval operations for interval-valued fuzzy sliding controllers. Int. J. Intell. Syst. **32**(3), 291–309 (2016)
26. S. Chanas, On the interval approximation of a fuzzy number. Fuzzy Sets Syst. **122**, 353–356 (2001)
27. C.W. Tao, J. Taur, J.H. Chang, S.-F. Su, Adaptive fuzzy switched swing-up and sliding control for the double-pendulum-and-cart system. IEEE Trans. Syst. Man Cybern. Part B Cybern. **40**, 241–252 (2010)

A Remark on the Caristi's Fixed Point Theorem and the Brouwer Fixed Point Theorem

S. Dhompongsa and P. Kumam

Dedicated to Prof. Hung T. Nguyen for his 75th anniversary

Abstract It is well-known that a partial order induced from a lower semi-continuous map gives us a clear picture of a proof of the Caristi's fixed point theorem. The proof utilized Zorn's lemma to guarantee the existence of a minimal element which turns out to be a desired fixed point. The proof cannot be carried over to prove the Brouwer fixed point theorem. We show that making an idea of ordering, we get a proof of the later one.

Mathematics Subject Classification: Primary 47H10 · Secondary 54H25

1 Introduction

A well-known fixed point theorem on the setting of complete lattice is Tarski's fixed point theorem which states that:

S. Dhompongsa (✉)
KMUTTFixed Point Research Laboratory, KMUTT-Fixed Point Theory and Applications Research Group, Department of Mathematics, Faculty of Science, King Mongkut's University of Technology Thonburi (KMUTT), 126 Pracha-Uthit Road, Bang Mod, Thrung Khru, Bangkok 10140, Thailand
e-mail: sompong.dho@kmutt.ac.th

P. Kumam
Center of Excellence in Theoretical and Computational Science (TaCS-CoE), Science Laboratory Building, King Mongkut's University of Technology Thonburi (KMUTT), 126 Pracha-Uthit Road, Bang Mod, Thrung Khru, Bangkok 10140, Thailand
e-mail: poom.kumam@mail.kmutt.ac.th

V. Kreinovich (ed.), *Statistical and Fuzzy Approaches to Data Processing, with Applications to Econometrics and Other Areas*, Studies in Computational Intelligence 892,
https://doi.org/10.1007/978-3-030-45619-1_7

93

Theorem 1 (Tarski's Fixed Point Theorem) *[12] If X is a complete lattice and $T : X \to X$ is monotone (order preserving), then the set of fixed points forms a non-empty complete lattice. That is a monotone self map on a complete lattice has a least fixed point.*

Tarski's theorem has a huge applications in various fields. For examples

(i) Tarski's $\Rightarrow$ Cantor-Schroder-Bernstein theorem, which asserts that whenever $f : B \to A$, $g : B \to A$ are both injective functions, then A and B are equivalent.

Proof The power set 2^A is a complete lattice under inclusion. Define $\varphi : 2^A \to 2^A$ by

$$X \mapsto A \setminus g(B \setminus f(X), \quad X \in 2^A.$$

φ is $\subset$-preserving:

$$\begin{aligned}
X \subset Y &\Rightarrow f(X) \subset f(Y) \\
&\Rightarrow B - f(X) \supset B - f(Y) \\
&\Rightarrow A - g(B - f(X)) \subset A - g(B - f(Y)) \\
&\Rightarrow \varphi(X) \subset \varphi(Y).
\end{aligned}$$

$$\begin{aligned}
\text{Tarski's} &\Rightarrow A \setminus X = g(B \setminus f(X)) \text{ for some } X \\
&\Rightarrow h := f_{|X} \cup f_{|g^{(-1)(A \setminus X)}} : A \to B \text{ is a desired bijection.}
\end{aligned}$$

(ii) Differential Equations.
(iii) Computer Science (it is used in the field of denotational semantics and abstract interpretation).
(iv) Game Theory.
(v) Graph Theory (a very nice application of Tarski is a generalization of Hall's Marriage Theorem).
(vi) Computational Geometry.
(vii) Tarski's FPT can be used to show the existence of fixed points of self maps on compact sets of probability measures equipped with stochastic ordering (i.e., existence of stationary distribution of dynamic economics) see [7] for more information on the applications of Tarski's theorem in economics.

For the setting of complete metric spaces, we refer to:

Theorem 2 (Caristi's Fixed Point Theorem) *Let (X, d) be a complete metric space and $\varphi : X \to \mathbb{R}$ l.s.c., and bounded from below. Assume that $T : X \to X$ satisfies*

$$d(x, Tx) \leq \varphi(x) - \varphi(Tx)$$

for each $x \in X$, then T has a fixed point.

Proof (cf. Khamsi [5]) Define an order $\prec_\varphi$ on X so that

$$x \prec_\varphi z \text{ if and only if } d(x, z) \leq |\varphi(z) - \varphi(x)| = \varphi(z) - \varphi(x) \in \mathbb{R}^1_+ = [0, \infty).$$

Claim: With respect to the order $\prec_\varphi$, $(X, \prec_\varphi)$ contains a minimal element. To see this, we first observe, by the boundedness from below of φ, that $\sum_{n=1}^\infty d(x_{n-1}, x_n) < \infty$ for any decreasing sequence $\{x_n\}$ in X.

Now if $\{x_\alpha\}_{\alpha \in \Gamma}$ is a decreasing chain in $(X, \prec_\varphi)$, take an increasing sequence $\{\alpha_n\}$ of elements from Γ such that $\lim_{n\to\infty} \varphi(x_{\alpha_n}) = \inf_{\alpha \in \Gamma}\{\varphi(x_\alpha)\}$. From the above observation, $\{x_{\alpha_n}\}$ converges to $\overline{x} \in X$. As φ is being l.s.c., $\varphi(\overline{x}) \leq \varphi(x_{\alpha_n})$ for all n. For a fixed n_0, $\varphi(x_{\alpha_{n_0}}) - \varphi(\overline{x}) \geq \varphi(x_{\alpha_{n_0}}) - \varphi(x_{\alpha_n}) \geq d(x_{\alpha_{n_0}}, x_{\alpha_n})$ for all $n \geq n_0$ which implies $\varphi(x_{\alpha_{n_0}}) - \varphi(\overline{x}) \geq d(x_{\alpha_{n_0}}, \overline{x})$. This shows that $\overline{x}$ is a lower bound of $\{x_{\alpha_n}\}$. Next suppose for some $\beta \in \Gamma$, $x_\beta \prec_\varphi x_{\alpha_n}$ for all n. This implies $d(x_\beta, x_{\alpha_n}) \leq \varphi(x_{\alpha_n}) - \varphi(x_\beta) \leq \varphi(x_{\alpha_n}) - \inf_{\alpha \in \Gamma}\{\varphi(x_\alpha)\}$. Taking $n \to \infty$ we see that $x_\beta = \overline{x}$. On the other hand, if $x_{\alpha_{n_0}} \prec_\varphi x_\beta$ for some n_0, we have $\overline{x} \prec_\varphi x_\beta$. In any case, this shows that $\overline{x}$ is a lower bound of $x_{\alpha \alpha \in \Gamma}$. Apply Zorn's lemma to conclude that $(X, \prec_\varphi)$ contains a minimal element as claimed. If $\overline{x}$ is a minimal element, $T\overline{x} \prec_\varphi \overline{x}$ would imply $T\overline{x} = \overline{x}$, i.e., $\overline{x}$ is a fixed point of T. $\qquad\square$

We extend the theorem to a higher dimensional space. Let (X, d) be a complete metric space, $K = \mathbb{R}^d_+ = [0, \infty)^d$, $\varphi : X \to \mathbb{R}^d$ is lower semi-continuous (l.s.c.) of [6, 8], i.e., at any $\overline{x} \in X$ and for any sequence (x_n) in X converging to $\overline{x}$, there exists a sequence $\{b_n\}$ in $\mathbb{R}^d$ converging to $\varphi(\overline{x})$ such that $b_n \leq \varphi(x_n)$ for all n. Here $a \leq b$ in $\mathbb{R}^d$ means $b - a \in K$. Following Tammer [11], φ is said to be bounded below if for some $y \in \mathbb{R}^d$, the range $\varphi(X) \subset y + K$. In the following, we may assume y to be 0.

Theorem 3 (Caristi's Fixed Point Theorem) *Let (X, d) be a complete metric space and $\varphi : X \to \mathbb{R}^d$ l.s.c. and bounded from below. Any map $T : X \to X$ satisfying*

$$\varphi(x) - \varphi(Tx) \in K \text{ and } d(x, Tx) \leq \|\varphi(x) - \varphi(Tx)\|$$

for all $x \in X$ has a fixed point.

Choose δ so that $\|\cdot\| \leq \delta\|\cdot\|_1$. Define an order $\prec_\varphi$ on X so that

$$x \prec_\varphi z \text{ if and only if } \varphi(z) - \varphi(x) \in K \text{ and } d(x, z) \leq \delta\|\varphi(z) - \varphi(x)\|_1.$$

$\prec_\varphi$ is a partial order: We only need to verify the transitivity of $\prec_\varphi$. This follows from the fact that $\|(a_1, \ldots, a_d)\|_1 = a_1 + \ldots + a_d$ if all $a_1, \ldots, a_d$ are nonnegative, which can apply to vectors $(a_1, \ldots, a_d)$ in K.

Proof As before, if $\{x_\alpha\}_{\alpha \in \Gamma}$ is a decreasing chain in $(X, \prec_\varphi)$, take an increasing sequence α_n of elements from Γ such that $\lim_{n\to\infty} \|\varphi(x_{\alpha_n})\|_1 = \inf_{\alpha \in \Gamma}\{\|\varphi(x_\alpha)\|_1\}$. Thus, x_{α_n} converges to some $\overline{x} \in X$. As φ is being l.s.c., there exists a sequence (b_n) in $\mathbb{R}^d$ converging to $\varphi(\overline{x})$ and $\varphi(x_{\alpha_n}) - b_n \in K$ for all n. For a fixed n_0, $\delta\|\varphi(x_{\alpha_{n_0}}) - $

$b_n\|_1 \geq \delta\|\varphi(x_{\alpha_{n_0}}) - \varphi(x_{\alpha_n})\|_1 \geq d(x_{\alpha_{n_0}}, x_{\alpha_n})$ for all $n \geq n_0$ which implies $\delta\|\varphi(x_{\alpha_{n_0}}) - \varphi(\overline{x})\|_1 \geq d(x_{\alpha_{n_0}}, \overline{x})$. Since $\varphi(x_{\alpha_{n_0}}) - b_n$ is a finite sum $\sum_{k=n_0}^{n-1} \varphi(x_{\alpha_k}) - \varphi(x_{\alpha_{k+1}}) + \varphi(x_{\alpha_n}) - b_n$ of elements in K, its limit $\varphi(x_{\alpha_{n_0}}) - \varphi(\overline{x})$ also belongs to K. Thus $\overline{x}$ is a lower bound of x_{α_n}. Next suppose for some $\beta \in \Gamma$, $x_\beta \prec_\varphi x_{\alpha_n}$ for all n. Observe that $\varphi(x_{\alpha_n})$ belongs to the cone $\varphi(x_\beta) + K$ and the norm $\|\varphi(x_\beta)\|_1$ is smallest in this cone. Therefore $\lim_{n\to\infty} \|\varphi(x_{\alpha_n})\|_1 = \|\varphi(x_\beta)\|_1$. Consequently, any convergent subsequence of $\varphi(x_{\alpha_n})$ must converge to $\varphi(x_\beta)$. This implies $\varphi(x_{\alpha_n})$ converges to $\varphi(x_\beta)$. From $d(x_\beta, x_{\alpha_n}) \leq \delta\|\varphi(x_{\alpha_n}) - \varphi(x_\beta)\|_1$, we see by taking $n \to \infty$ that $x_\beta = \overline{x}$.

As for another case when $x_{\alpha_{n_0}} \prec_\varphi x_\beta$ for some n_0, we have $\overline{x} \prec_\varphi x_\beta$ In conclusion, $\overline{x}$ is a lower bound of $x_{\alpha_{\alpha \in \Gamma}}$. Zorn's lemma then ensures that $(X, \prec_\varphi)$ contains a minimal element as claimed. Take a minimal element $\overline{x}$. Thus $T\overline{x} - \overline{x}$ and $\overline{x} - T\overline{x}$ both belong to K which implies $T\overline{x} = \overline{x}$, i.e., $\overline{x}$ is a fixed point.   □

We note in passing that Caristi's fixed point theorem and the following two famous theorems are equivalent (see [4, 9, 10]).

Theorem 4 (Ekeland) *Let (X, d) be a complete metric space and $\varphi : X \to \mathbb{R}$ l.s.c. and bounded from below. Let $\epsilon > 0$ and $u \in X$ satisfy $\varphi(u) < \inf_{x \in X} \varphi(x) + \epsilon$. Then there exists some point $v \in X$ such that*

(i) $\varphi(v) \leq \varphi(u)$,
(ii) $d(u, v) \leq 1$,
(iii) For each $w \neq v$, $\varphi(v) - \epsilon d(u, v) < \varphi(w)$.

Theorem 5 (Takahashi) *Let (X, d) be a complete metric space and $\varphi : X \to \mathbb{R}$ l.s.c. and bounded from below. Suppose that for each $u \in X$ with $\inf_{x \in X} \varphi(x)$, there exists $v \in X$, $v \neq u$ and $\varphi(v) + d(u, v) < \varphi(u)$. Then f attains its infimum.*

We note that Caristi's fixed point theorem generalizes the Banach fixed point theorem. To see this, if f is a contraction with the contraction coefficient $k \in (0, 1)$, then define $\varphi(x) = \frac{1}{1-k} d(x, f(x))$.

$$\Rightarrow \varphi(x) - \varphi(f(x)) = \frac{1}{1-k}[d(x, f(x)) - d(f(x), f^2(x))]$$
$$> \frac{1}{1-k}[d(x, f(x)) - kd(x, f(x))]$$
$$= d(x, f(x)).$$

Neither the contraction property nor the continuity of the self map required in the Caristi's fixed point theorem.

Of course, the most famous theorem in fixed point theory is the following:

Theorem 6 (Brouwer Fixed Point Theorem (BFT)) *For the unit cube $[0, 1]^d$ of the Euclidean space $\mathbb{R}^d$, let $T : [0, 1]^d \to [0, 1]^d$ be a continuous function. Then T has a fixed point, i.e., a point $x \in [0, 1]^d$ with $T(x) = x$.*

In some sense, both Tarski's and Caristi's theorems involve some orders. It is the purpose of this paper to present a proof of BFT by way of ordering.

2 Preliminaries

Consider $\| \cdot \| = \| \cdot \|_2$ and put $K = [0, \infty)^d$, $K^0 = int\,K$. Let $\{e_1 \ldots, e_d\}$ be the standard basis for $\mathbb{R}^d$, that is, based on the Kronocker delta δ_{ij}, $e_j = \{(\delta)_{ij}\}_{i=1}^d$. For each $u \in [0, \sqrt{d}]$, let $H_u \subset \mathbb{R}^d$ be the hyperplane passing the point $(u, \ldots, u)$ and having $\overline{1} = (1, \ldots, 1)$ as its normal vector. For $j = 1, \ldots, d$, let π_j be the projection along e_j from a subset of points $(x_1, \ldots, x_d)$ of H_u into $[0, 1]^d$, where $x_i \in [0, 1]$ for $i \neq j$. Let S_{uj} be the image of π_j and the simplex $\Delta_u = [0, 1]^d \cap H_u$.

Write the given continuous function $T : [0, 1]^d \to [0, 1]^d$ as $T = (f_1, \ldots, f_d)$ where $f_j : K \to [0, 1]$ is continuous for each j. For each u, draw the graph of f_j restricted to S_{uj} via the formula

$$
g_{uj} : (x_1, \ldots, x_{j-1}, 0, x_{j+1}, \ldots, x_d) \mapsto
$$
$$
(x_1, \ldots, x_{j-1}, f_j(x_1, \ldots, x_{j-1}, x_{uj}, x_{j+1}, \ldots, x_d), x_{j+1}, \ldots, x_d)
$$

for $(x_1, \ldots, x_{j-1}, x_{uj}, x_{j+1}, \ldots, x_d) \in S_{uj}$. Denote f_{uj} for the restriction $f_j|_{S_{uj}}$. Thus the graph of f_j at u means the set of points $(x_1, \ldots, x_{j-1}, f_j(x_1, \ldots, x_{j-1}, x_j, x_{j+1}, \ldots, x_d), x_{j+1}, \ldots, x_d), (x_1, \ldots, x_{j-1}, x_j, x_{j+1}, \ldots, x_d) \in S_{uj}$. Lastly, denote T_u for the restriction of T to $\cup_{j=1}^d S_{uj}$.

Our proof relies on this result (whose proof included for convenience).

Lemma 1 *[2, Lemma 3.1] For each u, the graphs of $f_1, \ldots, f_d$ at u intersect at a point. That is there are points $(a_1, \ldots, a_d), (x_1, \ldots, x_d) \in [0, 1]^d$, such that $f_j(a_1, \ldots, a_{j-1}, x_j, a_{j+1}, \ldots, a_d) = a_j$, and $(a_1, \ldots, a_{j-1}, x_j, a_{j+1}, \ldots, a_d) \in S_{uj}$ for each j.*

Proof Put $\pi_j : (x_1, \ldots, x_d) \mapsto \sum_{i=1, i \neq j}^d x_i e_i, (x_1, \ldots, x_d) \in [0, 1]^d$, and consider the function

$$
S := \pi_1 g_{ud} \pi_d \ldots g_{u3} \pi_3 g_{u2} \pi_2 g_{u1} : \pi_1([0, 1]^d) \to \pi_1([0, 1]^d).
$$

By the BFT (for $[0, 1]^{d-1}$), $S(v) = v$ for some v. Clearly $g_1(v)$ is a desired point of intersection.

In the following proof, instead of looking for x with $Tx - x \in K$, we look for x with $Tx - x \notin -K^0$.

3 A New Proof of the BFT (A Complementation of [1–3])

Proof Suppose 0 is not a fixed point of T. Thus obviously, $T(0) \notin -K^0$. We can see that for sufficiently small u, the following properties hold:

(i) For some j, some components of the set $\{x \in \Delta_u : f_{uj}(x) < x_j\}$ lie properly inside the intersection of the sets $\{x \in \Delta_u : f_{uj}(x) < x_i\}$ for all $i \neq j$. Or they have empty intersection with the intersection of the sets $\{x \in \Delta_u : f_{uj}(x) < x_i\}$ for all $i \neq j$.

(ii) Each component of the set $\{x \in \Delta_u : f_{uj}(x) < x_j\}$ for any j has an empty intersection with the intersection of the sets $\{x \in \Delta_u : f_{uj}(x) > x_i\}$.

Clearly, $T_u(x) - x \notin -K^0$ for all $x \in \Delta_u$, for all such u.

Suppose u_0 is the supremum of all u where $T_v(x) - x \notin -K^0$ for all $x \in \Delta_u$, for all $v < u$. Clearly, $u_0 > 0$ and if (i) holds, $f_{uj}(x) = x_j$, i.e., $T_u(x) - x$ lies in the boundary of $-K^0$. But when $u \geq u_0$ with $u - u_0$ is sufficiently small, and (i) holds, we would have $f_{uj}(x) < x_j$, i.e., $T_u(x) - x$ lies in $-K^0$. We deform f_{uj} to f'_{uj} so that $f'_{uj} = x_j$. This can be done by a projection along e_j onto Δ_u. Let put $T'_u = (f'_{u1}, \ldots, f'_{ud})$. In addition to (i) and (ii) above, u assumes as well the property:

(iii) T'_u produces no extra points of intersection $(a_1, \ldots, a_d < u)$ of $f'_{u1}, \ldots, f'_{ud}$ with $a_1 + \cdots + a_d \geq u$.

Let u_1 be the supremum of all u where for all $v < u$, v satisfies (i), (ii), (iii). If $u_1 = \sqrt{d}$, then $\bar{1}$ is a fixed point of T. This follows from Lemma 2.1 together with (ii) and (iii). Thus $0 < u_1 < \sqrt{d}$, and we claim that T has a fixed point in the set Δ_{u_1}. For if there is no such a fixed point, then we first observe that for $0 < u < u_1 < w < \sqrt{d}$ with small $w - u$, say $w - u < \delta$, so that, for some $\epsilon > 0$, $\|T(x) - x| > \epsilon$ for all $x \in U_{i=1,\ldots,d, u < v < w}, S_{vi}$, and the δ-neighborhoods of the disjoint sets in (i) and (ii) are disjoint. Here we follow the uniform continuity of T. Define for each j,

$$f''_{wj}(x_1, \ldots, x_{j-1}, x_{wj}, x_{j+1}, \ldots, x_d) = f_{wj}(x_1, \ldots, x_{j-1}, x_{wj}, x_{j+1}, \ldots, x_d)$$
$$+ f'_{uj}(x_1, \ldots, x_{j-1}, x_{uj}, x_{j+1}, \ldots, x_d)$$
$$- f_{uj}(x_1, \ldots, x_{j-1}, x_{uj}, x_{j+1}, \ldots, x_d) \quad (1)$$

where $(x_1, \ldots, x_{j-1}, x_{uj}, x_{j+1}, \ldots, x_d) \in S_{uj}$ and $(x_1, \ldots, x_{j-1}, x_{wj}, x_{j+1}, \ldots, x_d)$ $\in S_{wj}$. Clearly f''_{wj} is continuous. Consider the estimates:

$$|f_w j''(x_1, \ldots, x_{j-1}, x_{wj}, x_{j+1}, \ldots, x_d) - f'_{uj}(x_1, \ldots, x_{j-1}, x_{uj}, x_{j+1}, \ldots, x_d)|$$
$$= |f_{wj}(x_1, \ldots, x_{j-1}, x_{wj}, x_{j+1}, \ldots, x_d) - f_{uj}(x_1, \ldots, x_{j-1}, x_{uj}, x_{j+1}, \ldots, x_d)|$$
$$\leq T(x_1, \ldots, x_{j-1}, x_{wj}, x_{j+1}, \ldots, x_d) - T(x_1, \ldots, x_{j-1}, x_{uj}, x_{j+1}, \ldots, x_d)$$
$$\leq \sup_{\|x-y\| \leq w-u} \|T(x) - T(y)\|$$
$$\to 0 \text{ as } w - u \to 0. \quad (2)$$

(1) implies that $f''_{wj}(x_1, \ldots, x_{(j-1)}, x_{wj}, x_{(j+1)}, \ldots, x_d) = f_{wj}(x_1, \ldots, x_{(j-1)}, x_{wj}, x_{(j+1)}, \ldots, x_d)$ if and only if $f'_{uj}(x_1, \ldots, x_{(j-1)}, x_{uj}, x_{(j+1)}, \ldots, x_d) = f_{uj}(x_1, \ldots, x_{(j-1)}, x_{uj}, x_{(j+1)}, \ldots, x_d)$. Thus, together with (2), w satisfies (i), (ii), (iii) for all w with $w - u$ sufficiently small, contradicting to the definition of u_1. Therefore T has a fixed point in the set Δ_{u_1} and the proof is complete. $\square$

Acknowledgements The authors acknowledge the financial support provided by the Center of Excellence in Theoretical and Computational Science (TaCS-CoE), KMUTT. The first author also would like to thank the Excellence Center in Economics, Chiang Mai University for the support.

References

1. N. Chuensupantharat, P. Kumam, S. Dhompongsa, A graphical proof of the Brouwer fixed point theorem. Thai J. Math. **15**(3), 607–610 (2017)
2. S. Dhompongsa, J. Nantadilok, A simple proof of the Brouwer fixed point theorem. Thai J. Math. **13**(3), 519–525 (2015)
3. S. Dhompongsa, P. Kumam, An elementary proof of the Brouwer fixed point theorem. Thai J. Math. **17**(2), 539–542 (2019)
4. I. Ekeland, Sur les problèmes variationnels. C. R. Acad. Sci. Paris Sér. A-B **275**, A1057–A1059 (1972)
5. M.A. Khamsi, Remarks on Caristi's fixed point theorem. Nonlinear Anal. **71**(1–2), 227–231 (2009)
6. M.A. Mansour, A. Metrane, M. Théra, Lower semicontinuous regularization for vector-valued mappings. J. Glob. Optim. **35**(2), 283–309 (2006)
7. H. Nguyen, Lectures on where to apply fixed point theory to economics?: Measure theory, probability\statistics, fuzzy logics & research issues, delivered at department of mathematics, King Mongkut's University of Technology Thonburi, Bangkok, Thailand (2017)
8. J.P. Penot, M. Thera, Semi-continuous mappings in general topology. Archiv der Mathematik **38**(1), 158–166 (1982)
9. W. Takahashi, Existence theorems generalizing fixed point theorems for multivalued mappings, in *Fixed Point Theory and Applications* (Marseille, 1989). Pitman Research Notes in Mathematics Series, vol. 252 (Longman Scientific & Technical, Harlow, 1991), pp. 397–406
10. W. Takahashi, *Nonlinear Functional Analysis: Fixed Point Theory and Its Applications* (Yokohama Publishers, Yokohama, 2000)
11. C. Tammer, A variational principle and a fixed point theorem, in *System modelling and optimization* (Compiègne, 1993). Lecture Notes in Control and Information Sciences, vol. 197 (Springer, London, 1994), pp. 248–257
12. A. Tarski, A lattice-theoretical fixpoint theorem and its applications. Pac. J. Math. **5**(2), 285–309 (1955)

Thick Sets, Multiple-Valued Mappings, and Possibility Theory

Didier Dubois, Luc Jaulin, and Henri Prade

Abstract Carrying uncertain information via a multivalued function can be found in different settings, ranging from the computation of the image of a set by an inverse function to the Dempsterian transfer of a probabilistic space by a multivalued function. We then get upper and lower images. In each case one handles so-called *thick sets* in the sense of Jaulin, i.e., lower and upper bounded ill-known sets. Such ill-known sets can be found under different names in the literature, e.g., *interval sets* after Y. Y. Yao, *twofold fuzzy sets* in the sense of Dubois and Prade, or *interval-valued fuzzy sets*, ... Various operations can then be defined on these sets, then understood in a disjunctive manner (epistemic uncertainty), rather than conjunctively. The intended purpose of this note is to propose a unified view of these formalisms in the setting of possibility theory, which should enable us to provide graded extensions to some of the considered calculi.

Keywords Thick set · Interval analysis · Possibility theory · Inverse image · Uncertainty

1 Introduction

The links between interval calculus [21] and possibility theory [13, 33] are well-known, as well as the interest for interval calculus in robust control [20]. The need for guaranteed approximation has led B. Desrochers and L. Jaulin to propose an

D. Dubois · H. Prade
IRIT - CNRS, 118, route de Narbonne, 31062 Toulouse Cedex 09, France
e-mail: dubois@irit.fr

H. Prade
e-mail: prade@irit.fr

L. Jaulin (✉)
Lab-STICC, ENSTA-Bretagne, 2 rue François Verny, 29200 Brest, France
e-mail: lucjaulin@gmail.com

V. Kreinovich (ed.), *Statistical and Fuzzy Approaches to Data Processing, with Applications to Econometrics and Other Areas*, Studies in Computational Intelligence 892,
https://doi.org/10.1007/978-3-030-45619-1_8

101

original *thick set* and *thick interval* calculus [5, 6]. This preliminary research note initiates a study of links between the latter calculus with other works dealing with uncertainty due to incomplete knowledge, with a view to extend these notions to the settings of possibility and fuzzy set theories. Such issues have been addressed by Hung Nguyen in some of his previous works on ill-known sets, and fuzzy numbers.

2   Thick Sets and Other Related Notions

A thick set [6] $[\![\mathbb{A}]\!]$ on a referential U (in general $\mathbb{R}^n$) is an interval in the power set 2^U defined by a pair of sets (A_*, A^*) such that $A_* \subset A^*$, namely

$$[\![\mathbb{A}]\!] = [A_*, A^*] = \{A \in 2^U | A_* \subset A \subset A^*\} \tag{1}$$

This means that it is an ill-known set defined by lower and upper bounds.

Formally speaking, a thick set could be represented by a fuzzy set with a three-valued membership function $\mu : U \to \{0, 1/2, 1\}$ as, for instance, the so-called *ensembles flous* in the sense of Gentilhomme [18], who represented concepts by means of a central area A_* and a peripheral area $A^* \setminus A_*$. It includes ill-known sets called x *interval sets* in [30], based on Kleene logic, for which, in the peripheral area, relevant information for concluding to belonging or not is incomplete (1/2 means unknown). See also [30, 31]. In [24] they are called *set intervals*. This use of pairs of sets to enclose a family of subsets was actually already discussed in [14].

Another example is the case of *rough sets* [27] where uncertainty comes from a lack of attributes when exactly describing a set of objects given in extension. It differs from *twofold sets* [12] where uncertainty comes from a lack of information on the attribute values of objects. Clearly, rough sets are known in extension, but their intension is ill-known (due to the lack of a sufficient number of attributes for discriminating among elements), while the converse happens for twofold fuzzy sets: their intension is given but their extension is ill-known (the lack of information on attribute values prevent from deciding whether or not an element satisfies or not a prescribed set of properties).

Such generalized sets, at least viewed as a nested pair of sets, have been also introduced in the fuzzy set literature at different points in time. Let us mention *interval-valued fuzzy sets* (Zadeh [32], Sambuc [28]). They are thick fuzzy sets in the sense of Jaulin, namely, pairs $[\![\mathbb{F}]\!] = (F_*, F^*)$ of fuzzy sets that bracket an ill-known fuzzy set $F : \mu_{\mathbb{F}_*} \leq \mu_{\mathbb{F}} \leq \mu_{\mathbb{F}^*}$ (a particular case of type 2 fuzzy sets [32]). We can also mention *twofold fuzzy sets* [12], which are pairs of fuzzy sets that are strongly nested (the support of the former is included in the core of the other). They represent a fuzzy set of elements that belong more or less necessarily (certainly) to a crisp set A, itself included in a fuzzy superset of elements that belong more or less possibly to A, these two fuzzy sets being induced by the fact that the relevant information for deciding membership to A or not is incomplete. Besides, a single fuzzy set may also represent a set ill-located between the core and the support of the fuzzy set, see, e.g., [26].

In the following, we first insist that the meaning of a thick set depends on what the sets involved represent in the context where they are used, emphasizing the difference between ontic and epistemic viewpoints. Then we show that thick sets are at work in Dempster approach to belief function and in interval arithmetics. A concrete example is shown indicating the approximate nature of the thick set representation in some cases.

3 Epistemic Sets and Ontic Sets

To clarify the intended meaning of thick sets, it is important to understand what the sets involved in their definition represent two kinds of things. A set, be it classical or fuzzy, may represent [15]:

- either a complex entity made of the *conjunction* of its elements—we then speak of *ontic sets*. An ontic set either represents a real object, or more generally the set itself is the entity we are trying to identify.
- or a set of mutually exclusive possible values for a variable—we then speak of *epistemic sets*. It reflects an imprecise piece of information on the value of a variable. The characteristic function of the set is a possibility distribution.

This distinction is crucial for the proper handling of sets and thick sets in computations.

A thick set, as a set of sets, is often understood as an epistemic set of ontic ones, i.e., it represents an ill-known set, which is itself considered as ontic; for example, an area to be surely covered by an observation device [19] is an ontic set. Or a set of trajectories any of which ensures that a robot can go through between two obstacles [16, 17]. Note that one may consider epistemic sets of epistemic sets. For instance, I am not sure about what my neighbor knows about the state of affairs. Her knowledge can be represented by an epistemic set, but my knowledge about her (e.g., from what she said to me) is a family of possible epistemic states, one of which is the correct one. This kind of set of sets can be found in epistemic logic, a simple example of which is the MEL logic [1].

Note that the thick set representation is often only an approximation of a family of sets [14]. In the finite case the thick set representation is linear in the number of elements of the referential, while a general subset of sets described in extension is exponential (not all of them are intervals in the Boolean algebra of subsets). In some cases, the thick set representation will be insufficient. For instance, consider the ontic set of languages spoken by an individual. Suppose we know that this person can speak exactly two languages without knowing which ones. This piece of knowledge cannot be usefully represented by a thick set. Only the empty set is included in all possible 2-element subsets, and only the whole set of languages contain all of them.

4 Thick Sets and Dempster's Construction

An example of thick set is made by the pair of lower and upper sets of solutions
$S \subseteq \Omega$, A_* and A^* respectively, of the set equation

$$f(S) = A \subseteq V$$

where A is a set and $f : \Omega \to V$ is an ill-known function belonging to a set thereof,
represented by a multimapping Γ. This is a thick set inversion problem, the solutions
$S = f^{-1}(A)$ of which all satisfy $S \in [A_*, A^*]$, with

$$A_* = \{\omega : \forall f \in \Gamma, \exists a \in A \mid a = f(\omega)\} = \{\omega : \Gamma(\omega) \subseteq A\} = \bigcap_{f \in \Gamma} f^{-1}(A);$$

$$A^* = \{\omega : \exists f \in \Gamma, \exists a \in A \mid a = f(\omega)\} = \{\omega : \Gamma(\omega) \cap A \neq \varnothing\} = \bigcup_{f \in \Gamma} f^{-1}(A).$$

Dempster [3] uses this framework to define lower and upper probabilities from a
probabilistic space (Ω, P) and a multivalued mapping $\Gamma : \Omega \to V$. This mapping
represents the incomplete knowledge about a random variable, i.e., a function f
that relates a sample space to an observation space V. The value $v = f(\omega)$ is the
result of the measurement of a feature of ω. If a random experiment produces a
result ω, the corresponding observation is an ill-known value $v = f(\omega) \in \Gamma(u)$ if
the measurement tool that should yield $v = f(\omega)$ is imperfectly known. We are
facing an ill-observed random variable [2].

As a consequence of incomplete information, we do not know the precise value
of the probability $P_f(A) = P(f^{-1}(A))$ of the event $f(\omega) \in A$ on Ω. Only an
upper bound $P^*(A) = P(\{\omega : \Gamma(\omega) \cap A \neq \varnothing\})$ and a lower bound $P_*(A) = P(\{\omega : \Gamma(\omega) \subseteq A\})$ can be computed. The same construction can be made starting from a
possibilistic space [11], where the set Ω is equipped with a possibility distribution.

In this model, we are thus using the description of the inverse image $f^{-1}(A)$ of an
event $A \subset \Omega$ when the function f is ill-known. It is given by the thick subset $[A_*, A^*]$
of Ω. The interval $[P_*(A), P^*(A)] = [P(A_*), P(A^*)]$ is the "probability" of this
thick subset, and contains the set of possible values of the probability $P(f^{-1}(A))$
when $f \in \Gamma$. But not all values in the interval $[P_*(A), P^*(A)]$ are of the form
$P(f^{-1}(A))$ for some $f \in \Gamma$ [2]. Clearly if Ω is finite, the set $\{P(f^{-1}(A)) : f \in \Gamma\}$
is finite too.

Besides, the work of Denœux et al. [4] can be seen as an extension of thick sets
to belief functions, where variables are set-valued so that the range set V is itself a
power set.

5 Case of Interval Arithmetics

Another instantiation of what the previous framework is given by the pair of lower and upper sets of solutions X_* and X^* of the equation $x - u = v$ (and thus $x = u + v$) where $u \in M$, $v \in N$, M, N being intervals.

One may interpret the equation $x - u = v$ in an uncertain context in two ways:

- Looking for the set $X^* = \{x : \exists u \in M, v \in N,$ such that $x = u + v\}$.
- Or looking for the set $X_* = \{x : \forall u \in M, \exists v \in N,$ such that $x = u + v\}$.

These maximal and minimal sets are respectively given by two interval addition operations: namely Minkowski's subtraction and addition defined respectively by

$$X^* = M \oplus N = \{x : (x \ominus M) \cap N \neq \varnothing\} = \{u + v : u \in M, v \in N\}$$
$$X_* = M \boxplus N = \{x : (x \ominus M) \subseteq N\}$$

with $x \ominus M = \{x - u \mid u \in M\}$. $M \boxplus N$ is the largest subset S such that $\forall x \in S, \forall u \in M, x - u \in N$. In other words, it is the subset of x such that $-M$ translated by x is included in N.

For instance, if $M = [m, m']$ and $N = [n, n']$, we have $M \oplus N = [m + n, m' + n']$, and

$$M \boxplus N = \begin{cases} [m + n', m' + n] & \text{if } m + n' \leq m' + n \\ M \boxplus N = \varnothing & \text{otherwise.} \end{cases}$$

It can be checked that $M \boxplus N \subseteq M \oplus N$, and that the length of $M \boxplus N$ is the length of M reduced by the one of N, while the length of $M \oplus N$ is the length of M augmented by the one of N. The operation $\oplus$ is said to be pessimistic, and the operation $\boxplus$ is said to be optimistic [9, 10].

This interval analysis problem is a particular case of solving the equation $f(S) = A$ as used in Dempster's construction. Here, Ω is the domain of x, and V is the domain of v, $A = N$ and $x \ominus M$ plays the role of $\Gamma(x)$. Function $f_u(x) = x - u$ is determined by choosing a value $u \in M$. So,

$$\Gamma = \{f_u \mid u \in M\}$$

and the equation to be solved is to find intervals X such that $f_u(X) = X - u = N$ when all that is known is that $u \in M$. Then each candidate solution is of the form $X = f_u^{-1}(N) = \{u + v \mid v \in N\} = u \oplus N$. It follows that:

$$M \boxplus N = \bigcap_{f \in \Gamma} f^{-1}(N) = \bigcap_{u \in M} u \oplus N,$$
$$M \oplus N = \bigcup_{f \in \Gamma} f^{-1}(N) = \bigcup_{u \in M} u \oplus N.$$

In the notations of the previous section, one should write $M \boxplus N = N_*$ and $M \oplus N = N^*$.

But here we see that intervals $M \boxplus N$ and $M \oplus N$ do not solve the same problem. If M and N are epistemic sets representing ill-known values, $M \oplus N$ describes the uncertainty about x induced by the one on u and v. If N represents a tolerance interval to be respected, $M \boxplus N$ describes the values of x allowed for making sure that the uncertainty about $x - u$ remains bounded by N in spite of the poor knowledge about u, described by M.

But, let us suppose that M and N are ontic, and represent the positions on the real line of two rods. Then the length of $M \oplus N$ is the one of the rod obtained by concatenation of M and N. By contrast, $M \boxplus N$ is the set of points certainly covered by the rod M if it is translated by a length $v \in N$.

These two operations $\boxplus$ and $\oplus$ can be generalized when M and N are fuzzy intervals [9, 10, 29].

6   An Example: The Problem of the Two Goats

Let us consider two goats, each one is attached to a stake by a rope the length of which is 10 m. The position of the stakes, m_i for goat i, $i \in \{1, 2\}$ is ill-known. We only know that

$$m_1 \in [\mathbf{m}_1] = [0, 1] \times [2, 10]; \quad m_2 \in [\mathbf{m}_2] = [10, 16] \times [0, 1].$$

The area $A(i)$ grazed by goat i is an ontic set pervaded by uncertainty, due to partial information on the positions of the stakes. This is represented by the thick set

$$[\![\mathbb{A}(i)]\!] = [A_*(i), A^*(i)]$$

with

$$A_*(i) = [\mathbf{m}_i] \boxplus D, A^*(i) = [\mathbf{m}_i] \oplus D,$$

where D is the disc with centrer 0 and radius 10. The area A grazed by at least one goat belongs to the thick set

$$[\![\mathbb{A}]\!] = [\![\mathbb{A}(1)]\!] \cup [\![\mathbb{A}(2)]\!] = [A_*(1) \cup A_*(2), A^*(1) \cup A^*(2)].$$

The set A is an ontic set, while the rectangles $[\mathbf{m}_1]$, $[\mathbf{m}_2]$ are epistemic (in black on the figure). We are certain that none of the goats can reach the area in blue. Let us note that the grazed areas are not *all* the subsets between $A_*(1) \cup A_*(2)$ and $A^*(1) \cup A^*(2)$. For instance, A is always a connected set, while $[\![\mathbb{A}]\!]$ may contain sets with several disjoint connected components. The thick set is an encompassing approximation of the grazed areas that are effectively possible (Fig. 1).

Fig. 1 The grazed area
contains the set $\mathbb{A}_*$ (pink)
and is contained in $\mathbb{A}^*$
(yellow)

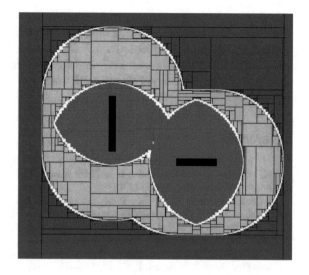

7 Conclusion

This note suggests the existence of links between several works having different
motivations. Thick sets are pairs of nested classical subsets. The framework of possibility theory should allow us to extend their calculus to the case of fuzzy thick sets,
thus permitting to introduce gradedness in the uncertainty. However, for this fuzzy
set extension, one may think of two approaches: i) looking for the solution of a fuzzy
equation of the form $f(S) = F$ where F is a fuzzy set and f is only known via a fuzzy
set-valued mapping $\tilde{\Gamma}$; ii) working in terms of alpha level-cuts, i.e., solve $f(S) = F_\alpha$
with $f \in \tilde{\Gamma}_\alpha$, for $0 < \alpha \leq 1$. However it is not clear that the two approaches give the
same results for the lower solution. For instance while $(M \oplus N)_\alpha = M_\alpha \oplus N_\alpha$ under
mild continuity conditions as shown very early by Nguyen [22], it is straightforward
to see that $(M \boxplus N)_\alpha \neq M_\alpha \boxplus N_\alpha$, because the latter intervals will generally not be
nested. This raises the question of the comparison between the two views, so as to
understand which one is better in which context.

8 Dedication

This short note is dedicated to Hung T. Nguyen. The first and the last authors of
this note had the privilege to meet Hung very early in the late seventies short after
he published an important paper discussing the expression of Zadeh's extension
principle (for extending a mapping to fuzzy arguments) in terms of alpha level-cuts
with application to fuzzy arithmetics [22]. Hung has been a continuous supporter
and contributor to fuzzy set theory, and the first and the last authors were fortunate

enough to collaborate with him, in a friendly manner, on two overview papers on important fuzzy set issues [7, 8]. Among all his contributions, let us also particularly mention his pioneering works on the relation between belief functions and random sets [23], and his works on interval- and fuzzy-valued probabilities [25], a topic clearly related to the issues of this research note.

References

1. M. Banerjee, D. Dubois, A simple logic for reasoning about incomplete knowledge. Int. J. Approx. Reason. **55**, 639–653 (2014)
2. I. Couso, D. Dubois, L. Sanchez, Random sets and random fuzzy sets as Ill-perceived random variables, in *SpringerBriefs in Computational Intelligence* (Springer, Berlin, 2014)
3. A.P. Dempster, Upper and lower probabilities induced by a multivalued mapping. Ann. Math. Statist. **38**, 325–339 (1967)
4. T. Denœux, Z. Younes, F. Abdallah, Representing uncertainty on set-valued variables using belief functions. Artif. Intell. **174**(7–8), 479–499 (2010)
5. B. Desrochers, L. Jaulin, Computing a guaranteed approximation of the zone explored by a robot. IEEE Trans. Automat. Contr. **62**(1), 425–430 (2017)
6. B. Desrochers, L. Jaulin, Thick set inversion. Artif. Intell. **249**, 1–18 (2017)
7. D. Dubois, H. T. Nguyen, H. Prade, M. Sugeno, Introduction: the real contribution of fuzzy systems, in *Fuzzy Systems: Modelling and Control*, ed. by H.T. Nguyen, M. Sugeno (Kluwer Academic Publishers, Boston, 1998), pp. 1–14
8. D. Dubois, H. T. Nguyen, H. Prade, Possibility theory, probability and fuzzy sets: misunder-standings, bridges and gaps, in *Fundamentals of Fuzzy Sets*, ed. by D. Dubois, H. Prade. The Handbooks of Fuzzy Sets Series (Kluwer, 2000), pp. 343–438
9. D. Dubois, H. Prade, Inverse operations for fuzzy numbers, in *PRE proceedings of IFAC Symposium on Fuzzy Information, Knowledge Representation and Decision Analysis*, Marseille, July 19–21 (1983), pp. 391–396. *Fuzzy Information, Knowledge Representation and Decision Analysis*, ed. by E. Sanchez, M.M. Gupta (Pergamon Press, 1984), pp. 399–404
10. D. Dubois, H. Prade, Fuzzy set-theoretic differences and inclusions and their use in the analysis of fuzzy equations. Control. Cybern. (Warsaw) **13**, 129–146 (1984)
11. D. Dubois, H. Prade, Evidence measures based on fuzzy information. Automatica **21**, 547–562 (1985). Preliminary version: Upper and lower possibilities induced by a multivalued mapping, in *PRE proceedings of IFAC Symposium on Fuzzy Information, Knowledge Representation and Decision Analysis*, Marseille, July 19–21 (1983), pp. 174–152. *Fuzzy Information, Knowledge Representation and Decision Analysis*, ed. by E. Sanchez (Pergamon Press, 1984)
12. D. Dubois, H. Prade, Twofold fuzzy sets and rough sets—some issues in knowledge represen-tation. Fuzzy Sets Syst. **23**(1), 3–18 (1987)
13. D. Dubois, H. Prade, *Possibility Theory* (Plenum Press, 1988)
14. D. Dubois, H. Prade, On incomplete conjunctive information. Comput. Math. Appl. **15**(10), 797–810 (1988)
15. D. Dubois, H. Prade, Gradualness, uncertainty and bipolarity: making sense of fuzzy sets. Fuzzy Sets Syst. **192**, 3–24 (2012)
16. H. Farreny, H. Prade, Uncertainty handling and fuzzy logic control in navigation problems, in *Proceedings of International Conference on Intelligent Autonomous Systems*, ed. by L.O. Hertzberger, F.C.A. Groen (Amsterdam, 1986), pp. 218–225, (North-Holland, 1987)
17. H. Farreny, H. Prade, Tackling uncertainty and imprecision in robotics, in *Proceedings of 3rd International Symposium of Robotics Research*, Gouvieux (Chantilly), ed. by O. Faugeras, G. Giralt (M.I.T. Press, 1985), pp. 85–91

18. Y. Gentilhomme, Les ensembles flous en linguistique. Cahiers de Linguistique Théorique and Appliquée (Bucarest) **5**, 47–63 (1968)
19. L. Jaulin, Solving set-valued constraint satisfaction problems. Computing **94**(2–4), 297–311 (2012)
20. L. Jaulin, M. Kieffer, O. Didrit, E. Walter, *Applied Interval Analysis: With Examples in Parameter and State Estimation, Robust Control and Robotics* (Springer, London, 2001)
21. R. Moore, *Interval Analysis* (Prentice-Hall, 1966)
22. H.T. Nguyen, A note on the extension principle for fuzzy sets. J. Math. Anal. Appl. **64**(2), 369–380 (1978)
23. H.T. Nguyen. On random sets and belief functions. J. Math. Anal. Appl. **65**(3), 531–542 (1978). Reprinted in Classic Works of the Dempster-Shafer Theory of Belief Functions, ed. by R. Yager, L. Liu (A.P. Dempster, G. Shafer, advisory eds.). Studies in Fuzziness and Soft Computing, 219, Chap. 5 (2008), pp. 105–116
24. H.T. Nguyen, V. Kreinovich, From numerical intervals to set intervals (Interval-related results presented at the first international workshop on applications and theory of random sets). Reliab. Comput. **3**(1), 95–102 (1997)
25. H.T. Nguyen, V. Kreinovich, B. Wu, G. Xiang, Computing statistics under interval and fuzzy uncertainty, in *Applications to Computer Science and Engineering. Studies in Computational Intelligence 393* (Springer, 2012)
26. H.T. Nguyen, V. Kreinovich, O. Kosheleva, Membership functions representing a number vs. representing a set: proof of unique reconstruction, in *Proceedings of IEEE International Conference on Fuzzy Systems (FUZZ-IEEE'16)*, Vancouver, July 24–29 (2016), pp. 657–662
27. Z. Pawlak, *Rough Sets. Theoretical Aspects of Reasoning about Data* (Kluwer Academic Publishers, Dordrecht, 1991)
28. R. Sambuc, Fonctions Φ-floues. Application à l'aide au diagnostic en pathologie thyroidienne. Thèse Université de Marseille (1975)
29. E. Sanchez, Solution of fuzzy equations with extended operations. Fuzzy Sets Syst. **12**, 237–248 (1984)
30. Y.Y. Yao, Interval sets and interval-set algebras, in *Proceedings of the 8th IEEE International Conference on Cognitive Informatics (ICCI'09)*, June 15–17, ed. by G. Baciu, Y. Wang, Y. Yao, W. Kinsner, K. Chan, L.A. Zadeh (Hong Kong, 2009), pp. 307–314
31. J.T. Yao, Y. Yao, V. Kreinovich, P. Pinheiro da Silva, S.A. Starks, G. Xiang, H.T. Nguyen, Towards more adequate representation of uncertainty: from intervals to set intervals, with possible addition of probabilities and certainty degrees, in *Proceedings of International Conference on Fuzzy Systems (FUZZ-IEEE408)*, June 1–6 (Hong Kong, 2008), pp. 983–990
32. L.A. Zadeh, Quantitative fuzzy semantics. Inf. Sci. **3**, 159–176 (1971)
33. L.A. Zadeh, Fuzzy sets as a basis for a theory of possibility. Fuzzy Sets Syst. **1**, 3–28 (1978)

A Tacit Assumption Behind Lewis Triviality That Is Not Applicable to Product Space Conditional Event Algebra

I. R. Goodman and Donald Bamber

Abstract For a given (generally, assumed atomic) nontrivial Boolean algebra B and any a, b in B with b ≠ Ø, a *conditional event* (a|b) is said to exist if for all probability measures P over B, there exists some mapping P^(not necessarily a probability measure) and some algebra B^ (not necessarily Boolean) containing (a|b) such that if P(b) > 0, then P^((a|b)) = P(a|b), ordinary conditional probability of a given b. But, in turn, Lewis' famous triviality result shows equivalently that if his *forcing hypothesis* holds—i.e., B and B^ at least overlap at B*(a, b, (a|b)), the Boolean subalgebra generated by a, b, and (a|b) (and for consistency P = P^)—then P *must be trivial* in the sense that a, b must be P-independent. The apparent informal implication of this, as gleaned from Lewis' own comments and that of the numerous papers that are based on his result, is that this spells the "death knell" for Stalnaker's thesis or even for any attempt at constructing a consistent nontrivial algebra of conditional events deriving from a Boolean algebra B. In this paper, we show two basic results contradicting this "conclusion". First, fully consistent and nontrivial Boolean (in fact, sigma-) algebras of conditional events can always be constructed relative to any given Boolean algebra B, based upon a standard product probability space construction. This avoids Lewis' forcing hypothesis—and thus Lewis triviality—yet satisfies natural isomorphic and homomorphic relations with B. Second, a key step in Lewis' own proof, involving conditioning, when generalized to a setting where his forcing hypothesis no longer is imposed, leads directly to a simple criterion determining the structure of consistent nontrivial Boolean conditional event algebras. This criterion is also directly related to the well-known property of *import/export* in AI (Artificial Intelligence) systems and conditional logic.

I. R. Goodman
San Diego, CA, USA

D. Bamber (✉)
Cognitive Sciences Department, University of California, Irvine, Irvine, USA
e-mail: dbamber@uci.edu

V. Kreinovich (ed.), *Statistical and Fuzzy Approaches to Data Processing, with Applications to Econometrics and Other Areas*, Studies in Computational Intelligence 892,
https://doi.org/10.1007/978-3-030-45619-1_9

Keywords Boolean algebra · Conditional event algebra · Lewis triviality theorem · Stalnaker thesis · Import/export · Lifting property · Product Space Conditional Event Algebra

1 Introduction

1.1 Preliminaries: Motivation, Aim and Scope of This Paper

There has been a number of times when the implications of a significant mathematical or logical breakthrough are misinterpreted by a number of researchers doing work in fields related to the discovery. Examples of this include: the conclusions that in general no "completely closed form" procedures can be established to solve polynomial equations of degree higher than four; James-Stein biased estimators for gaussian linear regression parameter vectors having dimension greater than two supposedly makes obsolete the use of best linear unbiased estimators or even Bayesian estimators; Lindley's extension [24] of the DeFinetti-Savage scoring game-theoretic characterization of admissibility for finitely additive probability measures among the class of all measures of uncertainty makes any measure of uncertainty inadmissible relative to finitely additive probability measures (but, see also [14]); Thom's—and later Ziemann's—introduction and over-application of "catastrophe theory" to the modeling of real-world phenomena; and any number of well-known "paradoxes", such as the Banach-Tarski rigid motion size-changing transformations (but all being nonmeasurable).

One prime instance of the above misinterpretation of correct results is connected with *Lewis' triviality result* [23], a consequence of Lewis' *forcing hypothesis* of proper *conditional events* and their (relative unconditional) *consequent* and *antecedent* events all made to lie in the same Boolean algebra. (Clarification of the above italicized terms are provided to some extent in greater detail, beginning with Sect. 1.2.) Such a belief is readily obtained from Lewis' own comments and the subsequent literature cited in Sect. 1.4, where even one paper [2] commented on the "growing cottage industry of papers" concerning Lewis' result. Two main targets of Lewis, as well as of many of the above mentioned papers, and as gleaned from personal correspondence, have been the non-Boolean conditional event algebra of Adams [1] and the conditional event algebras proposed by Stalnaker [31, 32]. At least in the case of Adams, consistency and nontriviality are explicitly demonstrated throughout his work cited above. In addition, it should be remarked that Lewis' forcing hypothesis and conclusion are not applicable in two ways to Adams' work: not only because of the non-Boolean structure of Adams' conditional event algebra, but also by the minimal role played in Adams forcing conditional events and unconditional events to lie in the same non-Boolean algebra (see [38]).

The aim of this paper is to address the seemingly unwarranted tacit conclusion of both Lewis and his "entourage" of cited papers in Sect. 1.4 that *consistent nontrivial Boolean conditional event algebras do not exist*. This is accomplished by showing: (i)

Lewis' basic *forcing hypothesis* yielding of the triviality result need not be satisfied by a fully consistent and nontrivial Boolean conditional event algebra and (ii) a key step in Lewis' proof when extended in a natural way to a more general setting where the forcing hypothesis is not imposed also need not hold in general; however when it does, it has the same effect as Lewis' original forcing: hypothesis: triviality. This property—or lack thereof—is also seen to tie in directly with the equally well-known potential property of import-export in AI and conditional logic.

Following this preliminary section, the subsequent Sections of this paper provide the rigorous setting for the relevant concepts discussed above, beginning with Sect. 1.2 concerning Boolean algebras, conditional events, and conditional event algebras, followed by Sect. 1.3 concerning Boolean conditional event algebras (BCEA), and Sect. 1.4, where an equivalent form of Lewis' forcing hypothesis and triviality result, together with proof are presented. In turn, Sect. 1.5 presents, in a now more rigorous setting, the central issues of this paper as stated above, where Eq. (27) introduces the *lifting property* and related concepts. Continuing this, Sect. 2 develops, again in a fully rigorous setting, connections between lifting property concepts and Lewis triviality or lack thereof for BCEA in general and PSCEA (Product Space Conditional Event Algebra) in particular. The conclusion is that in general the lifting property while holding for Lewis, because of his forcing hypothesis, no longer holds in general for a wide class of BCEAs including PSCEA.

1.2 Boolan Algebras, Conditional Events, Conditional Event Algebras

Consider a nontrivial Boolean algebra $\mathcal{B}$, identified as usual, with a tuple consisting of:

(i)   a set B of events a, b, c, d, ... (with, or without subscripts) together with specially designated events Ω (universal/unit/event), $\emptyset$ (null/zero event);

(ii) a set of closed operators &, $\vee$, $(.)'$, $\Rightarrow$, Δ, ...—respectively, conjunction/"and", disjunction/ "or", negation/complement/"not", material conditional "if(.), then (..)", equivalently, "not (.) or (..)",), symmetric difference/ "(((.) and (..)') or ((..) and (.)'))", ...—all acting on the events a, b, c, d, ...;

(iii) a set of relations <, $\leq$, $\geq$, =, $\neq$, $\perp$, ...—respectively, proper subevent, subevent, subevent in opposite direction, equality, non-equality, disjoint/non-overlapping, ... all acting on the events a, b, c, d, ...;

(iv) the set of the usual Boolean algebra axioms and properties/ theorems (including commutativity, associativity, idempotence, absorption, mutual distributivity of &,$\vee$, involution of $(.)'$, DeMorgan property of &, $\vee$, $(.)'$, etc.) as shown, e.g. in [37].

Symbolically, we can write the identification

$$\mathcal{B} \leftrightarrow (B; \Omega, \emptyset; \&, \vee, (.)', \Rightarrow, \Delta \ldots; <, \leq, \geq, =, \neq, \perp, \ldots) \qquad (1)$$

and simply use from now on—unless necessary to avoid ambiguity of concept—just B or $\mathcal{B}$.

In turn, we consider certain formal objects that play an algebraic/ set and numerical/ probabilistic role with respect to the events a, b, c, d, … in B, analogous to the numerical role played by fractions to whole numbers.

For any a, b in B, with b $\neq$ Ø, consider the formal object (a|b), to be read as "a, given b", or "if b, then a", or even "b is sufficient for a", "a is necessary with respect to b", or even further (depending on the context) "b, at least in part, causes a", "a is, at least in part, inferred by b", etc. In the above, b plays the role of *antecedent event* and a—or a&b—that of *consequent event* with respect to *simple conditional event* (a|b). Furthermore, to tie in the above formal *conditional operator* (.|..) with the usual arithmetic division operator (.÷..) (or (./..)), the former acting as a binary operator

$$(.|..) : B \times (B\neg\{\emptyset\}) \rightarrow (B|B), \tag{2}$$

where the set of all simple conditional events is designated as (B|B) (or more properly, but more cumbersome) (B|(B¬{Ø})), *we impose the constraint that*, denoting the set of all probability measures P over $\mathcal{B}$, or symbolically P: B → [0, 1] (real unit interval), as $\wp(\mathcal{B})$, or more simply $\wp(\mathcal{B})$, for each P in $\wp(\mathcal{B})$, there exists mapping P^:(B|B) → [0, 1] such that for all (a|b) in (B|B)

$$P^\wedge((a|b)) = P(a|b) =^d P(a \& b)/P(b), \text{ for only } P(b) > 0, \tag{3}$$

i.e., right-hand side of Eq. (3) is the ordinary conditional probability of a given b.

In order to show that Eq. (2) can always be satisfied in a fully consistent conditional-probabilistic way for any given Boolean algebra $\mathcal{B}$, we will additionally assume, without loss of generality, $\mathcal{B}$ to be atomic (see again, e.g. [37]), so that also $\wp(\mathcal{B})$ and (B|B) satisfy Eq. (3).

Furthermore, we also seek to establish an (conditional event) algebra $\mathcal{B}^\wedge$ (at first, not necessarily Boolean) generated by (B|B), that extends in some reasonable way $\mathcal{B}$ and all of its events, operators, and relations as symbolized in the right-hand sides of Eq. (1), using "hat" notation for each such extension. For example, consider the typical possible combination of extended &, ∨, (.)′, in the form of the *compound conditional event*

$$Q =^d \left((a|b)\&^\wedge(c|d)\right)^{'\wedge} \vee^\wedge (e|b \vee d)) \text{ in } B^\wedge, \tag{4}$$

where, for each P in $\wp(\mathcal{B})$, so that P(b), P(d) > 0, P^(Q) is a well-defined finite arithmetic combination of P acting on some subset of relative atoms generated by all of the antecedents and consequents of simple conditional events (a|b), (c|d), (e|f), such as below

$$P^\wedge(Q) = P^\wedge\left(((a|b)^{'\wedge} \vee^\wedge (c|d)^{'\wedge}) \vee^\wedge (e | b \vee d)\right)$$

$$= P^\wedge \left(\left(\left(a'|b \right) \vee^\wedge \left(c'|d \right) \right) \vee^\wedge \left(e\,|b \vee d \right) \right)$$

$$= P^\wedge \left(\left(\left(\left(\left(a'\&b \vee c'\&d \right) \times \Omega \right) \vee^\wedge \left(b'c\&d \right) \times \left(a'|b \right) \vee^\wedge \left(\left(d'\&a\&b \right) \times \left(c'|d \right) \right) \right) |b \vee d \right) \vee^\wedge \left(e|b \vee d \right) \right)$$
$$\qquad \vee^\wedge \left(e\,|b \vee d \right))$$

$$= P^\wedge \left(\left(\left(a'\&b \vee c'\&d \right) \times \Omega \; \vee^\wedge \; \left(b'\&c\&d \right) \times \left(a'|b \right) \vee^\wedge \left(d'\&a\&b \right) \times \left(c'|d \right) \right) \vee^\wedge e \times \Omega \,|\, b \vee d \right)$$

$$= P^\wedge \left(\left(\left(\left(a'\&b \vee c'\&d \right) \& e' \right) \times \Omega \; \vee^\wedge \; \left(b'\&c\&d\&e' \right) \times \left(a'|b \right) \vee^\wedge \left(d'\&a\&b\&e' \right) \times \left(c'|d \right) \right.$$
$$\qquad \left. \vee^\wedge \left(e\&(b \vee d) \times \Omega \right) |b \vee d \right)$$

$$= \left(P^\wedge \left(\left(a'\&b \vee c'\&d \right)\&e' \right) \times \Omega \right) + P^\wedge \left(b'\&c\&d\&e' \times \left(a'|b \right) \right) + P^\wedge \left(d'\&a\&b\&e' \times \left(c'|d \right) \right)$$
$$\qquad + P^\wedge \left(e\&(b \vee d) \times \Omega \right))) / P(b \vee d)$$

$$= \left(P\left(a'\&b\&e' \right) + P\left(c'\&d\&e' \right) - P\left(a'\&b\&c'\&d\&e' \right) \right.$$
$$\qquad + P\left(b'\&c\&d\&e' \right) \cdot P\left(\left(a'|b \right) + P\left(d'\&a\&b\&e' \right) \cdot P\left(c'|d \right) + P\left(e\&(b \vee d) \right) \right) / P(b \vee d). \qquad (5)$$

Note that the above example, if true, shows a conditional event algebra that has at least some Boolean algebra and relatively simple computational properties. This leads us to the next Section.

1.3   Boolean Conditional Event Algebras and Product Space Conditional Event Algebra

The above example in Sect. 1.2 exhibits an evaluation of the probability of a finite logical compound of simple conditional events in the form of a finite feasible computation of probabilities of relative atoms formed from the corresponding antecedents and consequents of those simple conditional events. In addition to imposing Boolean structure for such logical compounds, the resulting algebra can be required to be a *full Boolean algebraic extension of initial* $\mathcal{B}$, in the sense that not only does Eq. (1) hold and the "hat" analog of Eq. (1) also holds, i.e.,

$$\mathcal{B}^\wedge \leftrightarrow \left(\mathcal{B}^\wedge; \Omega^\wedge, \emptyset^\wedge; \&^\wedge, \vee^\wedge, (\cdot)^{\,\prime\wedge}, \Rightarrow^\wedge, \Delta^\wedge, \ldots; <^\wedge, \leq^\wedge, \geq^\wedge, \neq^\wedge, \perp^\wedge, \ldots \right). \quad (6)$$

as a Boolean algebra, but also the natural mapping holds:

$$(\cdot|\Omega): B \to (B|\Omega) \subseteq (B|B) \subseteq B^\wedge, (a|\Omega) =^d (\cdot|\Omega)(a), \text{ for any } a \text{ in } B, \quad (7.1)$$

as an *embedding of* $\mathcal{B}$ into $\mathcal{B}^\wedge$, i.e., an isomorphism that is $\left(P, P^\wedge \right)$—preserving as a special case of Eq. (3), where b is fixed to be Ω:

$$P^\wedge \left((a|\Omega) \right) = P(a), \text{ all } a \text{ in } B. \qquad (8)$$

Furthermore, it can also be required that for any b in B, $b \neq \emptyset$, denoting (B|b) as obviously the set of all (a|b) in (B|B) with b fixed in the modification of Eq. (7.1), replacing Ω by b, yields

$$(\cdot|b):B \to (B|b) \subseteq (B|B) \subseteq B^{\wedge}, (a|b) =^{d} (.|b)(a), \text{ for any a in B,} \qquad (7.2)$$

as a *surjective homomorphism* of B to (B|b) with the analogue of Eq. (8) being simply Eq. (3) again.

Such a (full) Boolean conditional event algebra $B^{\wedge}$ extending B—when it exists—of course has all of the usual properties of an ordinary Boolean algebra, such as mentioned earlier, but now applied to its simple conditional events and their componds. In addition, as a Boolean algebra admitting nontrivial probability measures, $B^{\wedge}$ would be part of the probability space triple

$$\left(\Omega^{\wedge}, B^{\wedge}, P^{\wedge}\right), \text{ for each P in } \wp(B). \qquad (9)$$

Analogous, to P being in $\wp(B)$, $P^{\wedge}$ being in $\wp(B^{\wedge})$, similarly, all properties of ordinary probability spaces are applicable to such *conditional event algebra probability spaces*, as e.g., *modularity*:

$$P^{\wedge}((a|b) \vee (c|d)) = P^{\wedge}((a|b)) + P^{\wedge}((c|d)) - P^{\wedge}((a|b)\&^{\wedge}(c|d))$$
$$= P(a|b) + P(c|d) - P^{\wedge}((a|b)\&^{\wedge}(c|d)), \qquad (10)$$

where $P^{\wedge}((a|b)\&^{\wedge}(c|d))$ could also be evaluated finitely, analogous to that in Eq. (5), but in general, requiring less computations, because of its simple probability conjunction structure.

With the above said, for completeness we now point out and sketch an outline of construction of a specific example of such a Boolean conditional event algebra, and the resulting probability spaces it produces. For background and history of conditional event algebras in general not necessarily Boolean and relations with the alternative Classical Logic/Boolean algebra material conditional operator $\Rightarrow$ see [1, 4–7, 16, 18, 29–32].

Let B be a given nontrivial Boolean Algebra (in general, assumed atomic) and consider for any P in $\wp(B)$, the probability space (Ω, B, P). In turn, following the standard approach to constructing a countable cartesian product probability space from individual factor/marginal probability spaces (see, e.g., [21, 28]), where in this case all factor probability spaces are identical to (Ω, B, P), produces the probability space denoted as $(\Omega^{\wedge}_{o}, B^{\wedge}_{o}, P^{\wedge}_{o})$, with the o-subscript added to the notation in Eq. (9) to indicate the above specific construction. To this we must indicate which events in $B^{\wedge}_{o}$ are simple conditional and, more generally, compound conditional events.

First note that the universal event for $B^\wedge_o$ is

$$\Omega^\wedge_o =^\wedge_o \Omega \times \Omega \times \Omega \times \ldots \text{ (countable infinite factors)}, \tag{11}$$

while by standard convention for cartesian products, the null event for $B^\wedge_o$ is

$$\emptyset^\wedge_o =^\wedge_o a_1 \times a_2 \times a_3 \times \ldots, \text{ all } a_j \text{ in } B, \text{ with at least some } a_j = \emptyset, j = 1, 2, 3, \ldots \tag{12}$$

Also, for any c in B, define here

$$(c|\Omega)_o =^\wedge_o c \times \Omega^\wedge_o, \tag{13}$$

so that also, compatible with Eqs. (12) and (13),

$$\emptyset^\wedge_o =^\wedge_o (\emptyset|\Omega)_o, \quad \Omega^\wedge_o =^\wedge_o (\Omega|\Omega)_o. \tag{14}$$

More generally, for any a, b in B, with $b \neq \emptyset$, noting that relative to $B^\wedge_o$, the product events $a\&b \times \Omega^\wedge_o, b' \times a\&b \times \Omega^\wedge_o, b' \times b' \times a\&b \times \Omega^\wedge_o, b' \times b' \times b' \times a\&b \times \Omega^\wedge_o,$ ... in $B^\wedge_o$ are all mutually disjoint ($\perp^\wedge_o$ relation). Define their disjoint disjunction as the conditional event

$$(a|b)_o =^\wedge_o \left(a\&b \times \Omega^\wedge_o\right)\vee^\wedge_o\left(b' \times a\&b \times \Omega^\wedge_o\right)\vee^\wedge_o\left(b' \times b' \times a\&b \times \Omega^\wedge_o\right)\vee^\wedge_o\ldots \tag{15}$$

In the above construction, for each P in $\wp(B)$, $P^\wedge_o$ is the usual countable infinite product probability measure generated by P as an identical factor a countable number of times. This leads to the evaluation, assuming $P(b) > 0$,

$$\begin{aligned}
P^\wedge_o((a|b)) &= P^\wedge_o\left(a\&b \times \Omega^\wedge_o\right) + P^\wedge_o\left(b' \times a\&b \times \Omega^\wedge_o\right) + P^\wedge_o\left(b' \times b' \times a\&b \times \Omega^\wedge_o\right) + \ldots \\
&= P(a\&b) + \left(P(b') \cdot P(a\&b)\right) + \left(P(b')^2 \cdot P(a\&b)\right) + \left(P(b')^3 \cdot P(a\&b)\right) + \ldots \\
&= P(a\&b) \cdot \left(1 + P(b') + P(b')^2 + \ldots\right) \\
&= P(a\&b) \cdot \left(1/(1 - P(b'))\right) \\
&= P(a|b),
\end{aligned} \tag{16}$$

showing Eq. (3) is satisfied.

Next, a number of useful properties and characterizations has been derived for probability spaces $(\Omega^\wedge_o, B^\wedge_o, P^\wedge_o)$, P in $\wp(B)$, which we dub *Product Space Conditional Event Algebra* (PSCEA). This is accomplished quite readily by using both the equivalent recursive form of Eq. (15)

$$(a|b)_o =^\wedge_o (a\&b|b)_o =^\wedge_o (a\&b \mid \Omega) \vee^\wedge_o \left(b' \times (a|b)\right), \tag{17}$$

and extending the definition of simple conditional events to include consequents arbitrary in $B^\wedge{}_0$, with antecedents still in B, analogous to the definition in Eq. (15), replacing everywhere among the conditional events in $B^\wedge{}_0$ (see e.g. [15, 17]). In particular, for any a, b, c, d in B, any P in $\wp(B)$, with P(b), P(d) > 0, the conjunction holds

$$(a|b)_o \,\&^\wedge{}_o\, (c|d)_o =^\wedge_o (C|b \vee d)_o, \tag{18}$$

$$C =^d (a\&b\&c\&d|\Omega)_o \vee^\wedge{}_o a\&b\&d' \times (c|d) \vee^\wedge{}_o c\&d\&b' \times (a|b), \tag{19}$$

so that

$$P^\wedge{}_o\big((a|b)_o \,\&^\wedge{}_o\, (c|d)_o\big) = P^\wedge{}_o((C \mid b \vee d)_o) = P^\wedge{}_o(C)/P(b \vee d), \tag{20}$$

$$P^\wedge{}_o(C) = P^\wedge{}_o (a\&b\&c\&d \mid \Omega)_o) + P^\wedge{}_o \big(a\&b\&d' \times (c|d)\big) + P^\wedge{}_o\big(c\&d\&b' \times (a|b)\big)$$
$$= P(a\&b\&c\&d) + P(a\&b\&d') \cdot P(c|d) + P(c\&d\&b') \cdot P(a|b). \tag{21}$$

Note that when b = d in the above computations, the natural simple reduction holds for Eqs. (18)–(21)

$$(a|b)\&^\wedge(c|b) =^\wedge{}_o(C|b), \; C =^d a\&b\&c\&b = a\&c\&b, \; P^\wedge\big((a|b)\&^\wedge(c|b)\big) = P(C|b), \tag{21.1}$$

compatible with the homomorphism property of Eq. (7.2). On the other hand, when $d = \Omega$ in the above computations, the also simple reduction holds for Eqs. (18)–(21)

$$(a|b)_o \,\&^\wedge{}_o(c|\Omega)_o =^\wedge{}_o (C|b \vee \Omega)_o =^\wedge{}_o(C|\Omega)_o, \; C =^d (a\&b\&c|\Omega)_o\vee^\wedge{}_o c\&b' \times (a|b),$$
$$P^\wedge\big((a|b)_o \,\&^\wedge{}_o (c|\Omega)_o\big) =P^\wedge((C|\Omega)_o)= P^\wedge(C) = P(a\&b\&c) + P\big(c\&b'\big) \cdot P(a|b). \tag{21.2}$$

(For more details, various properties, characterizations, and potential application of PSCEA see e.g. [12, 13, 15, 17, 35, 36], based in part on earlier work [34] involving deduction via only single conditional probability constraints without use of any CEA. For the earliest attempts at establishing the essential equivalent of the foundations of PSCEA, see the independent work of McGee [25] and Van Fraasen [33].)

1.4   Lewis' Main Triviality Result Revisited

With a fundamental sketch of the concept of BCEA and PSCEA established in Sects. 1.2 and 1.3, we can now make more specific the thrust of this paper via the following Sects. 1.4 and 1.5. First, in this Section we present one version and its succinct proof of the now well-known main Lewis triviality result [Lewis]. At first glance, because of the very short nature of proof, this may appear almost trivial itself to the large number of researchers [2, 3, 8–11, 19, 20, 22, 26, 27] who have taken interest in generalizing Lewis' negative result as described below. But, it is this very simple proof that contains an apparently overlooked relation that, when placed in a natural more general setting than Lewis assumes, reflects in general nontriviality of BCEA.

Theorem 1 *D. Lewis, 1976* [23] (reformulated)

Let B *(or* B—*see Eq. (1)) be a given nontrivial Boolean algebra. Then, for any events* a, b *in* B *with* $\emptyset < ab < b < \Omega$, *there cannot exist in the same Boolean algebra* B *conditional event* (a|b) *(as defined in Sect. 1.2), unless the space of probability measures* $\wp(B)$ *is so restricted that* a *and* b *are P-independent, for all P in* $\wp(B)$, $0 < P(ab) < P(b) < 1$.

Proof In this very short and correct proof, we add one additional valid step that is omitted in the original proof—since it is trivially true under *Lewis' forcing hypothesis* (especially, left-hand side of Eq. (22)) that assumes

$$a, b, (a|b) \text{ in } B, \emptyset < ab < b < \Omega, \text{ P in } \wp(B), 0 < P(ab) < P(b) < 1. \quad (22)$$

Equation (22) and the probability space properties of (Ω, B, P) imply also that

$$\text{for all } c \text{ in } B, \text{ with } P(c) > 0, (a|b)\&c \text{ in } B, P(\cdot|c) =^d P(\cdot\&c)/P(c) \text{ in } \wp(B), \quad (23)$$

$$P((a|b)\,|c)(= P((a|b)\&c\,|c)) = P(\cdot|c)((a|b)) = (P(\cdot|c)(a|b) = P(a\&b\&c)/P(b\&c)), \quad (24)$$

$$P(a|b) = P((a|b)) = P((a|b)\,\&c) + P((a|b)\&c') = P((a|b)|c) \cdot P(c) + P((a|b)|c') \cdot P(c'), \quad (25)$$

including the cases of c = a, a'.
In turn, evaluating the right-hand side of Eqs. (24), (25) for c = a, a' shows

$$P((a|b)\,|\,a) = 1, P((a|b)\,|a') = 0, P(a|b) = 1 \cdot P(a) + 0 \cdot P(a') = P(a), \quad (26)$$

which shows the P-independence of a and b.

                                                                                    ◆

1.5 Can Lewis Triviality Still Hold for BCEA When His Forcing Hypothesis Doesn't ?

Equation (24) above is the "obvious" omitted equation in Lewis' proof—again said, because under Lewis' forcing hypothesis (left-hand side of Eq. (22)), it is indeed obvious. However, a natural generalization of this relation when Lewis's forcing hypothesis no longer holds is not so obvious. Since the tacit conclusion drawn from the large number of papers extending Lewis' result appears to be that BCEA (and hence PSCEA) could not nontrivially exist in general, this paper is aimed toward responding to such conclusions. In brief we ask—and hopefully, answer, to some degree—the question of whether or not

$$P^{\wedge}((a|b) \,|(c|\Omega)) = P(\cdot|c)^{\wedge}((a|b)), \, a, b, c \text{ in B with } \emptyset < ab, bc < b < \Omega, \quad (27)$$

for especially $c = a, a'$, as is so in the case when the forcing hypothesis is in effect.

Clearly, if Eq. (27) does hold, for say $c = a, a'$, as in Lewis' proof, then a slightly modified version of the left-hand side of Eq. (26) does hold, which together with an also slight modification of Eq. (25), implies the right-hand side of Eq. (26), i.e., Lewis triviality holds in that a, b must be P-independent!

On the other hand, when Eq. (27) does not necessarily hold, it is not clear that Lewis triviality will hold, again, when the forcing hypothesis of the left-hand side of Eq. (22) is not imposed. In fact, the remainder of this paper is concerned with delineating as much as possible the relationship between consistent BCEA—and in particular PSCEA—and failure of the property equivalent to Eq. (27) to hold. To this end, it is demonstrated that a large class of consistent BCEA exist—including PSCEA—for which that property fails to hold, unlike that in Lewis' triviality proof.

2   Lifting Properties and Lewis Triviality—Or Lack of It—When Forcing Hypothesis Is No Longer Imposed

2.1   Introductory Remarks and Basic Definitions

First, from now on, for purpose of simplicity, when no ambiguity arises, we omit the Boolean "&" operator between events a and b, such as denoting ab for a&b, and we use other simplified notation where possible.

Suppose also from now on throughout this section, unless otherwise specified, that B (or $\mathcal{B}$) is an arbitrary given nontrivial (i.e. has at least, besides $\emptyset$, Ω, events a, b with $\emptyset \neq a < b \neq \Omega$) atomic Boolean algebra with all of the usual properties associated with Boolean algebra [37], bearing in mind the formal tuple identification of Eq. (1)

$$\mathcal{B} \leftrightarrow \left(B; \Omega, \emptyset; \&, \vee, (.)', \Rightarrow, \Delta \ldots; <, \leq, \geq, =, \neq, \perp, \ldots \right), \quad (1)$$

and as before $\wp(B)$ (or $\wp(\mathcal{B})$) denotes the set of all probability measures over B (or $\mathcal{B}$).

Also, as defined in Sect. 1.3, $B^{\wedge}$(or $\mathcal{B}^{\wedge}$) more, denotes any BCEA generated by/associated with B (or $\mathcal{B}$), and $P^{\wedge}$ in $\wp(B^{\wedge})$ being a probability measure extending

P in the sense described also in Sect. 1.3. Similarly, recall from the above Sections the definition of the conditional probability P(.|c) in $\wp(B)$, determined by conditioning P in $\wp(B)$ on fixed c in B, assuming P(c) > 0, and its extension P(.|c)^ in $\wp(B^\wedge)$ (where well-defined). Further, recall from Sect. 1.3 PSCEA, the special BCEA that represents a product space construction of B and related probability space (Ω, B, P), for any P in $\wp(B)$. Relative to PSCEA, for simplicity where possible, we omit the $_o$ subscript that indicates its particular construction.

Definitions For any a, b, c in B (with cb $\neq \emptyset$) and any P in $\wp(B)$, with P(cb) > 0, we say that B^ has the *lifting property* wrt (a, b, c, P) *iff* the following (simplified version of Eq. (27)) holds:

$$P^\wedge((a|b) \mid (c|\Omega)) = P(.|c)^\wedge((a|b)). \tag{28}$$

Also, say that B^ has the *import-export property* wrt ((a|b), c, P), assuming P(cb) > 0, *iff*

$$P^\wedge((a|b)|(c|\Omega)) = P(a|cb), \tag{29.1}$$

being the natural conditional event algebra analogue of the classical logic material conditional operator $\Rightarrow$, where

$$(c \Rightarrow (b \Rightarrow a)) = ((cb) \Rightarrow a). \tag{29.2}$$

Say that B^ has the *modus ponens property iff*

$$(a|b)\&^\wedge(b|\Omega)) =^\wedge (ab|\Omega), \text{ for all possible a, b in B, b} \neq \emptyset, \tag{30}$$

whence immediately, for all P in $\wp(B)$,

$$P^\wedge((a|b)\&^\wedge(b|\Omega)) = P(ab), \text{ for all possible a, b in B, P(b)} > 0, \text{P in } \wp(B). \tag{31}$$

Say that B^ has the *recursive expansion property iff* for all a, b in B, b $\neq \emptyset$,

$$(a|b) =^\wedge (ab|\Omega) \vee^\wedge ((b'|\Omega)\&^\wedge (a|b)). \tag{32}$$

2.2 Some Basic Relations Between Lifting Properties and Lewis Triviality in a General Context

Lemma 1 *Let B^ be a BCEA and a, b, c in B any P in $\wp(B)$, with P(cb) > 0. Then, B^ has the* lifting property *wrt (a, b, c, P) iff B^ has the* import-export property *wrt ((a|b), c, P).*

Proof Follows directly from the appropriate definitions.◆

Lemma 2 *Let* B^ *be a BCEA. Then,*
    B^ *has the modus ponens property iff* B^ *has the recursive expansion property.*

Proof Follows from again use of the definitions together with basic Boolean properties of B^.◆

Lemma 3 *Let* B^ *be a BCEA and* a, b, c *in* B, P *in* $\wp$(B), *with* P(cb) > 0. *Then*

(i)   B^ *has the lifting property* wrt (a,b,c,P)*iff* $P^\wedge((a|b)|(c|\Omega)) = P(a|cb)$
                                                                              iff $P^\wedge((a|b)\&^\wedge(c|\Omega)) \cdot P(b|c) = P(abc)$

(ii) *If* B^ *has modus ponens and* P(cb′) > 0, *then*
        (B^ *has the lifting property* wrt (a, b, c, P)
        iff $P(a|bc) = P^\wedge((a|b) \mid (c|\Omega)) = P((a|b) \mid (cb' \mid \Omega)) = P^\wedge((a|b) \mid (cb \mid \Omega)))$.

(iii) *If* ∅ ≠ c ≤ b, *then* B^ *has the lifting property* wrt (a, b, c, P).

Proof Again, all straightforward from basic Boolean properties applied to the definitions. ◆

Theorem 2 *PSCEA has the lifting property only for P-trivial combinations of events*

Suppose B^ *is PSCEA* wrt B *and* a, b, c, *in* B, P *in* $\wp$(B) *arbitrary such that* P(cb), P(cb′) > 0. *Then*
    B^ *has lifting property* wrt (a, b, c, P)

$$\textit{iff}\quad P(a|cb) = P(a|b) = P^\wedge((a|b) \mid (c|\Omega))$$

$$\textit{iff}\quad (a|b) \text{ and } (c|\Omega) \text{ are } P^\wedge\text{-independent.}$$

Proof First note from Eq. (21.2),

$$P^\wedge\big((a|b) \,\&^\wedge(c|\Omega)\big) = P(a\&b\&c) + P\big(c\&b'\big) \cdot P(a|b). \tag{21.2}$$

Then, applying the definition of the lifting property here and noting that PSCEA always has the modus ponens property and as characterized in Lemma 3(ii), one can show the desired relation is satisfied, using the basic properties of PSCEA, or directly, using the calculus of relations for PSCEA, as outlined in Sect. 1.3. ◆

Definition Let B^ be a given BCEA associated with B and let a, b, c in B and P in $\wp$(B), with P(cb), P(c′b) > 0. Then, define B^ to have the *bilifting property* wrt (a, b, c, P) *iff* B^ has both the lifting property wrt (a, b, c, P) and the lifting property wrt (a, b, c′, P).

Remark to Theorem 3 Suppose B^ is PSCEA wrt B and a, b, c, in B, P in $\wp$(B) arbitrary such that P(cb), P(cb′) > 0. Then, in general B^ cannot have the lifting property wrt (a, b, c, P), unless in the first place the trivial relation holds that a and b are P-conditional independent wrt c. Hence, in general, the possible extension of Lewis' triviality theorem cannot be applied to PSCEA.

Proof This is an immediate consequence of Theorem 1. ♦

Apropos to the comments for Eq. (27), Sect. 1.4, the definition of the bilifting property is readily seen to be the natural generalization of Lewis' key step. This leads to the following implication for any BCEA that would have the bilifting property wrt some (a, b, c, P):

Theorem 3 *Bilifting essentially implies triviality for a BCEA*

Let $B^\wedge$ be a BCEA such that there is some a, b, c in B, P in $\wp(B)$ with $P(cb)$, $P(c'b) > 0$. Then

If $B^\wedge$ has the bilifting property wrt (a, b, c, P), then either

(i) $P(a|b) = P(a|cb) = P(a|c'b)$, i.e., a, b are $P(.|c)$-independent, and, equivalently, $(a|b)$ and $(c|\Omega)$ are $P^\wedge$-independent,

 or

(ii) b and c are P-independent.

Proof First, apply the total probability expansion theorem, much like Eq. (25) in the proof of Lewis result, which is valid without any forcing assumption, but here P and $P^\wedge$ do not necessarily coincide

$$P(a|b) = P^\wedge((a|b)|(c|\Omega)) \cdot P(c) + P^\wedge\big((a|b)\big|(c'|\Omega)\big) \cdot P(c'). \tag{33}$$

In turn, applying the bilifting assumption to $P^\wedge((a|b)|(c|\Omega))$ and $P^\wedge\big((a|b)\big|(c'|\Omega)\big)$ in Eq. (33) and using Lemma 1 for c and c' separately, shows

$$P(a|b) = P(a|cb) \cdot P(c) + P(a|c'b) \cdot P(c'). \tag{34}$$

On the other hand, the total probability expansion theorem can be applied directly as

$$\begin{aligned} P(a|b) &= P(ac|b) + P\big(ac'|b\big) \\ &= P(a|cb) \cdot P(c|b) + P\big(a|c'b\big)\big) \cdot P(c'|b). \end{aligned} \tag{35}$$

Subtracting Eq. (35) from Eq. (34) shows

$$\begin{aligned} 0 &= P(a|cb) \cdot (P(c) - P(c|b)) + P(a|c'b) \cdot (P(c') - P(c'|b)) \\ &= P(a|cb) \cdot (P(c) - P(c|b)) + P\big(a|c'b\big) \cdot \big(P(c'|b) - P(c')\big) \\ &= \big(P(a|cb) - P\big(a|c'b\big)\big) \cdot (P(c) - P(c|b)). \end{aligned} \tag{36}$$

Finally, Eq. (36) produces the desired result. ♦

Corollary 2 *For PSCEA lifting and bilifting properties are equivalent.*
Let $B^\wedge$ be PSCEA wrt B and suppose a, b, c in B, P in $\wp(B)$, with $P(cb), P(c'b) > 0$. Then,

> (i) B^ *has the lifting property at* (a, b, c, P)
> *iff* (ii) B^ *has the bilifting property at* (a, b, c, P).

Proof This follows immediately from the symmetry of Theorem 3 in c and c′, for any given a, b and that (a|b) and (c|Ω) are P^-independent iff (a|b) and the negation $(c|\Omega)'^{\wedge}$ are P^-independent, from a basic property for any probability space—thus applicable here to probability space (Ω^, B^, P^). Finally, the equivalence of P^-independence for (a|b) and (c′ |Ω) to that of (a|b) and $(c|\Omega)'^{\wedge}$ is also a consequence of the natural isomorphism between B and (B|Ω)—so that actually,

$$\left(c'|\Omega\right) = (c|\Omega)'^{\wedge}.$$

◆

3 Conclusion

One could consider further potential lifting and bilifting properties of BCEA's or even PSCEA wrt events (a, b, c, P), where c = a or c = a′, for an even closer analogue of the setting for Lewis' triviality proof, but without his forcing hypothesis. But, it is clear from Corollary 2 and Theorem 3, that this is actually unnecessary, since essential triviality has already been established for the more general hypothesis that (single) lifting can hold.

Section 1.3 points out the existence of nontrivial BCEA's (through the construction of PSCEA) for which Lewis' forcing hypothesis is essentially never valid. In that section it is also demonstrated that regardless of that hypothesis, another key element in his proof (trivially obviously true under the forcing hypothesis, but not so in the more general setting of not imposing forcing) also cannot hold, unless in the very first place triviality of probability wrt the unconditional events is assumed (the paradox of assuming the conclusion of an argument).

References

1. E.W. Adams, *A Primer of Probability Logic* (CSLI Publications, Stanford, CA, 1998) (Supersedes in detail much of Adams' work on conditional event algebra since 1965)
2. R. Bradley, More triviality. J. Philos. Log. **28**, 129–139 (1999)
3. R. Bradley, Conditioning and non-monotonic probabilities. J. Logic Lang. Inform. **15**(1–2), 63–81 (2006)
4. P.G. Calabrese, The logic of quantum measurements in terms of conditional events. Log. Found. IGPL **14**(3), 415–435 (2006)
5. P.G. Calabrese, *Logic and Conditional Probability: A Synthesis* (College Publications, Norcross, GA, 2017)
6. A.H. Copeland, Postulates for the theory of probability. Am. J. Math. **63**(4), 741–762 (1941)

7. B. De Finetti, Foresight: its logical laws, its subjective sources, in *Studies in Subjective Probability*, ed. by H.E. Kyberg, H.E. Smokler (Wiley, New York, 1964), pp. 97–154. Translated from French of original paper in *Annales l'Institute Henri Poincare* 7: 1–68 (1937)
8. F.E. Doring, Probabilities of conditionals. Philos. Rev. **103**, 689–700 (1994)
9. G.J. Dorn, Popper's laws of the excess of the probability of the [material] conditional over the conditional probability. Conscriptus: Zeitschrift fur Philosophie **26**, 3–61 (1992–1993)
10. E. Eels, B. Skyrms (eds.), *Probabilities and Conditionals* (Cambridge University Press, Cambridge UK, 1995)
11. B. Fitelson, The strongest possible Lewisian triviality result. Wiley Online Library (1995)
12. I.R. Goodman, Toward a comprehensive theory of linguistic and probabilistic evidence: two new approaches to conditional event algebra. IEEE Trans. Man Mach. Cybern. **24**(12), 1685–1698 (1994)
13. I.R. Goodman, Applications of product space algebra of conditional events and one point random set coverage of fuzzy sets to the development of conditional fuzzy sets. Fuzzy Sets Syst. **69**(3), 257–278 (1995)
14. I.R. Goodman, H.T. Nguyen, On the scoring approach to admissibility of uncertainty measures in expert systems. J. Math. Anal. Appl. **159**(2), 550–594 (1991)
15. I.R. Goodman, H.T. Nguyen, A theory of conditional information for probabilistic inference in intelligent systems: part II, product space approach. Inf. Sci. **76**(1, 2), 13–42 (1994)
16. I.R. Goodman, H.T. Nguyen, E.A. Walker, *Conditional Inference and Logic for Intelligent Systems: A Theory of Measure-Free Conditioning* (North-Holland, Amsterdam, 1991)
17. I.R. Goodman, R.P. Mahler, H.T. Nguyen, *Mathematics of Data Fusion (Part 3)* (Springer, New York, 1997)
18. T. Hailperin, *Boole's Logic & Probability*, 2nd edn. (Elsevier Science, Amsterdam, 1986)
19. A. Hajek, Probabilities of conditionals—revisited. J. Philos. Log. **18**, 423–428 (1989)
20. A. Hajek, Probability, Logic, and Probability Logic, in *The Blackwell Companion to Logic*, ed. by L. Goble, Chap. 16 (Blackwell, Oxford, UK, 2001), pp. 362–384
21. P.R. Halmos, *Measure Theory* (Springer, New York, 1978)
22. J. Hawthorne, On the logic of nonmonotonic conditionals and conditional probabilities. J. Philos. Log. **25**(2), 185–218 (1996)
23. D. Lewis, Probabilities of conditionals and conditional probabilities. Philos. Rev. **85**(3), 297–315 (1976) (especially the main triviality result on pp. 299–300)
24. D.V. Lindley, Scoring rules and the inevitability of probability. Int. Stat. Rev. **50**(1), 1–11 (1982)
25. V. McGee, Conditional probabilities and compounds of conditionals. Philos. Rev. **98**(4), 485–541 (1989)
26. P. Milne, The simplest Lewis-style triviality proof yet? Analysis **63**(4), 300–303 (2003)
27. C.G. Morgan, E.R. Mares, Conditionals, probability and non-triviality. J. Philos. Logic **24**(5), 455–467 (1995)
28. J. Neveu, *Mathematical Foundations of the Calculus of Probability* (Holden-Day, San Francisco, 1984)
29. K. Popper, A set of independent axioms for probability. Mind **47**(186), 275–277 (1938)
30. G. Schay, An algebra of conditional events. J. Math. Anal. Appl. **24**(2), 334–344 (1968)
31. R.C. Stalnaker, A theory of conditionals, in *Studies in Logical Theory*, ed. by N. Rescher (Blackwell, Oxford, UK, 1968), pp. 98–112
32. R.C. Stalnaker, Probability and conditionals. Philos. Sci. **37**, 64–86 (1970)
33. B. Van Fraasen, Probabilities of conditionals, in *Foundations of Probability Theory, Inference, and Statistical Theories of Science*, ed. by W.L. Harper, C. Hooker (D. Reidel, Dordrecht, Holland, 1976), pp. 261–301
34. D. Bamber, Entailment with near surety of scaled assumptions of high conditional probability. J. Philos. Logic **29**(1), 1–74 (2000)
35. D. Bamber, I.R. Goodman, H.T. Nguyen, Deduction from conditional knowledge. Soft Comput. **8**(4), 247–255 (2004)
36. I.R. Goodman, H.T. Nguyen, Mathematical foundations of conditionals and their probability distributions. Int. J. Uncertain. Fuzziness Knowl. Based Syst. **3**(3), 247–339 (1995)

37. E. Mendelson, *Boolean Algebra and Switching Circuits* (McGraw Hill, New York, 1970)
38. D. Bamber, I.R. Goodman (to appear) Basic relations between Adams' conditional event algebra and Boolean conditional event algebras

Arrondir le cercle…

Emmanuel Haven

Abstract In this short contribution, I attempt to highlight a very small sample of the many works Professor Nguyen has contributed to multiple areas of mathematics, statistics and other areas of science.

1 My First Encounter with Professor Nguyen

I had the pleasure meeting Professor Nguyen, for the very first time, only about a year ago. Hung had been so kind to invite me at two events, one in Thailand and one in Vietnam. I will never forget the very first time I saw him. It was in the grand lecture theatre of Chiang Mai University where the inaugural presentations would take place. Upon this encounter with Hung, I could sense immediately that he had been, very actively, reading one of my papers. He had spotted an issue with one paper, which indeed was a weakness. Hung did mention this to me in a completely disarming way: friendly and firmly. After those first five minutes of us discussing, I felt a sense of elation and hope at the same time. There are still people in this world who live for the kick they get out of inquiring! Hung's 'type' of inquiring is of the sharp and rigorous type. This is an individual who still at a very respectable advanced age, is following a work regime that many people of my age (in the fifties) would have trouble emulating. Whilst we were in Thailand, Hung introduced us to his home and to his daily working schedule. I have wondered, several times since that visit, how he does it. I am afraid to ask him.

For the newcomer in the precious Hung Family, please let me try to expand on how I would view Hung's contributions. I shall certainly not pretend to be anywhere complete in my assessment. Furthermore, my views need to really be taken 'cum

E. Haven (✉)
Memorial University, St. John's, NF, Canada
e-mail: ehaven@mun.ca

IQSCS, Leicester, UK

© The Editor(s) (if applicable) and The Author(s), under exclusive license
to Springer Nature Switzerland AG 2021
V. Kreinovich (ed.), *Statistical and Fuzzy Approaches to Data Processing, with Applications
to Econometrics and Other Areas*, Studies in Computational Intelligence 892,
https://doi.org/10.1007/978-3-030-45619-1_10

127

(very large) grano salis'. I am not a mathematician nor a hard scientist of any sorts. Rather, I trained as a theoretical economist who then converted himself to finance.

2   A Very Brief Overview of Some of the Work I Read Hung Produced

As many readers of this Festschrift will know, Hung trained as a mathematician in France. He did all his degrees in France and topped his educational cycle with a Ph.D. from the University of Lille. He started his career in the very early 1980s at New Mexico State University, where he is now an Emeritus Professor in the Department of Mathematical Sciences. He had numerous stints at visiting positions notably at UC Berkeley and Harvard University but also outside of the US, in Belgium at the Free University of Brussels and in Japan at the Tokyo Institute of Technology. His publication record stretches over a wonderfully wide array of papers which appeared in top publications like *Comptes Rendus de l'Académie des Sciences de Paris; Biometrika* and *Annals of Statistics*. He published numerous books in a variety of areas such as statistics, fuzzy modelling and logic. A topic he is very interested in is 'uncertainty' and recently he has deployed an enormous amount of energy in using the idea of quantum probability to formalizing uncertainty, especially in areas like finance and economics. Hung has also contributed in very substantial ways to the problem of using/misusing p-values and hypothesis testing in the social sciences. This is an important debate which we can argue has led to explaining some of the problems associated with replicability in the social sciences.

I am most acquainted with the work Hung has recently contributed on the important issue of how some ideas from the formalism of quantum mechanics can be applied to issues in the social sciences. I shall say more on this in a next section of this paper.

2.1   The 1970s and the 1980s

It is known that Lotfi Zadeh, when he introduced the idea of fuzzy sets in the 1960s [1], encountered some formidable opposition. However, he persisted. It took until the 1990s, almost a quarter century after his first paper was published, for fuzzy sets to begin to be known to the wider public and showcase the enormous growth of applications in a variety of areas (going well beyond engineering applications only). As a graduate student, I did attend some of the IFSA and operations research meetings where fuzzy mathematics was being applied to many very interesting problems, including problems in economic theory. Hung wrote one of his first papers on fuzzy sets, about ten years after Zadeh's seminal paper. This was the paper which dealt with the extension principle in fuzzy sets [2]. His doctoral background in probability must have interested him in fuzzy sets. As a graduate student, I often got confused about membership functions and erroneously thought them to be probability distributions. I should have perused Hung's other 1978 paper [3], one of the first papers in the newly created *Fuzzy Sets and Systems,* the flagship journal of the fuzzy set community.

At the end of the 1970s and the early 1980s Hung reverts back to contributions to probability theory with papers delivering important contributions to martingales [4], Markov processes [5] and diffusion models [6].

It goes without saying that Hung's contributions to the edifice of fuzzy set theory has been extremely substantial. As an example, his book *A First Course in Fuzzy Logic* is currently in its 4th edition [7].

2.2   The 1990s and the Twenty-First Century

It is really interesting to see how Hung's academic creativity is unleashed in the 1990s. He contributes on a variety of topics in areas such as hazard rate models [8] and the uses of belief functions [9]. He also produces a multitude of important monographs ranging from stochastic processes [10] to computer science [11].

Hung continues actively working on fuzzy and random sets but in the twenty first century he definitely shows a very high interest in uncertainty modelling with the help of new probability tools. He continues the work others have also embarked on, such as David Trafimow's contributions [12] on how p values and hypothesis testing need to be approached in the social sciences.

3   Quantum Probability

One can get to know people via the intermediary of many things whether it is place or circumstance. As I mentioned in the beginning of this paper, the circumstance by which I discovered Hung in person, was via his discovery on how quantum probability had been applied in social science.

The idea of quantum probability has of course its roots in quantum mechanics via the well known double-slit experiment. The concept of probability waves was born out of that very experiment. The use of such probability and the interpretation linked to subtle issues which may have a connection to quantum mechanics proper, was first introduced by Andrei Khrennikov. Andrei produced a substantial body of work in probability not only geared towards the professional mathematician, but also with applications to social science in mind [13, 14]. Quantum probability is surely a generalization of probability but by no means the only generalization. A lot of the quantum-like work has focussed only applying it to the theoretical edifice of decision making. Hung wrote a very detailed document, entitled "Quantum Probability for Behavioral Economics" which was used for a short course he taught. In that very document he really poses, what I would dare to call some 'gem questions'. The document really does a great job in convincing a non-initiated audience to understanding why non-commutative probability can really be relevant to social science. This is important because it relates to a question that Hung often comes back on: how can uncertainty be captured or modelled via a non-commutative device? There is non-commutativity in the received information [15].

An important starting point in the discourse on using quantum probability in the social sciences, will be the so called decision making paradoxes which are very well known among economists and researchers who work in the area of decision making. The short course document Hung produced is really exceptionally well written. It contains the relevant references and really digs into some of the deep issues in the references Hung quotes.

One major amazement I experienced in getting to know Hung, was really how, in a very short period of time, he was able to write such a serious contribution. And this, it turned out was only the beginning! Much more additional work followed. We mention both of his contributions to the edited volumes by Vladik Khreinovich and co-editors [16, 17].

A very important question Hung posed was the one which queried whether behavioral economics should be using quantum probabilities? Maybe the answer should be that behavioral economics is to use probabilities which can respond to the violation of the law of total probability. That opens up, indeed, a wide set of possibilities which includes quantum probability.

Hung however has been adamant not to confine the quantum set-up to only economics and psychology. He also does believe that a peculiar interpretation to quantum mechanics, i.e. so called 'Bohmian mechanics' holds promise too. The origins of proposing Bohmian mechanics [18, 19] as a way to model information in financial processes goes back almost twenty years to the work of Khrennikov [20]. The idea here is that the quantum potential, emerging out of inserting the polar form of the wave function in the Schrödinger PDE, can be connected to a measure of information, known under the name of Fisher information. Whether the quantum potential is really a potential is a question that the foundations of quantum physics literature has been posing for some time. In further work on this very topic, especially in relation to finance, we do find additional pointers that the quantum potential does extract some 'type' of information from publicly available information. However, it is difficult to know what precisely it captures.

Svitek and co-authors [21] point out to the very interesting fundamental problem most of us face and that is the absence of sufficient data when investigating a problem. The authors make the interesting argument that a complex valued two dimensional system can often already do a very good job at approximating a very parameter rich environment. Kreinovich et al. [22] provide for an algebraic rationale why Bohmian mechanics may be palatable in a social science environment.

4 Arrondir le Cercle...

Here we come to the last section of this short contribution. In French language, 'arrondir le cercle' may not have meaning. It may (possibly) mean that one tries to smooth out the circle. But the circle is already that much round. Hence, what remains to be rounded out more? Maybe it means something else such as—'coming full circle'. Well, I leave it up to the reader to decide, but I do not think it is the second option which I denoted with that title.

Fuzzy logic which is one of the key topics Hung started his academic career with, seems to also be narrowly connected to the logic associated with quantum physics. However, even though the foundations of quantum physics have devoted work to this, it is really not clear how deep the resemblance between the two logics may be. I came across a beautiful presentation from Vladik Kreinovich; Ladislav Kohout and Eunjin Kim entitled "Square root of 'Not': a major difference between fuzzy and quantum logic" [23]. The central argument is that in classical logic the unary operation, to use the symbols the authors use, $s(a)$, yields two values for each of $s(0)$ and $s(1)$. The composition $s(s(a))$ is always, in classical logic, different from negation. But as the authors show, one can come up with an example, in quantum mechanics, which produces from the composition $s(s(a))$ a negation, $\neg a$. As the authors remark it is true that because of this result, one can factor large integers in so called polynomial time[1] (a huge advantage in terms of security protocols). In fuzzy logic however, for $s : [0, 1] \to [0, 1] : s(s(a)) = 1 - a$.

This brings me back to the oeuvre of Hung. He started his career, next to important mainstream work, with contributions to the mathematical edifice of multivalued logic to then converge to the very recent part of his career, with applications of quantum mechanics which indirectly involve quantum logic as an underpinning structure. Does it mean Hung has 'arrondit'[2] the circle or has he rather 'bouclé'[3] the circle? In any case, I wish Hung many more years of unbridled intellectual energy.

References

1. L.A. Zadeh, Fuzzy sets. Inf. Control., 338–353 (1965)
2. H.T. Nguyen, A note on the extension principle for fuzzy sets. J. Math. Anal. Appl. **64**, 369–380 (1978)
3. H.T. Nguyen, On conditional possibility distributions. Fuzzy Sets Syst. **1**(4), 299–309 (1978)
4. H.T. Nguyen, T.D. Pham, Loi des grands nombres pour les martingales et applications a la statistique. Comptes Rendus de l'Académie des Sciences de Paris **A-290**, 165–168 (1979)
5. H.T. Nguyen, Density estimation in a continuous time stationary Markov process. Ann. Stat. **7**(2), 341–348 (1979)

---

[1] As opposed to exponential time.

[2] As mentioned before, making more round would be one possible translation.

[3] This can be translated as closing the cercle. Or maybe also coming full circle.

6. H.T. Nguyen, T.D. Pham, Identification of non-stationary diffusion model by the method of sieves. SIAM J. Control. Optim. **20**(5), 603–611 (1982)
7. H.T. Nguyen, C.L. Walker, E.A. Walker, *A First Course in Fuzzy Logic*, 4th edn. (CRC Press, Chapman and Hall, 2019)
8. H.T. Hung, T.D. Pham, Strong consistency of maximum likelihood estimator in a change point hazard rate model. J. Stat. **2**(2), 203–216 (1990)
9. H.T. Hung, E.A. Walker, On decision making using belief functions, in *Advances in the Dempster-Shafer Theory of Evidence*, ed. by R. Yager et al. (1994)
10. H.T. Hung, D. Bosq, *A Course in Stochastic Processes: Stochastic Models and Statistical Inference* (Kluwer, 1996)
11. H.T. Hung, V. Kreinovich, *Applications of Continuous Mathematics to Computer Science* (Kluwer, 1997)
12. D. Trafimow et al., Manipulating the alpha level can not cure significance testing. Front. Psychol. **9**, 699 (2018)
13. A. Khrennikov, *Ubiquitous Quantum Structure: Drom Psychology to Finance* (Springer, Berlin, 2010)
14. E. Haven, A. Khrennikov, *Quantum Social Science* (Cambridge University Press, 2013)
15. H.T. Nguyen, *On Quantum Probability for Decision Analysis* (New Mexico State University and Chiang Mai University, 2019)
16. H.T. Nguyen, et al., On quantum probability calculus for modeling economic decisions, in *Structural Changes and their Econometric Modeling*, ed. by V. Kreinovich, S. Sriboonchitta. Studies in Computational Intelligence 808 (Springer, 2019)
17. H.T. Nguyen, et al. Beyond traditional probabilistic methods in econometrics, in *Beyond Traditional Probabilistic Models in Economics*, ed. by V. Kreinovich, et al. Studies in Computational Intelligence 809 (Springer, Cham, 2019)
18. D. Bohm, A suggested interpretation of the quantum theory in terms of hidden variables. Phys. Rev. **85**, 166–179 (1952)
19. D. Bohm, A suggested interpretation of the quantum theory in terms of hidden variables. Phys. Rev. **85**, 180–193 (1952)
20. A.Y. Khrennikov, Classical and quantum mechanics on information spaces with applications to cognitive, psychological, social and anomalous phenomena. Found. Phys. **29**, 1065–1098 (1999)
21. M. Svitek, O. Kosheleva, V. Kreinovich, T.N. Nguyen, Why quantum (wave probability) models are a good description of many non-quantum complex systems, and how to go beyond quantum models, in *Beyond Traditional Probabilistic Models in Economics*, ed. by V. Kreinovich, et al. Studies in Computational Intelligence 809 (Springer, Cham, 2019)
22. V. Kreinovich, O. Kosheleva, S. Sriboonchitta, Why Bohmian approach to quantum econometrics: an algebraic explanation. Department Technical Reports (University of Texas at El Paso, 2018)
23. V. Kreinovich, L. Kohout, E. Kim, Square root of 'Not': a major difference between fuzzy and quantum logic. Int. J. Gen. Syst. **40**(1), 111–127 (2011)

Beyond p-Boxes and Interval-Valued Moments: Natural Next Approximations to General Imprecise Probabilities

Olga Kosheleva and Vladik Kreinovich

Abstract To make an adequate decision, we need to know the probabilities of different consequences of different actions. In practice, we only have partial information about these probabilities—this situation is known as *imprecise probabilities*. A general description of all possible imprecise probabilities requires using infinitely many parameters. In practice, the two most widely used few-parametric approximate descriptions are p-boxes (bounds on the values of the cumulative distribution function) and interval-valued moments (i.e., bounds on moments). In some situations, these approximations are not sufficiently accurate. So, we need more accurate more-parametric approximations. In this paper, we explain what are the natural next approximations.

1 How Decision Theory Describes Human Preferences: A Brief Reminder

One of the main objectives of decision making theory is to help people make decisions. To be able to provide this help, we need to be able to take into account people's preferences.

A natural way to elicit preferences is to ask people to compare different alternatives and select the one which is, in their opinion, the most preferable. As a result, we get a large number of pairs of alternatives (A, A') in which the person preferred the alternative A to the alternative A'. However, directly dealing with such sets of pairs is difficult. One of the main reasons for this difficulty is that computers have been originally designed to process numbers, not discrete objects like sets of pairs, and computers are still much more efficient in processing numbers than in processing

O. Kosheleva · V. Kreinovich (✉)
University of Texas at El Paso, El Paso, TX 79968, USA
e-mail: vladik@utep.edu

O. Kosheleva
e-mail: olgak@utep.edu

© The Editor(s) (if applicable) and The Author(s), under exclusive license
to Springer Nature Switzerland AG 2021
V. Kreinovich (ed.), *Statistical and Fuzzy Approaches to Data Processing, with Applications to Econometrics and Other Areas*, Studies in Computational Intelligence 892,
https://doi.org/10.1007/978-3-030-45619-1_11

other types of data. Thus, to enhance computer processing of people's preferences, it is desirable to describe these preferences by numbers.

Such a description is indeed possible, it is known as the *utility* approach; see, e.g., [5, 8, 9, 13, 15]. In this approach, we select two fixed alternatives:

- a very bad alternative A_- which is worse than anything that we will actually encounter, and
- a very good alternative A_+ which is better than anything that we will actually encounter.

Then, for each number p from the interval $[0, 1]$, we can form a lottery $L(p)$ in which:

- we get A_+ with probability p and
- we get A_- with the remaining probability $1 - p$.

When $p = 0$, the lottery $L(0)$ coincides with the very bad alternative A_- and is, thus, worse than any actual alternative A: $L(0) < A$. When $p = 1$, the lottery $L(1)$ coincides with the very good alternative A_+ and is thus, better than the actual alternative A: $A < A_+$. The larger the probability p of the very good alternative, the better the lottery $L(p)$. So:

- if $L(p) < A$ and $p' < p$, then also $L(p') < A$;
- similarly, if $A < L(p)$ and $p < p'$, then also $A < L(p')$.

So, when we compare the alternative A with lotteries $L(p)$ corresponding to increasing values p, at some point, we switch from $L(p) < A$ to $A < L(p)$. Thus, for each alternative A, there exists a threshold value p_0 such that:

- for $p < p_0$, we have $L(p) < A$, while
- for $p > p_0$, we have $A < L(p)$.

This threshold value can be formally defined as $\sup\{p : L(p) < A\}$ or, equivalently, as $\inf\{p : A < L(p)\}$. This threshold value is called the *utility* of the alternative A; it is usually denoted by $u(A)$.

By definition, for every $\varepsilon > 0$, we have

$$L(u(A) - \varepsilon) < A < L(u(A) + \varepsilon).$$

Since the value ε can be arbitrarily small, and in practice, we do not notice very small differences, we can thus conclude that, from the practical purposes, the original alternative A is equivalent to the lottery $L(u(A))$. We will denote this practical equivalence by $A \equiv L(u(A))$.

So, if we have several alternatives, we should select the one whose utility is the largest.

Hence, to describe a person's preferences, we can use the utility function that assigns, to each alternative A, the utility $u(A)$ of this alternative to this person.

2 Which Action to Select: Discrete Case

How can we use the utility-based description of preferences when making a decision? In decision making, we need to select between several possible actions—including, sometimes, an "action" of doing nothing. The difficulty is that usually, we cannot exactly predict the consequences of each action: usually, several different consequences are possible.

For example, when we invest money in a new company, we may gain (if the company prospers) or we may lose (if the company fails). When a medical doctor prescribes a strong medicine to a patient, this may lead to the patient's recovery— or it may lead to the appearance of very bad side effects, which are, unfortunately, possible for strong medicines.

Usually, we cannot predict the exact consequence of each action a, but, based on the previous experience, we can estimate the probabilities $p_1, \ldots, p_n$ of different consequences $A_1, \ldots, A_n$. Let $u(A_i)$ denote the utility of each of the possible consequences. By definition of utility, this means that each alternative A_i is equivalent to a lottery $L(u(A_i))$ in which:

- we get A_+ with probability $u(A_i)$ and
- we get A_- with the remaining probability $1 - u(A_i)$.

Thus, the action a is equivalent to a complex two-stage lottery, in which:

- first, we select one of the alternatives A_i with the corresponding probability p_i, and then
- depending on which alternative A_i we selected on the first stage, we select either A_+ (with probability $u(A_i)$) or A_- (with probability $1 - u(A_i)$).

As a result of this two-stage lottery, we end up either with A_+ or with A_-. We can use the formula of complete probability to find the probability $u(a)$ of getting A_+:

$$u(a) = p_1 \cdot u(A_1) + \cdots + p_n \cdot u(A_n). \tag{1}$$

Thus, the action a is equivalent to lottery in which we get A_+ with probability $u(a)$ and A_- with the remaining probability. By definition of utility, this means that the utility of the action a is thus equal to $u(a)$.

When selecting an action, we need to select the action with the largest possible utility. The formula (1) for the action's utility is exactly the formula for the expected value of the utility. So, we can conclude that we need to select the action with the largest value of the expected utility.

3   Which Action to Select: General Case

In general, each action can have infinitely many possible consequences: e.g., if we give a strong medicine to a patient with fever, his resulting body temperature can take any value from normal to—in unfortunate cases—an even higher fever. In such situations, we still need to compare expected values, but we now have an integral instead of the sum:

$$u(a) = \int \rho(x) \cdot u(x) \, dx, \tag{2}$$

where $\rho(x)$ is the probability density and $u(x)$ is the utility of an alternative corresponding to the value x of the corresponding parameter (or parameters).

4   We Usually Know Probabilities with Some Uncertainty

Based on the finite sample of past experiences, we can only determine the corresponding distribution $\rho(x)$ with some uncertainty. In other words, instead of a single probability distribution with probability density $\rho(x)$, we have a whole class P of probability distributions which are consistent with our knowledge.

If two different distributions $\rho_1(x)$ and $\rho_2(x)$ are possible, this means that we can also have a case in which:

- we have the first distribution $\rho_1(x)$ with some probability α and
- we have the second distribution $\rho_2(x)$ with the remaining probability $1 - \alpha$.

This case corresponds to the probability density function $\alpha \cdot \rho_1(x) + (1 - \alpha) \cdot \rho_2(x)$, a *convex combination* of the two original distributions. Thus, with every two distributions, the class P contains their convex combinations—i.e., the class P is *convex*.

For each utility function $u(x)$, for different probability distributions from the class P, we have, in general, different values of the utility $u(a)$. In other words, instead a single value $u(a)$, we have a set of possible values. For a convex combination of two distributions $\rho_1, \rho_2 \in P$, the corresponding value $u(a)$ is also a convex combination of values u_1 and u_2 corresponding to the combined distributions. Thus, the set of possible value is convex—and is, therefore, an interval $[\underline{u}(a), \overline{u}(a)]$. (It should be mentioned that this interval can be infinite or semi-infinite.)

These intervals form a known alternative *prevision-based* representation of imprecise probabilities, when to each function $u(x)$, we assign the interval of possible values of $u(a)$:

$$[\underline{u}(a), \overline{u}(a)] = \left\{ \int \rho(x) \cdot u(x) \, dx : \rho \in P \right\}; \tag{3}$$

see, e.g., [2, 17].

5   How Can We Represent Imprecise Probabilities in Practice: Usual Approaches of Interval-Valued Moments and p-Boxes

Ideally, we should store intervals (3) corresponding to all possible utility functions $u(x)$. However, there are very many different functions, and it is not possible to store all the corresponding intervals in a computer. We should therefore limit ourselves to some few-parametric family of functions $u(x)$.

Which functions should we choose? We should select functions $u(x)$ corresponding to reasonable utilities. One class of such functions comes from the fact that usually, small changes in a parameter x leads to small changes in the alternative— and thus, to small changes in the corresponding utility. In mathematical terms, it is reasonable to describe this idea by saying that $u(x)$ smoothly depends on x. In many such cases, we can expand this function in Taylor series and keep only a few first terms in this expansion:

$$u(x) \approx u_0 + u_1 \cdot x + \cdots + u_k \cdot x^k.$$

For such functions, the expected value is simply a linear combination of the moments:

$$\int \rho(x) \cdot u(x) \, dx \approx u_0 + a_1 \cdot M_1 + \cdots + a_k \cdot M_k,$$

where $M_i \stackrel{\text{def}}{=} \int \rho(x) \cdot x^i \, dx$. Thus, in such situations, to represent the probability distribution, it is sufficient to represent its moments $M_1, \ldots, M_k$.

For imprecise probabilities, we do not know the exact moments, we only know moments with some uncertainty, i.e., for each i, we only know the intervals $[\underline{M}_i, \overline{M}_i]$. Such interval-valued moments are thus a natural way to present the corresponding uncertainty. For the case when we have several parameters $x = (x_1, \ldots, x_m)$, we similarly need to store intervals of possible values of the joint moments

$$M_{i_1 \ldots i_m} \stackrel{\text{def}}{=} \int \rho(x) \cdot x_1^{i_1} \cdot \ldots \cdot x_m^{i_m} \, dx.$$

Another important case is when we have an abrupt jump in utility for some value x_0 of the parameter x. For example, for a chemical plant:

- if the concentration of a potentially dangerous chemical in the air does not exceed the required very small threshold x_0, we are OK, but
- if $x > x_0$, the plant will suffer from fines and possible closure—which drastically decreases the corresponding utility.

In such cases, the utility function is approximately equal to a jump: $u(x) = u_-$ for $x < x_0$ and $u(x) = u_+$ for $x \geq x_0$. One can check that the resulting expected utility is equal to $u_+ + (u_- - u_+) \cdot F(x)$, where $F(x)$, as usual, denotes the corresponding value of the cumulative distribution function (cdf)—i.e., the probability that the corresponding random variable does not exceed x. In such cases, to determine the value of the expected utility, we need to know the values of the cdf $F(x)$.

In case of uncertainty, we do not know the exact values of $F(x)$. Instead, for each x, we only know the bounds $[\underline{F}(x), \overline{F}(x)]$ on $F(x)$. Such interval-valued cdf is known as the *probability box*, or a *p-box*, for short; see, e.g., [4].

In case of several variables, a natural idea is to describe similar bounds on the joint pdf

$$F(x_{01}, \ldots, x_{0m}) \stackrel{\text{def}}{=} \text{Prob}(x_1 \leq x_{01} \& \cdots \& x_m \leq x_{0m});$$

see, e.g., [14].

6 Data Processing Under Interval-Valued Moments and p-Boxes

In both approaches, the information about the actual (unknown) probability density function $\rho(x)$ comes in the form of bounds on the corresponding integrals:

$$\underline{v}_i \leq \int \rho(x) \cdot f_i(x) \, dx \leq \overline{v}_i, \tag{4}$$

where:

- for moments, $f_i(x) = x^i$ and
- for p-boxes, $f_i(x) = 1$ for $x < x_i$ and $f_i(x) = 0$ for all other x.

Based on this information, we need to estimate the value of the expected utility:

$$\int \rho(x) \cdot u(x) \, dx. \tag{5}$$

In other words, we need to find the smallest and the largest values of the objective function (5) under constraints (4).

In some cases, there are explicit formulas for the corresponding smallest and largest values; see, e.g., [4]. In general, we need to optimize the objective function (5) (which is linear in terms of the unknowns $\rho(x)$) under constraints (4) which are linear inequalities. Such problems of optimizing linear objective functions under linear inequality constraints are known as *linear programming* problems. There exist efficient algorithms for solving such problems; see, e.g., [3, 10, 16].

7 Need to Go Beyond p-Boxes and Interval-Valued Moments

p-boxes and interval-valued moments provide a reasonable first approximation to the actual description of imprecise probabilities. In some cases, however, this approximation is not accurate enough. Let us explain why this can happen, on a simple p-box example.

Indeed, suppose that we know that the probability distribution has the from $F(x) = F_0(x, c)$ for some family of functions $F_0(x, c)$ and some (unknown) value of the parameter c. As the simplest possible example, we can consider the family $F_0(x, c) = \widetilde{F}(x) + c$, where $\widetilde{F}(x)$ is a given function and the parameter c can take any value from some reasonably narrow interval $[-\varepsilon, \varepsilon]$.

In this case, for each x, the smallest possible value of the corresponding cdf is $\underline{F}(x) = \widetilde{F}(x) - \varepsilon$ and the largest possible value is $\overline{F}(x) = \widetilde{F}(x) + \varepsilon$. Thus, in the p-box representation, the above uncertainty will be represented by the p-box

$$[\underline{F}(x), \overline{F}(x)] = [\widetilde{F}(x) - \varepsilon, \widetilde{F}(x) + \varepsilon].$$

Let us now consider the following natural task: selecting a symmetric confidence interval corresponding to a given confidence level $1 - \alpha$. In terms of the cumulative density function, the requirement that a symmetric interval $[-x_0, x_0]$ is a confidence interval with given confidence level means that

$$\text{Prob}(x \in [-x_0, x_0]) = F(x_0) - F(-x_0) \geq 1 - \alpha.$$

The narrowest interval with this property is the interval for which

$$F(x_0) - F(-x_0) = 1 - \alpha.$$

In our case, we know that the cdf has the form $F(x) = \widetilde{F}(x) + c$. Thus,

$$F(x_0) - F(-x_0) = \widetilde{F}(x_0) - \widetilde{F}(-x_0),$$

and the desired requirement takes the form

$$\widetilde{F}(x_0) - \widetilde{F}(-x_0) = 1 - \alpha. \tag{6}$$

But what will happen if we only use the information from the p-box? In this case, the only information that we have about $F(x_0)$ is that

$$F(x_0) \in \left[\widetilde{F}(x_0) - \varepsilon, \widetilde{F}(x_0) + \varepsilon \right]$$

and the only information that we have about $F(-x_0)$ is that

$$F(-x_0) \in \left[\widetilde{F}(-x_0) - \varepsilon, \widetilde{F}(-x_0) + \varepsilon \right].$$

We have no information about the dependence between the values $F(x_0)$ and $F(-x_0)$; thus, based on the p-box information, the only conclusion that we can make about the difference $F(x_0) - F(-x_0)$ is that this difference is somewhere in the interval

$$\left[\left(\widetilde{F}(x_0) - \varepsilon \right) - \left(\widetilde{F}(-x_0) + \varepsilon \right), \left(\widetilde{F}(x_0) + \varepsilon \right) - \left(\widetilde{F}(-x_0) - \varepsilon \right) \right] =$$

$$\left[\left(\widetilde{F}(x_0) - \widetilde{F}(x_0) \right) - 2\varepsilon, \left(\widetilde{F}(x_0) - \widetilde{F}(x_0) \right) + 2\varepsilon \right];$$

see, e.g., [6, 11, 12].

The only way to guarantee that $F(x_0) - F(-x_0) \geq 1 - \alpha$ is to require that the smallest possible value of this difference is larger than or equal to $1 - \alpha$, i.e., that

$$\left(\widetilde{F}(x_0) - \widetilde{F}(x_0) \right) - 2\varepsilon \geq 1 - \alpha$$

or, equivalently, that

$$\widetilde{F}(x_0) - \widetilde{F}(-x_0) \geq 1 - (\alpha - 2\varepsilon).$$

The narrowest possible symmetric interval for which we can provide such a guarantee is the interval for which

$$\widetilde{F}(x_0) - \widetilde{F}(-x_0) = 1 - (\alpha - 2\varepsilon). \tag{7}$$

By comparing the formula (7) that comes from the p-box approximation with the actual formula (6), we see that because of this approximation, we have to take an interval that is much wider than necessary—namely, an interval corresponding to a much larger confidence level $1 - (\alpha - 2\varepsilon)$.

Such examples show that we need to go beyond p-boxes and interval-valued moments. What are the natural next approximation?

8 Case of Interval-Valued Moments: What Are the Natural Next Approximations?

In the moments method, we approximate the actual utility function by a polynomial $a_0 + a_1 \cdot x + \ldots + a_k \cdot x^k$. Thus, to get an adequate description of decision making under the corresponding imprecise probability, we need to know, for each of these polynomials, the range of the corresponding integrals.

Strictly speaking, there is no need to try all possible polynomials. First, the term a_0 just adds a constant to the expected utility, so it is sufficient to consider the polynomials $a_1 \cdot x + \ldots + a_k \cdot x^k$.

Second, if we know the integral $\int u(x) \cdot \rho(x)\,dx$ corresponding to a function $u(x)$, then, for every real number c, we can determine the integral

$$\int u_c(x) \cdot \rho(x)\,dx$$

corresponding to the function $u_c(x) = c \cdot u(x)$ as $c \cdot \int u(x) \cdot \rho(x)\,dx$. Thus, if we know the range $[\underline{u}, \overline{u}(a)]$ of possible values of the integral $\int u(x) \cdot \rho(x)\,dx$, then:

- for each number $c > 0$, we can compute the range $[\underline{u}_c, \overline{u}_c]$ for the function $u_c(x)$ as $[\underline{u}_c, \overline{u}_c] = [c \cdot \underline{u}, c \cdot \overline{u}]$, and
- for each number $c < 0$, we can compute the range $[\underline{u}_c, \overline{u}_c]$ for the function $u_c(x)$ as $[\underline{u}_c, \overline{u}_c] = [c \cdot \overline{u}, c \cdot \underline{u}]$;

see, e.g., [6, 11, 12].

By multiplying each polynomial by an appropriate constant, we can always get $a_1 = 1$. Thus, it is sufficient to consider polynomials of the type

$$x + a_2 \cdot x^2 + \ldots + a_k \cdot x^k.$$

So, a natural next approximation is to select some value k and to consider, for all possible values of the parameters $a = (a_2, \ldots, a_k)$, the bounds $\underline{M}(a_2, \ldots, a_k)$ and $\overline{M}(a_2, \ldots, a_k)$ on the integral

$$\int \rho(x) \cdot \left(x + a_2 \cdot x^2 + \ldots + a_k \cdot x^k \right) dx.$$

It is worth mentioning that for $k = 2$, we get linear combinations of the first and second moments; a similar idea—known as the Elastic Net (EN) method—is very successful in data processing; see, e.g., [1, 7, 18].

From the computational viewpoint, we are still Ok, since the corresponding constraints

$$\underline{M}(a_2, \ldots, a_k) \leq \int \rho(x) \cdot \left(x + a_2 \cdot x^2 + \ldots + a_k \cdot x^k \right) dx \leq \overline{M}(a_2, \ldots, a_k)$$

are still linear in terms of the unknowns $\rho(x)$ and thus, we can still use efficient linear programming algorithms to solve the corresponding computational problems.

9 Case of p-Boxes: What Are the Natural Next Approximations?

The use of p-boxes means we use bounds on the integrals $\int \rho(x) \cdot u(x)\,dx$ of the step-functions $u(x) = H(x - a)$ corresponding to different thresholds a, where $H(x)$ is the *Heaviside function* which is:

- equal to 0 for negative x and
- 1 for non-negative x.

A natural idea is thus to consider bounds on the above integral for linear combinations of such step-functions, i.e., for functions of the type

$$u(x) = c_1 \cdot H(a_1 - x) + c_2 \cdot H(a_2 - x) + \cdots + c_k \cdot H(a_k - x).$$

Without losing generality, we can assume that $a_1 < a_2 < \cdots < a_k$. Such functions are piece-wise constant:

- for $x \leq a_1$, we have $u(x) = c_1 + c_2 + \cdots + c_k$;
- for $a_1 < x \leq a_2$, we have $u(x) = c_2 + \cdots + c_k$;
- ...
- for $a_{k-1} < x \leq a_k$, we have $u(x) = c_k$; and
- for $x > a_k$, we have $u(x) = 0$.

Vice versa, every piece-wise constant function can be represented in this form.

In particular, for $k = 2$, $c_1 = -1$, and $c_2 = 1$, we get the characteristic function of the interval $[a_1, a_2]$—thus, by considering such functions, we can find the exact confidence interval and hence, solve the problem with the p-boxes that we mentioned in the previous sections.

Every function can be thus approximated. The larger k we take, the more accurate the resulting description—and we can get any desired accuracy by selecting an appropriate values k.

Similarly to the previous section, we can multiply each such function by $1/c_1$ and thus, get the case when $c_1 = 1$. Thus, we need to consider bounds

$$\underline{F}(a_1, c_2, a_2, \ldots, c_k, a_k) \text{ and } \overline{F}(a_1, c_2, a_2, \ldots, c_k, a_k)$$

on the integrals

$$\int \rho(x) \cdot (H(a_1 - x) + c_2 \cdot H(a_2 - x) + \cdots + c_k \cdot H(a_k - x)) \, dx =$$

$$(c_1 + c_2 + \cdots + c_k) \cdot \text{Prob}(x \leq a_1) + (c_2 + \cdots + c_k) \cdot \text{Prob}(a_1 < x \leq a_2) + \cdots +$$

$$c_k \cdot \text{Prob}(a_{k-1} < x \leq a_k).$$

The resulting constraints

$$\underline{F}(a_1, c_2, a_2, \ldots, c_k, a_k) \leq$$

$$\int \rho(x) \cdot (H(a_1 - x) + c_2 \cdot H(a_2 - x) + \cdots + c_k \cdot H(a_k - x)) \, dx \leq$$

$$\overline{F}(a_1, c_2, a_2, \ldots, c_k, a_k)$$

are still linear in terms of the unknowns $\rho(x)$. So, when processing this information, we can still use efficient linear programming algorithms to solve the corresponding computational problems.

Acknowledgments This work was supported in part by the National Science Foundation grants 1623190 (A Model of Change for Preparing a New Generation for Professional Practice in Computer Science) and HRD-1242122 (Cyber-ShARE Center of Excellence).

References

1. H. Alkhatib, I. Neumann, V. Kreinovich, C. Van Le, Why LASSO, EN, and CLOT: invariance-based explanation, in *Data Science for Financial Econometrics*, ed. by N.D. Trung, N.N. Thach, V. Kreinovich (Springer, Cham, Switzerland, 2020) to appear
2. T. Augustin, F. Coolen, G. de Cooman, M. Troffaes (eds.), *Introduction to Imprecise Probabilities* (Wiley, New York, 2014)
3. T.H. Cormen, C.E. Leiserson, R.L. Rivest, C. Stein, *Introduction to Algorithms* (MIT Press, Cambridge, Massachusetts, 2009)
4. S. Ferson, V. Kreinovich, L. Ginzburg, D.S. Myers, K. Sentz, *Constructing Probability Boxes and Dempster-Shafer Structures*. Sandia National Laboratories, Report SAND2002-4015, January (2003)
5. P.C. Fishburn, *Utility Theory for Decision Making* (Wiley, New York, 1969)
6. L. Jaulin, M. Kieffer, O. Didrit, E. Walter, *Applied Interval Analysis, with Examples in Parameter and State Estimation, Robust Control, and Robotics* (Springer, London, 2001)
7. B. Kargoll, M. Omidalizarandi, I. Loth, J.-A. Paffenholz, H. Alkhatib, An iteratively reweighted least-squares approach to adaptive robust adjustment of parameters in linear regression models with autoregressive and t-distributed deviations. J. Geod. **92**(3), 271–297 (2018)
8. V. Kreinovich, Decision making under interval uncertainty (and beyond), in *Human-Centric Decision-Making Models for Social Sciences*, ed. by P. Guo, W. Pedrycz (Springer, Berlin, 2014), pp. 163–193
9. R.D. Luce, R. Raiffa, *Games and Decisions: Introduction and Critical Survey* (Dover, New York, 1989)
10. D.G. Luenberger, Y. Ye, *Linear and Nonlinear Programming* (Springer, Cham, Switzerland, 2016)
11. G. Mayer, *Interval Analysis and Automatic Result Verification* (de Gruyter, Berlin, 2017)
12. R.E. Moore, R.B. Kearfott, M.J. Cloud, *Introduction to Interval Analysis* (SIAM, Philadelphia, 2009)
13. H.T. Nguyen, O. Kosheleva, V. Kreinovich, Decision making beyond Arrow's 'impossibility theorem', with the analysis of effects of collusion and mutual attraction. Int. J. Intell. Syst. **24**(1), 27–47 (2009)
14. R. Pelessoni, P. Vicig, I. Montes, E. Miranda, Bivariate p-boxes. Int. J. Uncertain. Fuzziness Knowl. Based Syst. **24**(2), 229–263 (2016)
15. H. Raiffa, *Decision Analysis* (McGraw-Hill, Columbus, Ohio, 1997)
16. R.J. Vanderbei, *Linear Programming: Foundations and Extensions* (Springer, New York, 2014)
17. P. Walley, *Statistical Reasoning with Imprecise Probabilities* (Chapman and Hall, London, UK, 1991)
18. H. Zou, T. Hastie, Regularization and variable selection via the elastic net. J. R. Stat. Soc. B **67**, 301–320 (2005)

How to Reconcile Maximum Entropy Approach with Intuition: E.g., Should Interval Uncertainty Be Represented by a Uniform Distribution

Vladik Kreinovich, Olga Kosheleva, and Songsak Sriboonchitta

Abstract In many practical situations, we only have partial information about the probabilities; this means that there are several different probability distributions which are consistent with our knowledge. In such cases, if we want to select one of these distributions, it makes sense not to pretend that we have a small amount of uncertainty—and thus, it makes sense to select a distribution with the largest possible value of uncertainty. A natural measure of uncertainty of a probability distribution is its entropy. So, this means that out of all probability distributions consistent with our knowledge, we select the one whose entropy is the largest. In many cases, this works well, but in some cases, this Maximum Entropy approach leads to counterintuitive results. For example, if all we know is that the variable is located on a given interval, then the Maximum Entropy approach selects the uniform distribution on this interval. In this distribution, the probability density $\rho(x)$ abruptly changes at the interval's endpoints, while intuitively, we expect that it should change smoothly with x. To reconcile the Maximum Entropy approach with our intuition, we propose to limit distributions to those for which the probability density's rate of change is bounded by some a priori value—and to limit the search for the distribution with the largest entropy only to such distributions. We show that this natural restriction indeed reconciles the Maximum Entropy approach with our intuition.

V. Kreinovich (✉) · O. Kosheleva
University of Texas at El Paso, El Paso, TX 79968, USA
e-mail: vladik@utep.edu

O. Kosheleva
e-mail: olgak@utep.edu

S. Sriboonchitta
Faculty of Economics, Chiang Mai University, Chiang Mai, Thailand
e-mail: songsakecon@gmail.com

V. Kreinovich (ed.), *Statistical and Fuzzy Approaches to Data Processing, with Applications to Econometrics and Other Areas*, Studies in Computational Intelligence 892,
https://doi.org/10.1007/978-3-030-45619-1_12

1 Formulation of the Problem

1.1 Interval Uncertainty Is Ubiquitous

Most of the information about the physical world comes from measurements. Measurements are never 100% accurate. Because of this, the measurement result $\widetilde{x}$ is, in general, different from the actual (unknown) value x of the desired quantity; see, e.g., [5].

In many practical situations, the only information that we have about the measurement error $\Delta x \stackrel{\text{def}}{=} \widetilde{x} - x$ is the upper bound Δ on its absolute value: $|\Delta x| \leq \Delta$. In such situations, once we have performed the measurement, the only information that we have about the actual value x is that this value belongs to the interval $[\widetilde{x} - \Delta, \widetilde{x} + \Delta]$; we do not know the probabilities of different values from this interval. In principle, many different distributions are possible—namely, all distribution which are located on this interval with probability 1.

There exist many techniques for dealing with such uncertainty; they are known as *interval computations*; see, e.g., [1, 3, 4].

1.2 Sometimes, We Need to Select a Single Distribution

While there exist many techniques for dealing with interval uncertainty, interval computations is a reasonably new field, its founding papers appeared only in mid-1950s. In contrast, traditional probabilistic methods have been in existence for several centuries. Not surprisingly, there are much more techniques for processing probabilistic uncertainty—when we know the distribution of measurement errors—than for processing interval uncertainty.

In many data processing situations, interval methods are still being developed, while there exist well-tested efficient probabilistic techniques. In such situations, we have no choice but to apply these techniques. However, to apply them, we need to select a single probability distribution from all the distributions located on the given interval.

A similar problem appears in more general situations, when we have partial information about the probabilities—and thus, to apply the existing probabilistic techniques, we need to select one of the possible distributions. Which distribution should we select?

1.3 Laplace Indeterminacy Principle and Maximum Entropy Approach

In the interval case, we do not have any reason to believe that some values from the given interval are more probable than others. Thus, it is reasonable to assume that all the points from the interval are equally probable, i.e., that we have a uniform distribution on this interval, with the constant probability density function $\rho(x) = \text{const}$. This idea is known as *Laplace Indeterminacy Principle*. It is a particular case of a more general *Maximum Entropy Principle* (see, e.g., [2]), according to which, from all possible probability distributions, we should select the one with the largest possible value of the *entropy*

$$S \overset{\text{def}}{=} - \int \rho(x) \cdot \ln(\rho(x)) \, dx. \tag{1}$$

1.4 For Interval Uncertainty, Maximum Entropy Approach Selects a Uniform Distribution

Let us apply the Maximum Entropy approach to the situation of interval uncertainty, when all we know is that $\rho(x) = 0$ for all x outside the given interval $[\underline{x}, \overline{x}]$. In this situation, among all the functions $\rho(x)$ located on this interval and that satisfy the condition that

$$\int \rho(x) \, dx = 1, \tag{2}$$

we must select the one for which the entropy S is the largest possible.

By applying the Lagrange multiplier technique to this constraint optimization problem of maximizing S, we get unconstrained optimization problem of maximizing the expression

$$- \int \rho(x) \cdot \ln(\rho(x)) \, dx + \lambda \cdot \left(\int \rho(x) \, dx - 1 \right)$$

for some parameter λ. Explicitly differentiating this expression with respect to $\rho(x)$ and equating the derivative to 0, we conclude that $- \ln(\rho(x)) - 1 + \lambda = 0$, hence $\ln(\rho(x)) = \text{const} = \exp(\lambda - 1)$ and therefore, $\rho(x)$ is also a constant. This constant can be found from the condition that $\int_{\underline{x}}^{\overline{x}} \rho(x) \, dx = 1$, i.e., that $(\overline{x} - \underline{x}) \cdot \rho(x) = 1$. Thus,

$$\rho(x) = \frac{1}{\overline{x} - \underline{x}}.$$

As the result, we get a uniform distribution on the interval $[\underline{x}, \overline{x}]$.

1.5   Why This Conclusion Is Counter-Intuitive

Intuitively, we expect that when the two events are close, their probabilities should also be close. In particular, we expect that when the values x and x' are close to each other, then the corresponding values $\rho(x)$ and $\rho(x')$ should also be close to each other.

Since we know that the probability distribution is located on the interval $[\underline{x}, \overline{x}]$ and thus, $\rho(\underline{x} - \varepsilon) = 0$ for all $\varepsilon > 0$, we thus expect to have $\rho(\underline{x}) = 0$, and similarly $\rho(\overline{x}) = 0$—and we expect the probability density function to be continuously rising from 0 to some value and then decreasing again to 0 as we reach the right endpoint $\overline{x}$ of the given interval.

In contrast to this natural intuition, for the uniform distribution—coming from using the Maximum Entropy approach—the value of the probability density function:

- jumps abruptly from 0 to $\dfrac{1}{\overline{x} - \underline{x}}$ as we cross into the interval,
- remains the same throughout this interval, and then
- abruptly drops back to 0 as we leave this interval.

1.6   This Is Not the only Case When Maximum Entropy Approach Leads to Counterintuitive Results

Similar counterintuitive results happen in more complex situations as well.

For example, suppose that, in addition to the bounds $\underline{x} \leq x \leq \overline{x}$ on the random variable x, we also know its first moment

$$E = \int x \cdot \rho(x)\, dx. \tag{3}$$

Then, the maximum entropy approach means that we maximize entropy S under constraints (2) and (3). For this constraint optimization problem, the Lagrange multiplier methods leads to the unconstrained problem of maximizing the expression

$$-\int \rho(x) \cdot \ln(\rho(x))\, dx + \lambda \cdot \left(\int \rho(x)\, dx - 1 \right) + \lambda_1 \cdot \left(\int x \cdot \rho(x)\, dx - E \right)$$

for some values λ and λ_1. Differentiating this expression with respect to $\rho(x)$ and equating the derivative to 0, we conclude that $- \ln(\rho(x)) - 1 + \lambda + \lambda_1 \cdot x = 0$. So, $\ln(\rho(x)) = a + \lambda_1 \cdot x$, where we denoted $a \overset{\text{def}}{=} \lambda - 1$ and thus, $\rho(x) = \exp(a + \lambda_1 \cdot x)$. This expression is positive on both endpoints $x = \underline{x}$ and $x = \overline{x}$ and thus, on each of the endpoints, there is a discontinuous jump to 0.

Similarly, if, in addition to the first moment E, we also know the second moment

$$M_2 = \int x^2 \cdot \rho(x) \, dx, \tag{4}$$

the Lagrange multiplier method leads to the unconstrained problem of maximizing the expression

$$-\int \rho(x) \cdot \ln(\rho(x)) \, dx + \lambda \cdot \left(\int \rho(x) \, dx - 1 \right) + \lambda_1 \cdot \left(\int x \cdot \rho(x) \, dx - E \right) +$$

$$\lambda_2 \cdot \left(\int x^2 \cdot \rho(x) \, dx - M_2 \right)$$

for which we get $\rho(x) = \exp(a + \lambda_1 \cdot x + \lambda_2 \cdot x^2)$, i.e., a truncated normal distribution—which also has jumps from 0 to positive values on each of the two endpoints $\underline{x}$ and $\overline{x}$ of the original interval.

1.7 How Can We Reconcile Maximum Entropy Approach with Intuition?

It is desirable to modify the Maximum Entropy approach so that it will be reconciled with our intuition. In this paper, we propose a natural way to do it.

2 Reconciling Maximum Entropy and Intuition: Idea and Consequences

2.1 How to Formalize Our Intuition

Let us first try to describe our intuition in precise terms. Intuitively, we do not expect the values of the probability density functions to jump. Moreover, we do not expect them to change too abruptly—this would be similar to jumping. Thus, our intuition means that there is an upper bound B of the rate with which the probability density function can change, i.e., that:

$$|\rho'(x)| \leq B \text{ for all } x. \tag{5}$$

This formalization naturally leads us to the following idea.

2.2 How to Reconcile the Maximum Entropy Approach and Our Intuition: Main Idea

In view of the above discussion, a reasonable idea is to add the inequality (5) as an additional constraint when maximizing the entropy. In other words, we select the distribution with the largest possible value of the entropy among all distributions which are consistent with our knowledge and which satisfy the additional constraint (5).

2.3 How to Actually Implement This Idea

In general, if we do not have any inequality constraints, if we simply want to maximize an objective function $F(y)$, then, from the fact that at the optimizing point y_{opt}, small changes of y do not increase the value of the objective functions, we conclude that all the partial derivatives of the objective function should be equal to 0 at this point. Indeed:

- if one of partial derivatives $\dfrac{\partial F}{\partial y_i}(y_{opt})$ was positive, then a small increase in the component y_i will increase the value of the objective function beyond the largest possible value $F(y_{opt})$, and

- if one of partial derivatives $\dfrac{\partial F}{\partial y_i}(y_{opt})$ was negative, then a small decrease in the component y_i will increase the value of the objective function beyond the largest possible value $F(y_{opt})$.

The only remaining option is that all the derivatives are zeros—this is exactly the usual calculus-based criterion that we used in the previous section to find the corresponding maxima.

When maximize a function $F(y)$ under an inequality constraint $G(y) \geq 0$, then for the optimizing point y_{opt}, we either have $G(y_{opt}) > 0$ or $G(y_{opt}) = 0$. In the first case, when we make small changes to y, the condition $G(y) > 0$ will still be satisfied. Thus, all small changes are allowed—and since these small changes cannot lead to an increase in the value of the objective function, we make the conclusion that all the partial derivatives will be 0.

In our examples, this means that for each x:

- we either have a strict inequality $|\rho'(x)| < B$, in which case all derivatives are 0s and we can derive the same formulas as before,
- or we have the equality $|\rho'(x)| = B$.

Let us describe what this means for each of the above three problems:

- when we only know that the random variable is located in an interval $[\underline{x}, \overline{x}]$,
- when we also know the first moment E, and
- when, in addition to the interval, we also know the first moment E and the second moment M_2.

2.4 Case When We only Know that the Random Variable Is Located on the Interval $[\underline{X}, \overline{X}]$.

In this case, the above argument leads to the following optimal function $\rho(x)$:

- near the left endpoint $\underline{x}$, where $\rho(\underline{x} - \varepsilon) = 0$, we cannot have a constant function $\rho(x)$ (which comes from equating the partial derivative with respect to $\rho(x)$ to 0), and we cannot have a function which is decreasing ($\rho'(x) = -B$); so, the only remaining choice is an increasing function with $\rho'(x) = B$ and thus,

$$\rho(x) = B \cdot (x - \underline{x});$$

- after the increasing function reaches a certain height, we get

$$\rho(x) = \rho_0$$

for some constant ρ_0;
- after that, when we are close to the right endpoint $\overline{x}$, the only remaining option is $\rho'(x) = -B$, for which

$$\rho(x) = B \cdot (\overline{x} - x).$$

So, in this case, we have a trapezoid probability density function, as shown in the following picture.

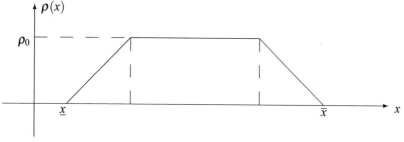

The value ρ can be determined from the condition (1), i.e., from the condition that the area under this graph be equal to 1. With the rate of rising equal to B, the rise from 0 to ρ_0 requires an x-length $\dfrac{\rho_0}{B}$. The joint areas of the ρ-increasing and the

ρ-decreasing triangles is thus equal to $\rho_0 \cdot \dfrac{\rho_0}{B}$. At the remaining part $L - 2 \cdot \dfrac{\rho_0}{B}$ of the interval, where we denoted $L \overset{\text{def}}{=} \overline{x} - \underline{x}$, the pdf $\rho(x)$ is equal to ρ_0, so its integral on this part of the original interval is equal to $\rho_0 \cdot \left(L - 2 \cdot \dfrac{\rho_0}{B} \right)$. Thus, the overall integral of $\rho(x)$ over the whole interval $[\underline{x}, \overline{x}]$ is equal to

$$\rho_0 \cdot \frac{\rho_0}{B} - \rho_0 \cdot \left(L - 2 \cdot \frac{\rho_0}{B} \right) = 1.$$

If we open the parentheses, combine similar terms, multiply all the terms by B, and move all the terms to the right-hand side, we get the following quadratic equation for ρ_0:

$$\rho^2 - L \cdot B \cdot \rho_0 + B = 0,$$

from which we can determine ρ_0 as

$$\rho_0 = \frac{L \cdot B - \sqrt{L^2 \cdot B^2 - 4B}}{2}. \tag{6}$$

Comment The usual uniform distribution corresponds to the case when there is no limitation on the rate of change, i.e., when $B \to \infty$. Let us show that in this limit, the expression (6) indeed leads to the usual formula $\rho_0 = \dfrac{1}{L}$. Indeed, we have

$$L^2 \cdot B^2 - 4B = L^2 \cdot B^2 \cdot \left(1 - \frac{4B}{L^2 \cdot B^2} \right) = L^2 \cdot B^2 \cdot \left(1 - \frac{4}{L^2 \cdot B} \right),$$

thus

$$\sqrt{L^2 \cdot B^2 - 4B} = \sqrt{L^2 \cdot B^2 \cdot \left(1 - \frac{4}{L^2 \cdot B} \right)} = L \cdot B \cdot \sqrt{1 - \frac{4}{L^2 \cdot B}}.$$

Since the ratio $z \overset{\text{def}}{=} \dfrac{4}{L^2 \cdot B}$ is, for large B, close to 0, we can use the fact that

$$\sqrt{1 - z} = 1 - \frac{z}{2} + O(z^2).$$

Then, we get

$$\sqrt{L^2 \cdot B^2 - 4B} = L \cdot B \cdot \left(1 - \frac{2}{L^2 \cdot B} + O\left(\frac{1}{B^2} \right) \right) = L \cdot B - \frac{2}{L} + O\left(\frac{1}{B} \right).$$

Thus,

$$\rho_0 = \frac{L \cdot B - \sqrt{L^2 \cdot B^2 - 4B}}{2} = \frac{L \cdot B - \left(L \cdot B - \frac{2}{L} + O\left(\frac{1}{B}\right)\right)}{2} = \frac{1}{L} + O\left(\frac{1}{B}\right).$$

So, for large B, we indeed get $\rho_0 \approx \frac{1}{L}$.

2.5 Discussion

In the above analysis, we assumed that we know the value B corresponding to our intuition. What if we do not know this value? In this case, it makes sense to select this bound B to be as small as possible—to guarantee the smallest possible rate of change of the corresponding probability density function. As we decrease B, the periods of increase and decrease become longer—until we reach the point when these two periods fill the whole interval and further decrease in B is not possible. For the resulting smallest possible value B, the probability density function becomes triangular.

Since the increase of $\rho(x)$ from 0 to the largest value ρ_0 is happening at the same rate as the decrease back to 0, the maximum value ρ_0 of $\rho(x)$ is attained at the midpoint $\tilde{x} = \frac{\underline{x} + \overline{x}}{2}$ of the original intervals. So, the resulting rate of increase B can be obtained by dividing the maximum value ρ_0 by the half-width $\Delta = \frac{\overline{x} - \underline{x}}{2} = \frac{L}{2}$ of the original interval: $B = \frac{\rho_0}{\Delta}$. In this case:

- for $x \leq \tilde{x}$, we have $\rho(x) = B \cdot (x - \underline{x})$, and
- for $x \geq \tilde{x}$, we have $\rho(x) = B \cdot (\overline{x} - x)$.

The area under this triangular curve is equal to $\frac{1}{2} \cdot L \cdot \rho_0 = 1$, thus $\rho_0 = 2L$ and hence,

$$B = \frac{\rho_0}{L/2} = \frac{2/L}{L/2} = \frac{4}{L^2}.$$

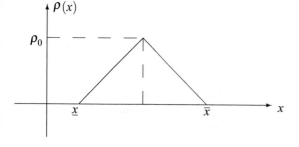

2.6 What If We Also Know the First Moment

In this case, similar to the previous case:

- we first have a linear increase $\rho(x) = B \cdot (x - \underline{x})$ until some value x_-;
- then we have an exponential distribution $\rho(x) = \exp(a + \lambda_1 \cdot x)$ until we reach some other value x_+;
- after that, we have a linear decrease $\rho(x) = B \cdot (\overline{x} - x)$.

Once we know a and λ_1, we can determine the transition values x_- and x_+ from the condition that the probability density function be continuous, i.e., from the conditions that $B \cdot (x - \underline{x}_-) = \exp(a + \lambda_1 \cdot x_-)$ and that $\exp(a + \lambda_1 \cdot x_+) = B \cdot (\overline{x} - x_+)$. The values a and λ_1 must be determined from the conditions (2) and (3).

2.7 What If We Also Know the First and the Second Moment

In this case, similar to the previous two cases:

- we first have a linear increase $\rho(x) = B \cdot (x - \underline{x})$ until some value x_-;
- then we have a (truncated) Gaussian distribution $\rho(x) = \exp(a + \lambda_1 \cdot x + \lambda_2 \cdot x^2)$ until we reach some other value x_+;
- after that, we have a linear decrease $\rho(x) = B \cdot (\overline{x} - x)$.

Once we know a, λ_1, and λ_2, we can determine the transition values x_- and x_+ from the condition that the probability density function be continuous, i.e., from the conditions that

$$B \cdot (x - \underline{x}_-) = \exp(a + \lambda_1 \cdot x_- + \lambda_2 \cdot x_-^2)$$

and that

$$\exp(a + \lambda_1 \cdot x_+ + \lambda_2 \cdot x_+^2) = B \cdot (\overline{x} - x_+).$$

The values a, λ_1, and λ_2 must be determined from the conditions (2), (3), and (4).

2.8 Multi-D Case

In the multi-D case, if all we know is that each variable $x_1, \ldots, x_n$ is located on the corresponding interval $[\underline{x}_i, \overline{x}_i]$, then the Maximum Entropy approach leads to a uniform distribution on the corresponding box

$$\mathscr{B} = [\underline{x}_1, \overline{x}_1] \times \cdots \times [\underline{x}_n, \overline{x}_n],$$

a distribution with the constant probability density

$$\rho(x) = \rho_0 = \frac{1}{(\overline{x}_1 - \underline{x}_1) \cdot \ldots \cdot (\overline{x}_n - \underline{x}_n)}.$$

This conclusion is also counterintuitive since on the border of this box, the probability density function changes abruptly from 0 to ρ_0, while intuitively, it should be continuous—and moreover, it should not change too fast.

In this case, it is reasonable to require a similar limitation on the rate of change. In the 1-D case, the limitation $|\rho'(x)| \leq B$ means that if, for some $\varepsilon > 0$, the two values x and y are ε-close (in the sense that $|x - y| \leq \varepsilon$), then the corresponding values of the probability density should not differ by more than $B \cdot \varepsilon$, i.e., that we should have $|\rho(x) - \rho(y)| \leq B \cdot \varepsilon$. Similarly, in the multi-D case, it is reasonable to require that if, for some $\varepsilon > 0$, the points $x = (x_1, \ldots, x_n)$ and $y = (y_1, \ldots, y_n)$ are ε-close—in the sense that all their components are ε-close, i.e., that $|x_i - y_i| \leq \varepsilon$ for all i—then we should have $|\rho(x) - \rho(y)| \leq B \cdot \varepsilon$.

The ε-closeness of two points can be equivalently described as $d_{\max}(x, y) \leq \varepsilon$, where we denoted

$$d_{\max}((x_1, \ldots, x_n), (y_1, \ldots, y_n)) \stackrel{\text{def}}{=} \max(|x_1 - y_1|, \ldots, |x_n - y_n|).$$

In these terms, the above requirement takes the form

$$|\rho(x) - \rho(y)| \leq B \cdot d_{\max}(x, y). \tag{7}$$

Similarly to the 1-D case, we propose, when looking for a distribution with the largest entropy, to limit ourselves only to probability distributions that satisfy this condition (7) for all x and y.

Under this restriction, in the situation when all we know is that the distribution is located in a box, then we only have $\rho(x)$ equal to some constant ρ_0 for all the points $x = (x_1, \ldots, x)$ whose max-distance $d_{\max}(x, \partial\mathcal{B}) \stackrel{\text{def}}{=} \min_{y \in \partial\mathcal{B}} d_{\max}(x, y)$ to the box's border $\partial\mathcal{B}$ does not exceed the ratio $\dfrac{\rho_0}{B}$. For points x which are closer to the border, we have

$$\rho(x) = B \cdot d_{\max}(x, \partial\mathcal{B}).$$

Acknowledgments This work was supported in part by the National Science Foundation grants 1623190 (A Model of Change for Preparing a New Generation for Professional Practice in Computer Science) and HRD-1242122 (Cyber-ShARE Center of Excellence).

References

1. L. Jaulin, M. Kiefer, O. Didrit, E. Walter, *Applied Interval Analysis, with Examples in Parameter and State Estimation, Robust Control, and Robotics* (Springer, London, 2001)
2. E.T. Jaynes, G.L. Bretthorst, *Probability Theory: The Logic of Science* (Cambridge University Press, Cambridge, UK, 2003)
3. G. Mayer, *Interval Analysis and Automatic Result Verification* (de Gruyter, Berlin, 2017)
4. R.E. Moore, R.B. Kearfott, M.J. Cloud, *Introduction to Interval Analysis* (SIAM, Philadelphia, 2009)
5. S.G. Rabinovich, *Measurement Errors and Uncertainties: Theory and Practice* (Springer, New York, 2005)

Time Series: How Unusual Local Behavior Can Be Recognized Using Fuzzy Modeling Methods

Vilém Novák and Viktor Pavliska

Abstract In this paper, we address the problem of automatic recognition of structural breaks in time series. The former are unexpected shifts of the course or sudden change of the volatility of time series. Structural breaks can be caused, e.g., by changes in the organization of a company, global or local economic development, global shifts in capital and labor, various kinds of outer influences such as discovery or depletion of natural resources, etc. Structural breaks in time series are usually detected using statistical methods. In this paper, we suggest using special non-statistical techniques of fuzzy modeling. We will employ two classes of them, namely the fuzzy transform (F-transform) and selected methods of Fuzzy Natural Logic (FNL). The fuzzy transform enables us to estimate the average slope of time series in an area characterized by a fuzzy set. The slope is then evaluated by evaluative linguistic expressions, which enables us to identify intervals with monotonous behavior and, consequently, identify structural breaks. Our method is simple, transparent, and computationally effective.

1 Introduction

This paper addresses the problem of automatic recognition of structural breaks in time series. These are specific unexpected shifts of their course or sudden change of their volatility. Such changes can be caused, e.g., by changes in the organization of a company, global or local economic development, global shifts in capital and labor, changes in resource availability due to outer influences, discovery or depletion of natural resources, etc. Recall that this problem belongs among the class of problems studied in the area of mining information from time series (cf. [4]).

V. Novák (✉) · V. Pavliska
Institute for Research and Applications of Fuzzy Modeling, University of Ostrava, NSC
IT4Innovations, 30. dubna 22, 70103 Ostrava 1, Czech Republic
e-mail: vilem.novak@osu.cz

V. Pavliska
e-mail: viktor.pavliska@osu.cz

V. Kreinovich (ed.), *Statistical and Fuzzy Approaches to Data Processing, with Applications to Econometrics and Other Areas*, Studies in Computational Intelligence 892,
https://doi.org/10.1007/978-3-030-45619-1_13

Structural breaks in time series are usually detected using statistical methods, for example, [2, 3, 23]. In this paper, we suggest using special techniques of fuzzy modeling for detection of structural breaks instead of the statistical ones. We will focus on two classes of methods, namely the *fuzzy transform* (F-transform) and some methods of *Fuzzy Natural Logic* (FNL). The detailed description of these techniques can be found in the book [19] and in many papers. We will describe how the methods mentioned above can be used for detection of structural breaks. This problem has been addressed already in [18] where they are identified using the estimation of the average tangent in an area characterized by a fuzzy set. For this purpose, we applied the first-degree F-transform and evaluated it through evaluative linguistic expressions (their theory is a part of FNL). In this paper, we extend this idea and identify the structural breaks by modification of the algorithm for identification of intervals with monotonous behavior described, e.g., in [16]. Our method works well for determination both of structural breaks in time series as well as in its volatility. The primary outcomes of our method are its *relative simplicity, transparency* and *computational effectiveness* because the complexity of the F-transform is linear.

2 F-Transform and Fuzzy Natural Logic in Time Series Analysis

2.1 Time Series and Its Decomposition

By a time series, we will understand a real-valued stochastic process

$$X : \mathbb{T} \times \Omega \longrightarrow \mathbb{R}$$

where Ω is a set of elementary random events and $\mathbb{T} = \{0, \ldots, q\} \subset \mathbb{N}$ is a finite set of numbers interpreted as time moments (cf. [1, 5]). Our fundamental assumption is that the time series can be decomposed into four components:

$$X(t, \omega) = Tr(t) + C(t) + S(t) + R(t, \omega), \qquad t \in \mathbb{T}, \omega \in \Omega \qquad (1)$$

where Tr is a *trend*, C is a *cyclic* component, S is a *seasonal* component and R is a random *noise*. Note that Tr, C, S are ordinary functions not having stochastic character. The model (1) is often simplified by considering a *trend-cycle* component TC which means that it includes two components from (1):

$$TC(t) = Tr(t) + C(t).$$

It is challenging to specify properties of the trend-cycle TC precisely. In general, it is a function with no clear periodicity or its periodicity is long, and we may not have enough data to specify the latter more exactly. Therefore, we will assume that the trend-cycle is a function with medium-long periodicity and very small modulus of continuity $\omega(h, TC)$.

Furthermore, we assume that the seasonal component $S(t)$ is a sum of periodic functions

$$S(t) = \sum_{j=1}^{r} P_j \left(\sin(\lambda_j t + \varphi_j) + \cos(\lambda_j t + \varphi_j) \right) \tag{2}$$

for some finite r where λ_j are frequencies, φ_j phase shifts and P_j are amplitudes. Recall that $\lambda = \frac{2\pi}{T}$ where T is the periodicity.

The noise R is assumed to be a sequence of (possibly independent) random variables $R(t)$ such that for each $t \in \mathbb{T}$, the $\mathbf{E}(R(t)) = 0$ and $\mathbf{Var}(R(t)) = \sigma^2 < +\infty$. Moreover, the covariance $\mathbf{Cov}(R(t), R(s))$ is assumed to depend on the time difference only and not on the concrete time t, i.e. the following holds for the covariance function for all $t \in \mathbb{T}$:

$$\gamma(\tau) = \mathbf{Cov}(R(t + \tau), R(t)).$$

We will suppose that R fulfills one of the following conditions:

$$\lim_{h \to \infty} \frac{1}{h} \int_0^h |\gamma(\tau)| d\tau = 0, \tag{3}$$

$$\gamma(\tau) = \sum_{j=1}^{M} G_j e^{i\lambda_j \tau} \tag{4}$$

for some $M \in \mathbb{N}$ where $G_j \in \mathbb{R}$ and $\lambda_j \in \mathbb{R}$.

In the sequel, we will often work with the time series X restricted to a certain subset of time moments $\bar{\mathbb{T}} \subset \mathbb{T}$. In this case, we will write $X|\bar{\mathbb{T}}$. Let $\bar{\mathbb{T}}$ be an interval $\bar{\mathbb{T}} = \{t_{j_1}, \ldots, t_{j_m}\} \subseteq \mathbb{T}$. By *length* of $\bar{\mathbb{T}}$ we mean $|\bar{\mathbb{T}}| = m$. We will also consider the operations of *left* XL (*right* XR) *extension of* $\bar{\mathbb{T}}$ by p time points. This is the interval

$$\mathrm{XL}_p(\bar{\mathbb{T}}) = \{t_{j_1-p}, \ldots, t_{j_1-1}, t_{j_1}, \ldots, t_{j_m}\},$$
$$\mathrm{XR}_p(\bar{\mathbb{T}}) = \{t_{j_1}, \ldots, t_{j_m}, t_{j_m+1}, \ldots, t_{j_m+p}\}.$$

Finally, by $l(\bar{\mathbb{T}})$ ($r(\bar{\mathbb{T}})$) we denote the first (the last) time moment, i.e., in our case, $l(\bar{\mathbb{T}}) = t_{j_1}$ ($r(\bar{\mathbb{T}}) = t_{j_m}$).

2.2 Fuzzy Transform and Time Series

2.2.1 Fuzzy Transform

The fuzzy transform (shortly, F-transform) is a procedure designed to deal with a bounded real continuous functions $f : [a, b] \longrightarrow [c, d]$ where $[a, b]$, $[c, d] \subset \mathbb{R}$. Its author is I. Perfilieva. The detailed description of the F-transform can be found in the literature, for example in [6, 10, 11, 19, 20].

The fundamental concept in the theory of F-transform is that of a *fuzzy partition*. There is a general definition of it that is based on the concept of a continuous and even generating function $A : [-1, 1] \longrightarrow \mathbb{R}$. In this paper, we will consider a more straightforward concept of a uniform Ruspini fuzzy partition as follows (cf. [19]).

Definition 1 Let $n \in \mathbb{N}$, $n \geq 2$, and $h \in \mathbb{R}$ be given. Let $c_0 < \cdots < c_n$ be fixed equidistant nodes within $[a, b]$, such that $c_0 = a, c_n = b$ and $c_k = a + kh$, $k = 0, \ldots, n$ where $h = \frac{b-a}{n}$. A set of continuous fuzzy sets $\mathcal{A}_h = \{A_0, \ldots, A_n\}$ on $[a, b]$ forms a *fuzzy partition of* $[a, b]$ if:

1. for all $k = 1, \ldots, n - 1$ and $x \in [x_k, x_{k+1}]$:

   - $A_k(c_k - x) = A_k(c_k + x)$,
   - $A_k(x) = A_{k-1}(x - h)$,
   - $A_{k+1}(x) = A_k(x - h)$.

2. $A_k(c_k) = 1$;
3. $A_k(x) = 0$ for $x \notin (c_{k-1}, c_{k+1})$, $k = 0, \ldots, n$ where we formally put $c_{-1} = a$ and $c_{n+1} = b$;
4. A_k strictly increases on $[c_{k-1}, c_k]$ and strictly decreases on $[c_k, c_{k+1}]$;
5. For all $x \in [a, b]$

$$\sum_{k=0}^{n} A_k(x) = 1. \tag{5}$$

Note the support of the fuzzy sets $A_0, \ldots, A_n$ has the width $2h$. In the theory of F-transform they are called *basic functions*. We will assume in this paper that they have a triangular shape.

Definition 2 Let a fuzzy partition $\mathcal{A}_h$ due to Definition 1 be given and consider a real, real-valued bounded function $f : [a, b] \longrightarrow [c, d]$. An $(n + 1)$-tuple

$$\mathbf{F}^m[f | \mathcal{A}_h] = (F_0[f | \mathcal{A}_h], \ldots, F_n[f | \mathcal{A}_h])$$

is called m-th degree *direct fuzzy transform* of f if

$$F_k^m[f | \mathcal{A}_h](x) = \beta_k^0[f | \mathcal{A}_h] + \beta_k^1[f | \mathcal{A}_h](x - c_k) + \cdots + \beta_k^m[f | \mathcal{A}_h](x - c_k)^2, \tag{6}$$

for all $k = 0, \ldots, n$. We call $F_k^m[f | \mathcal{A}_h]$ due to (6) *components* of the fuzzy transform.

Precise determination of the components (6) is in detail described in [10].

The following theorem is important for this paper.

Theorem 1 ([7]) *Let $\mathcal{A}_h$ be a fuzzy partition due to Definition 1 and let $f : [a, b] \longrightarrow \mathbb{R}$ be a real function that is four-times continuously differentiable on $[a, b]$. Let $F_k^m[f | \mathcal{A}_h], k \in \{1, \ldots, n - 1\}$, be the component (6). Then*

$$\beta_k^0[f | \mathcal{A}_h] = f(c_k) + O(h^2), \tag{7}$$

$$\beta_k^1[f | \mathcal{A}_h] = f'(c_k) + O(h^2), \tag{8}$$

$$\beta_k^2[f|\mathcal{A}_h] = \frac{f''(c_k)}{2} + O(h^2).\tag{9}$$

It follows from this theorem that each coefficient β_k^j provides a weighted average of values as well as of derivatives of the function f in the area characterized by the fuzzy set $A_k \in \mathcal{A}_h$.

Remark 1 It should be noted that only the nodes $c_1, \ldots, c_{n-1}$ should be considered when dealing with the F-transform. The reason is that if the fuzzy partition $\mathcal{A}_h$ is restricted to the interval $[a, b]$ then the basic functions at both left and right edge of it are incomplete and, therefore, the components are distorted.   □

Definition 3 Let $\mathcal{A}_h$ be a fuzzy partition due to Definition 1 and let $\mathbf{F}^m[f|\mathcal{A}_h]$ be the direct $\mathbf{F}^m$-transform of f. The function $\mathbf{I}^m[f|\mathcal{A}_h]$ defined by

$$\mathbf{I}^m[f|\mathcal{A}_h](x) = \sum_{k=0}^{n} F_k[f|\mathcal{A}_h] \cdot A_k(x), \quad x \in [a, b]\tag{10}$$

is called the *inverse F-transform*[1] of f.

The following property of the inverse F-transform has been proved.

Lemma 1 ([21, 22]) *Let $\mathcal{A}_h$ be a fuzzy partition due to Definition 1 and $f :$ $[a, b] \longrightarrow \mathbb{R}$ be a continuous function. Then*

$$\max_{x \in [a,b]} |f(x) - \mathbf{I}^m[f|\mathcal{A}_h](x)| \leq 2\omega(h, f)\tag{11}$$

where

$$\omega(h, f) = \max_{\substack{|x-y|<h \\ x,y \in [a,b]}} |f(x) - f(y)|$$

is the modulus of continuity of f.

Thus, the function $\mathbf{I}[f|\mathcal{A}_h]$ in (10) approximates the original function f with arbitrary precision depending on the choice of h in the fuzzy partition $\mathcal{A}_h$.

2.2.2   Application of the F-Transform to Time Series

For our approach to the analysis of time series, the following theorems are of crucial importance (for their proofs, see [8, 9, 20]).

Theorem 2 *Let S be the seasonal component (2) whose members have periodicities $T_1, \ldots, T_r$. Furthermore, let an h-uniform fuzzy partition $\mathcal{A}_h$ be formed over the*

---

[1] We will use the term "inverse fuzzy transform" in two meanings: (a) as the procedure for obtaining estimation of f and (b), as the function (10) approximating f.

equidistant set of nodes due to Definition 1 where the distance between the nodes is $h = \bar{d}\,\bar{T}$ *and* $\bar{T}$ *is the longest periodicity among all* $T_1, \ldots, T_r$. *Then the following holds for all* $t \in [c_1, c_{n-1}]$ *and* $m \in \{0, 1\}$:

$$\lim_{d \to \infty} |\mathbf{I}^m[S|\mathcal{A}_h](t)| = 0, \tag{12}$$

Theorem 3 *Let R be the noise of a time series due to the decomposition (28), which is stationary process with zero mean and finite variance. Furthermore, we suppose that the covariance function satisfies the assumption (3) or (4). Then, for any* $t \in \mathbb{R}$ *and* $m \geq 0$, *it holds that*

$$\lim_{h \to \infty} \mathbf{Var}\,(\mathbf{I}^m[R|\mathcal{A}_h](t)) = 0.$$

It follows from the theorems above that, by proper setting of the distance h between the nodes, the F-transform makes it possible to "wipe out" part or the whole of the seasonal component S of the time series and significantly reduce its noise. As a consequence, we can prove the following theorem.

Theorem 4 *Let* $X(t)$ *be a realization of the time series (28). Let* $\mathcal{A}_h$ *be an h-uniform fuzzy partition due to Theorem 2. Then there is a number* $D(m, h)$, $m \in \mathbb{N}$, *such that* $\lim_{h \to \infty} D(m, h) = 0$ *and*

$$|\mathbf{I}^m[X|\mathcal{A}_h](t) - TC(t)| \leq 2\omega(h, TC) + D(m, h), \qquad t \in [c_1, c_{n-1}]. \tag{13}$$

2.2.3 Computing the Trend-Cycle

Because the trend-cycle TC is, by the assumption, a smooth function, its modulus of continuity is very small, and we may conclude that the F-transform enables us to extract the trend-cycle TC from the time series $X(t)$ with high fidelity. Of course, the result depends on the estimation of the number r of periodic functions forming the seasonal component S and their periodicities. For this task, very helpful is the *periodogram* (see, e.g., [1] and elsewhere)

$$J(\lambda) = \frac{1}{2\pi|\mathbb{T}|}\left|\sum_{t \in \mathbb{T}} X(t)e^{-i\lambda t}\right|^2. \tag{14}$$

Function (14) attains significantly high values for frequencies (equivalently, period-icities $T = \frac{2\pi}{\lambda}$) detected in the time series X.

To specify the width $2h$ of basic functions in a fuzzy partition to be able to determine the trend-cycle TC due to Theorem 4, we may stem from the general OECD specification:

> *The trend-cycle is a component that represents variations of low or medium frequency while high fluctuations have been filtered out.*

Thus, we proceed as follows: let $\{T_1, \ldots, T_m\}$ be an ordered list of periodicities found using (14). Let us choose a periodicity $T_{mid} \in \{T_1, \ldots, T_m\}$ from the middle of this list and form a fuzzy partition $\mathcal{A}_h$ for $h = T_{mid}/2$. The estimation of the trend-cycle is then

$$TC = \mathbf{I}^m[X|\mathcal{A}_{(T_{mid}/2)}]. \tag{15}$$

The degree m of the F-transform can be set to $m \in \{0, 1\}$.

2.2.4 FT-Deviation

Finally, we introduce the concept of FT-deviation that is analogy to the statistical concept of the standard deviation.

Definition 4 Let $f : [a, b] \longrightarrow [c, d]$ be a bounded continuous function and $\mathbf{I}^0[f|\mathcal{A}_h]$ be its inverse F-transform w.r.t. the fuzzy partition $\mathcal{A}_h$. Then the FT-deviation of f is

$$\sigma_{FT} = \sqrt{\frac{1}{b-a} \int_a^b (f(x) - \mathbf{I}^m[f|\mathcal{A}_h](x))^2 dx} \tag{16}$$

The σ_{FT} well characterizes oscillation or volatility of f independently on its general course. In finite case, if the function f is given in a finite number of arguments $f(x_1), \ldots, f(x_n)$ then (16) reduces to

$$\sigma_{FT} = \sqrt{\frac{1}{n} \sum_{i=1}^n (f(x_i) - \mathbf{I}^m[f|\mathcal{A}_h](x_i))^2}. \tag{17}$$

Application of the F-transform to the time series analysis has several outcomes: it provides analysis of the time series, makes it possible to extract its trend and/or trend-cycle with high fidelity, and enables to estimate average derivatives of the time series (taken as a function) over an imprecisely specified area, while volatility of the time series in this area is ignored. These features enable us to apply methods of the fuzzy natural logic to various tasks, such as forecasting, automatic generation of linguistic description of specific parts of it (cf. [16, 17]), and to mining particular kinds of information.

2.3 Evaluation of Sizes and Local Trend of Time Series

2.3.1 Evaluative Linguistic Expressions

One of the tasks in mining information from time series is linguistic characterization of their trend, i.e., its direction and steepness. This is done using sentences of the form

$$Trend \text{ is } \langle \text{direction} \rangle \tag{18}$$

where

$$\langle \text{direction} \rangle := \textsf{stagnating} | \langle \text{special hedge} \rangle \langle \text{sign} \rangle, \tag{19}$$

$$\langle \text{sign} \rangle := \textsf{increasing} | \textsf{decreasing} \tag{20}$$

and

$$\langle \text{special hedge} \rangle := \emptyset | \textsf{negligibly} | \textsf{slightly} | \textsf{somewhat} | \textsf{clearly} | \textsf{roughly} |$$
$$\textsf{sharply} | \textsf{quite largely} | \textsf{fairly largely} | \textsf{hugely} | \textsf{significantly}.$$

Mathematical characterization of the trend is the tangent that can be computed using the first derivative. Hence, linguistic characterization of the trend means the evaluation of the magnitude of the corresponding tangent. We will use special expressions of natural language that belong to the class of the so-called, *evaluative linguistic expressions*. These are expressions of natural language using which people characterize position on a bounded ordered scale. The latter may consist of real measuring units such as meters, degrees, etc., but quite often, it is just an abstract scale consisting of unspecified units. Important for the semantics of evaluative expressions are order properties of these scales. Detailed presentation of their theory can be found in [12–14]. Note that the expressions due to (19) also belong among evaluative ones.

Evaluative expressions may have a complicated structure but for the purpose of this paper, it will suffice to consider only the *simple evaluative linguistic expressions* (possibly with signs) that have the general form

$$\langle \text{linguistic hedge} \rangle \langle \text{TE-adjective} \rangle. \tag{21}$$

The TE-adjectives form a special class of expressions that include gradable adjectives, evaluative adjectives, but also adjectives such as *left, middle, medium*, etc. They often form a *fundamental evaluative trichotomy* that consists of two antonyms and a middle member, for example *low, medium, high; clever, average, stupid; good, normal, bad*, etc. The triple of adjectives *small, medium, big* is taken as canonical. Note that evaluative linguistic expressions are quite often confused with fuzzy categories; for the difference, see [15].

To determine the semantics of evaluative expressions, we must first specify the *context*, i.e., to specify what does it mean "small, medium", or "big" in a given situation.

Definition 5 Let $(U, \leq)$ be a universe of discourse where $\leq$ is a linear ordering.[2] Let $v_L, v_S, v_R \in U$ be three elements such that $v_L < v_S < v_R$. Then a *context* is a set

---

[2] Without lack of generality, we may assume that $U \subset \mathbb{R}$.

⟨linguistic hedge⟩⟨small⟩   ⟨linguistic hedge⟩⟨medium⟩   ⟨linguistic hedge⟩⟨big⟩

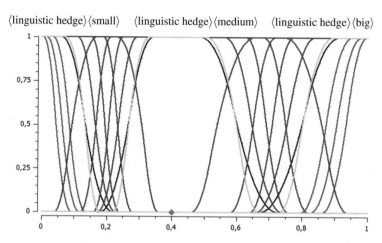

Fig. 1 Typical shapes of fuzzy sets (extensions) of simple evaluative linguistic expressions in a context $w = \langle v_L, v_S, v_R \rangle$

$$w = \{x \mid v_L \leq x \leq v_S\} \cup \{x \mid v_S \leq x \leq v_R\} \subset U. \tag{22}$$

The meaning of an evaluative expression is then a function[3] $W \longrightarrow \mathcal{F}(U)$ where W is a set of all contexts and $\mathcal{F}(U)$ is a set of all fuzzy sets over U.

We will work with a canonical set of simple evaluative expressions consisting of the TE-adjectives *small* (*Sm*), *medium* (*Me*), *big* (*Bi*) and special adjective *zero* (*Ze*). These can be completed by the hedges *extremely* (*Ex*), *significantly* (*Si*), *very* (*Ve*), *rather* (*Ra*), *more or less* (*ML*), *roughly* (*Ro*), *quite roughly* (*QR*), *very roughly* (*VR*). Experimental mathematical model of their meaning can be found, e.g., in [19]. Typical shapes of their extensions (fuzzy sets) in a given context are depicted in Fig. 1.

If a context w is known then an evaluative linguistic expression characterizing a given value y w.r.t. the context w can be determined using a function of *local perception*

$$LPerc\,(y, w). \tag{23}$$

Definition of such a function can be found in [19].

2.3.2 Determination of the Context

By Definition 5, the context is an interval of reals with a distinguished central point v_S. Informally, the $v_L \in U$ is the *most meaningful left bound*, $v_S \in U$ is the *most typical central point*, and $v_R \in U$ is the *most meaningful right bound*. Note that $v_S \in U$ can lay on any place inside the interval (v_S, v_R); not necessarily in its precise

---

[3] Such a function is called *intension*.

Table 1 Translation table between special and canonical evaluative expressions

Special hedge, direction of trend	$Ev[X|\bar{T}]$ from (24)
stagnating	extremely small or zero (Ex Sm or Ze)
negligibly	significantly small (Si Sm)
slightly	very small (Ve Sm)
somewhat	rather small (Ra Sm)
clearly	very roughly medium (VR Me)
roughly	very roughly big (VR Bi)
fairly large	roughly big (Ro Bi)
quite large	rather big (Ra Bi)
large	big (Bi)
sharply	very big (Ve Bi)
significantly	significantly big (Si Bi)
huge	extremely big (Ex Bi)

center. The interval $[v_L, v_S] \subset U$ contains all small values and $[v_S, v_R] \subset U$ all big ones. We will usually write the context as $w = \langle v_L, v_S, v_R \rangle$. Clearly, $x \in w$ means that x belongs to the set (22).

The "most meaningful" is, of course, vague specification. In practice, however, we are usually able to specify the context in a known concrete situation. For example, when speaking about high or small building in the Czech Republic, we know that there is no building taller than 100 m and the standard height of buildings is 30–40 m. Therefore, it is reasonable to set the context for them to, say, $w = \langle 5, 30, 80 \rangle$. In New York, on the other hand, it can be $w = \langle 40, 100, 300 \rangle$.

The following are few rules how the context should be specified.

(a) The *standard context* is the interval [0, 1] with the middle point specified using the golden ratio φ. Namely we set $v_L = 0$, $v_S = 1 - \frac{1}{\varphi}$ and $v_R = 1$. The reason is that the middle value v_S should be nearer to v_L than to v_R because people are able to distinguish better small values than big ones. Since φ is an irrational number, for practical reasons we may round $1 - \frac{1}{\varphi}$ to 0.4 since precise value is unnecessary. The standard context should be used when the evaluation is abstract (e.g., degrees of attractiveness) and no further specific information is at disposal.

(b) If data $\{d_1, \ldots, d_m\}$ are given then put $v_L = \min\{d_1, \ldots, d_m\}$, $v_R = \max\{d_1, \ldots, d_m\}$ and $v_S = v_L + (v_R - v_L) * 0.4$.[4]

(c) The context for trend of time series is equivalent to the context for tangent $w_{tg} = \langle v_L, v_S, v_R \rangle$. Because we must distinguish also between increase and decrease of the time series, we must specify positive context $w_{tg}^+ = \langle v_L^+, v_S^+, v_R^+ \rangle$ and negative one $w_{tg}^- = \langle v_R^-, v_S^-, v_L^- \rangle$. We usually put $v_L^- = v_L^+ = 0$ and write the

---

[4]Or, precisely, $v_S = v_L + (v_R - v_L) * (1 - \frac{1}{\varphi})$.

whole context as $w_{tg}^- \sqcup w_{tg}^+$ where $\sqcup$ denotes join of the negative and positive parts.

The right bounds $v_R^+ = \frac{M^+}{\bar{\mathbb{T}}}$ and $v_R^- = \frac{M^-}{\bar{\mathbb{T}}}$ where M^+, M^- are certain extremal values that can be reached by the time series during a given time period $\bar{\mathbb{T}}$. Both M and $\bar{\mathbb{T}}$ are usually specified by the user who knows the meaning of the time series. The v_S is then determined as in (b).

(d) The context $\langle v_L, v_S, v_R \rangle$ can be specified freely by the user on the basis of his/her knowledge of the concrete evaluated situation (cf. the discussion above).

2.3.3 Evaluation of Trend of Time Series over a Given Area

Let us consider a time interval $\bar{\mathbb{T}} \subset \mathbb{T}$. The following definition specifies evaluation of trend of the given time series X over the interval $\bar{\mathbb{T}}$, i.e., the trend of $X|\bar{\mathbb{T}}$.

Definition 6 Let $\bar{\mathbb{T}} \subset \mathbb{T}$ be a given time interval and $\mathcal{A}_h$ be a fuzzy partition with $h = |\bar{\mathbb{T}}|/2$. Furthermore, let $A_k \in \mathcal{A}_h$ be a basic function with the support $\text{Supp}(A_k) = \bar{\mathbb{T}}$ and let $\beta_k^1[(X|\bar{\mathbb{T}})|\mathcal{A}_h]$ be a value of the tangent (slope) of X over the area determined by A_k computed using formula (8). Then sentence (18) is construed as

$$\text{Trend of } X|\bar{\mathbb{T}} \text{ is } \pm Ev[X|\bar{\mathbb{T}}] \tag{24}$$

where $Ev[X|\bar{\mathbb{T}}]$ is an evaluative linguistic expression obtained using the function of local perception

$$\pm Ev[X|\bar{\mathbb{T}}] := LPerc(\beta_k^1[(X|\bar{\mathbb{T}})|\mathcal{A}_h], w_{tg}^- \sqcup w_{tg}^+). \tag{25}$$

By this definition, (25) assigns a linguistic expression $\pm Ev[X|\bar{\mathbb{T}}]$ to the value of the tangent $\beta_k^1[(X|\bar{\mathbb{T}})|\mathcal{A}_h]$ determined in (8) with respect to the context $w_{tg}^- \sqcup w_{tg}^+$.

Remark 2 The evaluative expressions characterizing slope of time series are specified in (19) and (20). It is possible to define their semantics. However, because we have already established the semantics of the canonical evaluative expressions, we can simplify the problem as follows. In which situation is the trend "sharply increasing"? A natural answer is: when the tangent is "very big". Similarly, we can also interpret the other expressions using which we evaluate the trend. This idea suggests to replace the expressions (19), (20) by the canonical evaluative expressions as suggested in Table 1. Of course, this is a simplification, and more exact model of the semantics of the former based on thorough linguistic analysis is necessary. For practical purposes, however, this approach may be sufficient.[5] □

---

[5]Let us remind the Zadeh's concept of *precisiated natural language* [25] that is considered as a technical simplification that works well in most practical situations.

3   Detection of Unusual Local Behavior of Time Series

3.1   *Structural Breaks in the Course of Time Series*

Structural breaks are sudden, huge changes in the ordinary course of the time series X. In [18], we suggested a method based on checking slope within two subsequent intervals determined by two adjacent basic functions A_i, $A_{i+1} \in \mathcal{A}_h$ for a particular fuzzy partition. In this paper, we modify this idea by employing a sophisticated method for finding intervals with the monotonous trend. This means that we decompose the time domain $\mathbb{T}$ into a set of intervals

$$\mathcal{T} = \{\mathbb{T}_i \mid i = 1, \ldots, s\}, \qquad \bigcup \mathcal{T} = \mathbb{T} \qquad (26)$$

such that the time series $X|\mathbb{T}_i$ has a monotonous trend evaluated due to Definition 6 and each two adjacent time intervals $\mathbb{T}_i$, $\mathbb{T}_{i+1}$ have a common time point.

Construction of intervals T_i proceeds starting with a *base interval* $\bar{\mathbb{T}}$ *of the width* $|\bar{\mathbb{T}}| = b$ that is step by step extended by an *increment of the length* p. The width b should be such a length of a time subinterval within which we can recognize the huge jump of values of time series. The latter is then considered as a structural break. We will determine b from the periodicity T_B given by the fundamental frequency of the time series; for example, $T_B = 12$ for monthly time series, $T_B = 30$ for daily time series, etc. Then we define[6]:

$$b = 0.4\, T_B. \qquad (27)$$

Algorithm 1 proceeds from right to left starting from the last known time moment of the time series X. Let an interval $\mathbb{T}_i = \{t_{j_i}, \ldots, t_{j_m}\}$ be processed being evaluated by the expression $\pm Ev[X|\mathbb{T}_i]$ from (25). First, we consider an interval $\mathrm{XL}_p(\{t_{j_i}\})$ and compute $\pm Ev[X|\mathrm{XL}_p(\{t_{j_i}\})]$. If the latter is equal to the former then we extend $\mathbb{T}_i$ to the left by p time moments to obtain a new interval $\mathrm{XL}_p(\mathbb{T}_i)$. We again compute its evaluation $\pm Ev[X|\mathrm{XL}_p(\mathbb{T}_i)]$. If the latter is equal to the original evaluation $\pm Ev[X|\mathbb{T}_i]$ then we put $\mathbb{T}_i := \mathrm{XL}_p(\mathbb{T}_i)$ and repeat the previous procedure. More precisely is this idea expressed in Algorithm 1.

The result of Algorithm 1 is the set of subsequent time intervals.

Definition 7 Let a realization $\{X(t) \mid t \in \mathbb{T}\}$ of the time series X be given and w_{tg} be a context for the trend of X. Let $\mathcal{T}$ be the set (26) of time intervals with monotonous trend detected using Algorithm 1. The interval $\mathbb{T}_i \in \mathcal{T}$ is an area of a *structural break* in the course of X if $Ev[X|\mathbb{T}_i] = \pm Ex\,Bi$ (i.e., the trend of X in the interval $\mathbb{T}_i$ is hugely increasing (decreasing)).

By this definition, only intervals with the hugely increasing (decreasing) trend can be considered as detection of structural breaks. The other cases, e.g., when the

---

[6]We consider here again the golden ratio.

Algorithm 1 Finding intervals with monotonous trend of time series X.

Input: $X, b, p, w_{tg}^-, w_{tg}^+$;
Require: $b > 2, 0 < p \le b, r(\mathbb{T}) > b$; // The time series must be sufficiently long
$i := 1; g := r(\mathbb{T})$;
repeat
    $\mathbb{T}_i := \mathrm{XL}_b(\{g\})$;
    $Ev[X|\mathbb{T}_i] := LPerc(\beta^1[X|\mathbb{T}_i], w_{tg}^- \sqcup w_{tg}^+)$;
    $e := \min(p, l(\mathbb{T}_i) - 1)$; // extension cannot exceed bounds of $\mathbb{T}$
    while $e > 0$ **do**
        $Y = \mathrm{XR}_{b-e}(\mathrm{XL}_e(\{l(\mathbb{T}_i)\}))$; // size of Y is equal to b
        $Ev[X|Y] := LPerc(\beta^1[X|Y], w_{tg}^- \sqcup w_{tg}^+)$;
        if $\pm Ev[X|Y] \neq \pm Ev[X|\mathbb{T}_i]$ **then**
            break; // extension has different evaluation
        end if

        $Ev[X|\mathrm{XL}_e(\mathbb{T}_i)] := LPerc(\beta^1[X|\mathrm{XL}_e(\mathbb{T}_i)], w_{tg}^- \sqcup w_{tg}^+)$;
        if $\pm Ev[X|\mathrm{XL}_e(\mathbb{T}_i)] \neq \pm Ev[X|\mathbb{T}_i]$ **then**
            break; // extended interval has different evaluation
        end if

        $\mathbb{T}_i := \mathrm{XL}_e(\mathbb{T}_i)$;
        $e := \min(p, l(\mathbb{T}_i) - 1)$; // next extension size
    end while
    print *Trend of $X|\mathbb{T}_i$ is* $\pm Ev[X|\mathbb{T}_i]$;

    $g := l(\mathbb{T}_i)$;
    $i := i + 1$;
until $g < b$
$s := i - 1$

---

trend is *largely* or *sharply* increasing (decreasing) can be treated as suspicious areas where structural breaks may occur.

Definition 7 can be applied provided that we correctly specify the context w_{tg} for the tangent. This can be obtained by comparing current values $X(t)$ with values occurring during the periods of the normal behavior of the time series, i.e., historical periods in which no exceptional behavior has been encountered.

3.1.1 Setting the Context

(a) Let a normal value of the tangent be known and equal to s. Then we put $v_L = 0$, $v_S = s$ and $v_R = s/0.4$.
(b) Let $X|\mathbb{T}_N = \{X(t) \mid t \in \mathbb{T}_N\}$ be a partial realization of the time series over the set $\mathbb{T}_N \subset \mathbb{T}$ of time moments when the time series behaved normally. Finally, let $\mathbf{I}^m[(X|\mathbb{T}_N)|\mathcal{A}_h]$ be estimation of the trend-cycle of $X|\mathbb{T}_N$ according to Sect. 2.2.3. Then we put $v_S = \frac{\sigma_{FT}}{0.4 \cdot T_B}$ where σ_{FT} is the FT-deviation (17). Finally, we put $v_R = v_S/0.4$.

3.2 Structural Breaks in Volatility of Time Series

However, we can also meet another kind of structural break that manifests itself in an abrupt change of the volatility of the time series (cf. Fig. 5). The volatility is usually characterized using standard deviation computed over a shifting time horizon. In this paper, we will estimate volatility using the F-transform. Our approach is a modification of the idea introduced by Troiano et al. in [24].

Definition 8 Let X be a time series and $\mathcal{A}_h$ a fuzzy partition with with the distance between nodes h that provides estimation of the trend-cycle due to (15). Put

$$V(t) = (X(t) - \mathbf{I}^m[X|\mathcal{A}_h](t))^2, \qquad t \in \mathbb{T}.$$

Then the *volatility* of X is a function $\hat{V}$ computed as the inverse F-transform of V

$$\hat{V}(t) = \mathbf{I}^m[V|\mathcal{A}_h](t), \qquad t \in \mathbb{T}. \tag{28}$$

Let w_V^+, w_V^- be positive and negative contexts for the tangent of V. Then, for a given $A_k \in \mathcal{A}$, we can find an evaluative expression

$$\pm Ev[V]_k := LPerc\left(\beta_k^1[V|\mathcal{A}_h], w_V^- \sqcup w_V^+\right) \tag{29}$$

that linguistically characterizes the slope of V in the area determined by $A_k \in \mathcal{A}_h$.

Definition 9 Let $\mathcal{A}_h$ be a fuzzy partition from Definition 8. The structural breaks in the volatility are all areas characterized by fuzzy sets $A_k \in \mathcal{A}_h$, $k \in \{1, \ldots, n\}$, for which $Ev[V]_k = \pm\langle hedge\rangle Bi$ where $Ev[V]_k$ is the evaluative expression (29).

To apply Definition 9, we can specify the context as follows: $v_L = 0$,

$$v_R = \frac{\max\{\hat{V}(t) \mid t \in \mathbb{T}\} - \min\{\hat{V}(t) \mid t \in \mathbb{T}\}}{2h} \tag{30}$$

and $v_S = 0.4\, v_R$.

4 Demonstration

In this section, we will demonstrate the above-described methods for the detection of structural breaks in time series. To see clearly how the methods work, we decided to demonstrate them on artificial time series, because in this case we precisely know where the structural breaks occur and so, we can convincingly check correctness of the algorithm. The time series were obtained by a combination of few real time

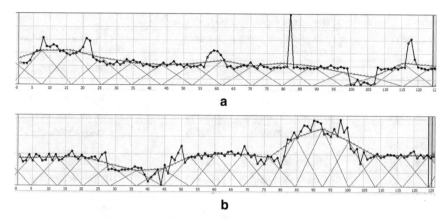

Fig. 2 Two artificial time series with various kinds of structural breaks. The pictures contain also trend-cycle estimated using the F-transform on the basis of a depicted fuzzy partition. The width of the basic functions is the following: **a** $2h = 18$, **b** $2h = 16$

series taken from the INDUSTRY subset of time series on a monthly basis from the M3-Competition published on the Internet.[7]

We created three artificial *monthly time series*. The first two manifest big structural breaks in their course (cf. Fig. 3) and the third one manifests structural breaks in volatility (see Fig. 5). The demonstration was prepared using the experimental software LFL-Forecaster[8] that has been developed to demonstrate our theory for the analysis and forecasting of time series, and also for mining information from them (cf. [19, Chap. 9]).

4.1 Structural Breaks in the Course of Time Series

The algorithm for detection of structural breaks in the course of time series is demonstrated on two kinds of artificial time series. They are depicted together with the estimation of their trend-cycle and fuzzy partition in Fig. 2. The trend-cycle was obtained using the F-transform based on the fuzzy partition based on uniformly distributed nodes with the distance h between them. Hence, the width of the basic functions is $2h$.

The distance h has been set on the basis of the detected periodicities using periodogram: time series (a) has periodicities 7.4, 12.6, 17.9, 26.5, 35.7, 41.7, 57.9; (b) has periodicities 2.4, 7.1, 12.6, 15.8, 25, 34.9, 50, 61.9. According to the theory, the proper value of h should be set on the basis of one of the periodicities selected from

---

[7]https://forecasters.org/resources/time-series-data/m3-competition/.

[8]The software was developed in the Institute for Research of Applications of Fuzzy Modeling of the University of Ostrava, Czech Republic.

Table 2 Intervals with monotonous trend detected in time series (a)

Interval	Linguistic evaluation of trend	Interval	Linguistic evaluation of trend
[121, 125]	Somewhat increasing	[63, 67]	Clearly decreasing
[117, 121]	**Hugely decreasing**	**[59, 63]**	**Hugely decreasing**
[113, 117]	**Hugely increasing**	**[53, 59]**	**Hugely increasing**
[109, 113]	Clearly increasing	[49, 53]	Clearly decreasing
[105, 109]	**Hugely increasing**	[45, 49]	Stagnating
[101, 105]	Clearly decreasing	**[41, 45]**	**Hugely increasing**
[97, 101]	**Hugely decreasing**	[37, 41]	Fairly large decrease
[93, 97]	Roughly decreasing	[33, 37]	Roughly decreasing
[87, 93]	Clearly increasing	[29, 33]	Roughly increasing
[83, 87]	Clearly decreasing	**[21, 29]**	**Hugely decreasing**
[79, 83]	**Hugely increasing**	**[15, 21]**	**Hugely increasing**
[71, 79]	Clearly increasing	**[7, 15]**	**Hugely decreasing**
[67, 71]	Quite largely increasing	**[1, 7]**	**Hugely increasing**

the middle of the corresponding list (cf. Theorems 2–4). Hence, we put $h = 9$ for the time series (a), and $h = 8$ for (b).

To find the structural breaks using Algorithm 1, we must specify the context, width b of the base interval, and increment p. According to Sect. 3.1, the context can be defined on the basis of the known *normal course*. The latter was taken from the flat parts of the respective time series as can be seen in Fig. 2. Of course, in reality, the ordinary course is determined from the knowledge of the real meaning of the time series.

The parameters for the respective time series were set as follows.

(A) The FT-deviation of the normal course of this time series is $\sigma_{FT} = 2.31$. The basic periodicity is set to $T_B = 12$. Hence, we obtain the context $w_{tg} = \langle -5.77/5, -2.31/5, 0 \rangle \sqcup \langle 0, 2.31/5, 5.77/5 \rangle$.

The base interval $b = 5$ ($\approx 0.4 \cdot 12$) and $p = 2$.

(B) The FT-deviation of the normal course of this time series is $\sigma_{FT} = 812$. The basic periodicity is set to $T_B = 12$. Hence, we obtain the context $w_{tg} = \langle -2031/5, -812/5, 0 \rangle \sqcup \langle 0, 812/5, 2031/5 \rangle$.

The base interval $b = 5$ ($\approx 0.4 \cdot 12$) and $p = 3$.

Table 3 Intervals with monotonous trend detected in time series (b)

Interval	Linguistic evaluation	Interval	Linguistic evaluation
[121, 125]	Stagnating	[52, 56]	Clearly increasing
[117, 121]	Roughly increasing	**[48, 52]**	**Hugely decreasing**
[113, 117]	Somewhat decreasing	**[44, 48]**	**Hugely increasing**
[105, 113]	Clearly increasing	[40, 44]	fairly large increase
[98, 105]	**Hugely decreasing**	**[36, 40]**	**Hugely decreasing**
[94, 98]	**Hugely increasing**	[32, 36]	Clearly increasing
[90, 94]	**Hugely decreasing**	[28, 32]	Clearly decreasing
[86, 90]	Clearly increasing	[24, 28]	Slightly decreasing
[79, 86]	**Hugely increasing**	[20, 24]	Somewhat decreasing
[75, 79]	Somewhat decreasing	**[16, 20]**	**Hugely decreasing**
[68, 75]	**Hugely decreasing**	[8, 16]	Clearly increasing
[60, 68]	Clearly decreasing	[4, 8]	Somewhat increasing
[56, 60]	Somewhat increasing	[1, 4]	Clearly increasing

One can see from Fig. 3 that all structural breaks in the respective time series are well detected. Moreover, we also immediately know whether it jumps up or down (in correspondence with the sign of β^1).

(A) Algorithm 1 detected altogether 26 intervals of monotonous trend—see Table 2. Among them 12 intervals are areas of structural break due to Definition 7. These intervals are in Table 2 printed in bold-face and graphically marked in Fig. 3.

(B) Algorithm 1 detected altogether 26 intervals of monotonous trend—see Table 3. Among them 9 intervals are areas of structural break due to Definition 7. These intervals are in Table 3 printed in bold-face and graphically marked in Fig. 3.

We compared our methods with classical Chow test for structural breaks. The results are in Fig. 4. One can see that unlike our methods, Chow test was not able to detect all the existing structural breaks.

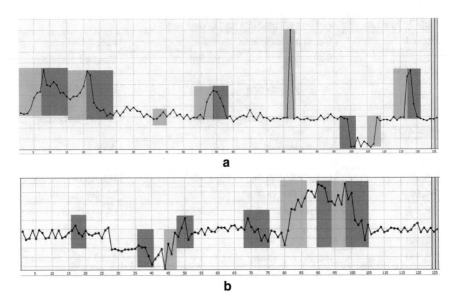

Fig. 3 Artificial time series (**a**) and (**b**) with marked structural breaks detected in their course using Algorithm 1

4.2   Structural Breaks in Volatility

The method for detecting this kind of structural breaks has been described in Sect. 3.2. We will demonstrate it on the time series depicted in Fig. 5 where also the trend-cycle is depicted with the width of basic functions $2h = 18$. This was chosen on the basis of the following detected periodicities using periodogram: 3.8, 7.5, 10.1, 15.6, 18, 22.3, 29.7, 41.3, 45.9, 61.9.

With respect to Definition 8, we compute volatility $\hat{V}$ using F-transform based on the fuzzy partition with the distance between nodes $h = 12.5$. The latter was determined on the basis of the periodicities 2.2, 13.8, 24.8, 41.4, 49, 61.9 detected in the function V. Finally, we compute the context: $v_R = \pm 3130$ using formula (30), $v_L = 0$ and $v_S = \pm 1252$. In Table 4 are estimations (8) of the local tangents.[9]

---

[9]Only full inner components are considered.

Fig. 4 Results of Chow test applied to time series (**a**) and (**b**)

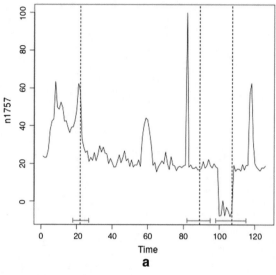

a

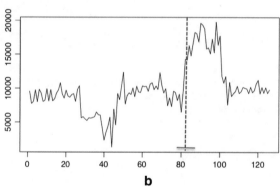

b

Table 4 Estimations of local tangents over areas characterized by basic functions of the width equal to $2h = 25$. The detected areas of structural breaks are typeset in bold face and graphically depicted in Fig. 5

k	$\beta_k^1[V|\mathcal{A}_h]$	Evaluation	k	$\beta_k^1[V|\mathcal{A}_h]$	Evaluation
1	43.75	ExSm	6	−504.49	RoSm
2	−9.52	Ze	7	**−2111.74**	**−VRBi**
3	404.71	RoSm	7	**−2213.04**	**−QRBi**
4	943.10	RaMe	9	−44.12	−ExSm
5	**3769.69**	**ExBi**			

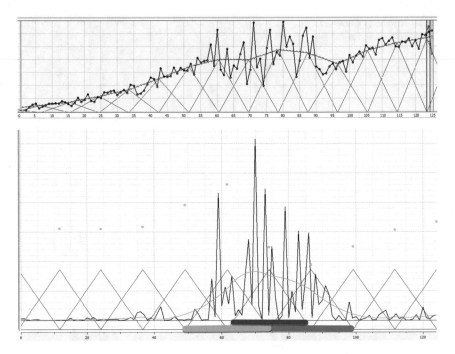

Fig. 5 *Up:* The artificial time series with structural break in volatility together with estimation of its trend-cycle. *Down:* The function V with its trend-cycle and marked structural breaks in volatility. The thick lines emphasize areas in which values of the tangent of volatility is big, i.e., the volatility is abruptly changing

5   Conclusions

In this paper, we suggested two methods of how structural breaks in time series can be detected using methods of fuzzy modeling. The first method detects structural breaks in the course of time series and the second one structural breaks in their volatility.

The methods are based on the application of the fuzzy transform and fuzzy natural logic, specifically its theory of evaluative linguistic expressions. Using them, we may automatically generate an evaluation of the trend of time series in an imprecisely determined area. The basic idea is to detect intervals with the monotonous trend. The intervals in which the trend is hugely increasing (decreasing) are probably areas with structural breaks.

The methods are robust w.r.t. the starting position of the fuzzy partition. It is also important to emphasize that the detection is extremely fast because the *time complexity of the fuzzy transform is linear*.

Acknowledgments The paper has been supported by the grant 18-13951S of GAČR, Czech Republic.

References

1. J. Anděl, *Statistical Analysis of Time Series* (SNTL, Praha, 1976). (in Czech)
2. S. De Wachter, D. Tzavalis, Detection of structural breaks in linear dynamic panel data models. Comput. Stat. Data Anal. **56**(11), 3020–3034 (2012)
3. P. Fischer, A. Hilbert, Fast detection of structural breaks, in *Proceedings of 21th International Conference on Computational Statistics*, Lisbon, Portugal (2014), pp. 9–16
4. T.-C. Fu, A review on time series data mining. Eng. Appl. Artif. Intell. **24**, 164–181 (2011)
5. J. Hamilton, *Time Series Analysis* (Princeton, Princeton University Press, 1994)
6. M. Holčapek, L. Nguyen, T. Tichý, Polynomial alias higher degree fuzzy transform of complex-valued functions. Fuzzy Sets Syst. **342**, 1–31 (2018)
7. V. Kreinovich, I. Perfilieva, Fuzzy transforms of higher order approximate derivatives: a theorem. Fuzzy Sets Syst. **180**, 55–68 (2011)
8. L. Nguyen, M. Holčapek, Suppression of high frequencies in time series using fuzzy transform of higher degree, in *Information Processing and Management of Uncertainty in Knowledge-Based Systems: 16th International Conference, IPMU 2016*, vol. 2, ed. by J. Carvalho, et al. (Springer, Cham, 2016), pp. 705–716
9. L. Nguyen, M. Holčapek, Higher degree fuzzy transform: application to stationary processes and noise reduction, in *Advances in Fuzzy Logic and Technology 2017*, vol. 3, ed. by J. Kacprzyk, et al. (Springer, Cham, 2018), pp. 1–12
10. L. Nguyen, M. Holčapek, V. Novák, Multivariate fuzzy transform of complex-valued functions determined by monomial basis. Soft Comput., 3641–3658 (2017)
11. L. Nguyen, V. Novák, Filtering out high frequencies in time series using F-transform with respect to raised cosine generalized uniform fuzzy partition, in *Proceedings of International Conference FUZZ-IEEE*, IEEE Computer Society (CPS, Istanbul, 2015), p. 2015
12. V. Novák, Mathematical fuzzy logic in modeling of natural language semantics, in *Fuzzy Logic—A Spectrum of Theoretical & Practical Issues*, ed. by P. Wang, D. Ruan, E. Kerre (Elsevier, Berlin, 2007), pp. 145–182
13. V. Novák, A comprehensive theory of trichotomous evaluative linguistic expressions. Fuzzy Sets Syst. **159**(22), 2939–2969 (2008)
14. V. Novák, On modelling with words. Int. J. Gen. Syst. **42**, 21–40 (2013)
15. V. Novák, Evaluative linguistic expressions vs. fuzzy categories? Fuzzy Sets Syst. **281**, 81–87 (2015)
16. V. Novák, Linguistic characterization of time series. Fuzzy Sets Syst. **285**, 52–72 (2016)
17. V. Novák, Mining information from time series in the form of sentences of natural language. Int. J. Approx. Reason. **78**, 192–209 (2016)
18. V. Novák, Detection of structural breaks in time series using fuzzy techniques. Int. J. Fuzzy Logic Intell. Syst. **18**(1), 1–12 (2018)
19. V. Novák, I. Perfilieva, A. Dvořák, *Insight into Fuzzy Modeling* (Wiley, Hoboken, New Jersey, 2016)
20. V. Novák, I. Perfilieva, M. Holčapek, V. Kreinovich, Filtering out high frequencies in time series using F-transform. Inf. Sci. **274**, 192–209 (2014)
21. I. Perfilieva, Fuzzy transforms: theory and applications. Fuzzy Sets Syst. **157**, 993–1023 (2006)
22. I. Perfilieva, M. Daňková, B. Bede, Towards a higher degree F-transform. Fuzzy Sets Syst. **180**, 3–19 (2011)
23. P. Preuss, R. Puchstein, H. Detter, Detection of multiple structural breaks in multivariate time series. J. Am. Stat. Assoc. **110**, 654–668 (2015)
24. L. Troiano, E. Mejuto, P. Kriplani, An alternative estimation of market volatility based on fuzzy transform, in *Proceedings of IFSA-SCIS*, Otsu, Japan (2017)
25. L.A. Zadeh, Precisiated natural language. AI Mag. **25**, 74–91 (2004)

Fuzzy-Based Methods in Data Analysis with the Focus on Dimensionality Reduction

Irina Perfilieva

Abstract We analyze a space with a fuzzy partition and show how it determines a measure of closeness. In the space with closeness, we characterize the corresponding Laplace operator and its eigenvectors. The latter serve as projection vectors to reduce the dimension of the original space. We show that the F-transform technique can be naturally explained in the language of dimensionality reduction.

Keywords Fuzzy partition · Laplace operator · F-transform · Dimensionality reduction

1 Introduction

Fuzzy modeling in data analysis relaxes requirements on rigorous data representation. Instead, it uses forms where real data are replaced by corresponding fuzzy sets. Despite of this obvious imprecision, this approach is very useful in numerical computations. It allows to be rough within areas with non-relevant data and optimally precise at localities of interest. However, the latter should be specified in the step of data preprocessing. For this purpose, we propose to apply the technique based on the theory of Laplace operators.

Informally, the Laplace (Laplacian) operator gives the difference between the average value of a function in the neighboring of a point, and its value at that point.

Why Laplacians? In PDE, Laplacian operator is used in the heat (diffusion) equation and characterizes a spacial change of the unknown function. In the classical

I. Perfilieva (✉)
Institute for Research and Applications of Fuzzy Modeling, University of Ostrava, NSC IT4Innovations, 30. dubna 22, 70103 Ostrava 1, Czech Republic
e-mail: irina.perfilieva@osu.cz

© The Editor(s) (if applicable) and The Author(s), under exclusive license
to Springer Nature Switzerland AG 2021
V. Kreinovich (ed.), *Statistical and Fuzzy Approaches to Data Processing, with Applications to Econometrics and Other Areas*, Studies in Computational Intelligence 892,
https://doi.org/10.1007/978-3-030-45619-1_14

case, this change is expressed using a difference of Gaussians. Therefore, a solution to the heat equation is expressed using the Gaussian function.

There are many applications where the use of the Laplacian operator is principal. Some quantum mechanics problems, although not diffusive in nature, are modeled by a mathematical analog of the heat (diffusion) equation. The latter is also used to model some phenomena arising in finance, like the Black-Scholes or the Ornstein-Uhlenbeck processes. In image processing and computer vision, the Laplacian operator appears in various tasks such as blob and edge detection, inpainting, etc. An approach based on Laplacian eigenmaps was proposed in [1] for a dimensionality reduction. The principal idea is to establish a projection of a higher dimensional data onto the closest manifold whose dimension is lower. The approach is focused on extraction principal components from data supplied with closeness.

In [2], Laplacian eigenmaps have been analyzed in the space where closeness is determined by a fuzzy partition. It has been shown that those eigenmaps can be selected as elements of a basis of the corresponding Hilbert space. Moreover, projections on those basis elements coincide with the F-(fuzzy) transform components [3, 4]. This fact shows that the method of F-transforms being one of fuzzy modeling techniques can be interpreted and classified as a conventional method of data analysis.

The aim of this contribution is to analyze a space with a fuzzy partition, consider the latter as a certain measure of closeness and characterize the corresponding Laplace operator and its eigenmaps. Then, we plan to show that in a space with a fuzzy partition, the F-transform technique can be explained in the language of dimensionality reduction.

In [5], an overview of mathematical properties and foundations of different dimensionality reduction techniques is given. In [1], an approach based on Laplacian eigenmaps, that characterizes the local neighborhood in terms of measure of closeness is proposed. The optimization problem focused on a low-dimensional representation of the data set is formulated, and the cost function is introduced. The proposed solution is formulated in terms of eigenmaps of the graph Laplacian.

In the fuzzy literature, the most relevant dimensionality reduction technique is the F-transform (shortly FzT) [3, 6]. It is based on a granulation of a domain (fuzzy partition) and gives a tractable image of an original data. The main characteristics with respect to input data: size reduction, noise removal, invariance to geometrical transformations, knowledge transfer from conventional mathematics, fast computation.

2 Preliminaries

2.1 Laplace Operator in Data Analysis

The Laplace operator or Laplacian is the name for many operators that are connected by the similar properties. In general, it can be characterized as a divergence of the gradient of a function on a certain (more often Euclidean) space. It is a linear operator on a functional space. Its generalization to functions defined on surfaces in Euclidean space or (even more general) on Riemannian and pseudo-Riemannian manifolds is known by the name Laplace-Beltrami operator.

The Laplace operator occurs in partial differential equations that describe many physical phenomena, such as electric and gravitational potentials. In particular, the Laplacian is used in the diffusion equation for heat and fluid flow, wave propagation, and quantum mechanics. In mathematical finance, it is the core of the Black-Scholes equation that characterizes the price evolution of a European call under the Black-Scholes model. In image processing and computer vision, the Laplacian is used in e.g., blob and edge detection.

In data analysis and especially in computer vision, the Laplace Beltrami operator on manifolds and close to it Laplacian of a graph are used more often. Below, we provide details for why fuzzy structured spaces are useful for efficient and reduced data representation.

We say that data $\mathbf{x}_1, \ldots, \mathbf{x}_k$ from a high-dimensional space $\mathcal{M}$—a manifold embedded in $\mathbb{R}^\ell$—have *efficient and reduced representation* $\mathbf{y}_1, \ldots, \mathbf{y}_k$ in the space $\mathbb{R}^m$, where $m \ll \ell$, if the following cost function (case $m = 1$)

$$\sum_{i,j=1}^{k} (y_i - y_j)^2 W_{ij}, \tag{1}$$

achieves a minimal value. In (1), W_{ij} is the measure of closeness between the prototypes $\mathbf{x}_i$ and $\mathbf{x}_j$.

Let us show that the problem of minimization the cost function in (1) follows from the requirement to preserve local measures of closeness between the given data points in $\mathcal{M}$ and in the image space $\mathbb{R}^m$. In other words, if a function $f : \mathcal{M} \to \mathbb{R}^m$ realizes the dimensionality reduction in the sense that

$$f(\mathbf{x}_i) = y_i, \ i = 1, \ldots, k, \tag{2}$$

then it should preserve local distances between the points $\mathbf{x}_1, \ldots, \mathbf{x}_k$ in $\mathcal{M}$, and their images $y_1, \ldots, y_k$ from the space $\mathbb{R}^m$, where $m \ll \ell$. This will be done in the following three steps.

1. As a first step, we retranslate (1) to the case where the original dataset $\{\mathbf{x}_1, \ldots, \mathbf{x}_k\}$ is identified with the set of graph vertices, and positive weights W_{ij} are assigned to edges between $\mathbf{x}_i$ and $\mathbf{x}_j$. The corresponding graph will be denoted as $G(V, E)$.

For $m = 1$, the problem of efficient and reduced representation of the dataset reduces to find the mapping $f : G \to \mathbb{R}$, such that connected points stay as close together as possible. This demand together with (2) leads to (1), see details in [7]. Moreover, (1) corresponds to $\int_{\mathcal{M}} \|\nabla f(x)\|^2$ where ∇f is a gradient of f, i.e., a vector in the tangent space $T\mathcal{M}_x$, such that given another vector $v \in T\mathcal{M}_x$, $df(v) = (\nabla f(x), v)_{\mathcal{M}}$.

2. On the other hand, we show that if f preserves local distances between the points $\mathbf{x}_1, \ldots, \mathbf{x}_k$ in $\mathcal{M}$, and their images $y_1, \ldots, y_k$, then f minimizes

$$\underset{\|\nabla f(x)\|_{L^2(\mathcal{M}=1)}}{\operatorname{argmin}} \int_{\mathcal{M}} \|\nabla f(x)\|^2. \tag{3}$$

Indeed, if $\mathbf{x}, \mathbf{u} \in \mathcal{M}$ are two neighboring points, then

$$|f(\mathbf{x}) - f(\mathbf{u})| \leq \|\nabla f(x)\| \|\mathbf{x} - \mathbf{u}\|_{\mathbb{R}^\ell} + o(\|\mathbf{x} - \mathbf{u}\|_{\mathbb{R}^\ell}),$$

where we made use of the assumption that $\mathcal{M}$ is isometrically embedded in $\mathbb{R}^\ell$. Therefore, $\|\nabla f(x)\|$ connects two estimates of closeness in spaces $\mathcal{M}$ and $\mathbb{R}$.

3. From the above given arguments, it follows that the best map that preserves local distances between the points $\mathbf{x}_1, \ldots, \mathbf{x}_k$ in $\mathcal{M}$, and their images $y_1, \ldots, y_k$, should fulfill (3) or (after discretization) (1). Therefore, the images $y_1, \ldots, y_k$ that fulfill (1) are indeed reduced and efficient representation of data $\mathbf{x}_1, \ldots, \mathbf{x}_k$.

What is still left is to show how the reduced representation is connected with a Laplace operator. Below, we explain in more detail. We will show that minimization of (1) leads to minimization of a quadratic form with a certain Laplace operator on vectors, and that a solution is given by a suitable eigenvector.

Let $W \in \mathbb{R}^{k \times k}$ be a weight matrix where entries W_{ij} measure closeness between data points $\mathbf{x}_i$ and $\mathbf{x}_j$. We define

$$L = D - W, \tag{4}$$

where D is the diagonal weight matrix such that

$$D_{ii} = \sum_{j=1}^{k} W_{ij}.$$

If W is a weight matrix of a graph, then the corresponding to it matrix L in (4) is known as a graph Laplacian. Let us show that for every vector $\mathbf{y} \in \mathbb{R}^k$,

$$\mathbf{y}^T L \mathbf{y} = \frac{1}{2} \sum_{i,j=1}^{k} (y_i - y_j)^2 W_{ij}.$$

Indeed,

$$\sum_{i,j=1}^{k} (y_i - y_j)^2 W_{ij} = \sum_{i,j=1}^{k} (y_i^2 + y_j^2 - 2y_i y_j)^2 W_{ij}$$

$$= \sum_{i=1}^{k} y_i^2 D_{ii} + \sum_{j=1}^{k} y_j^2 D_{jj} - 2 \sum_{i,j=1}^{k} y_i y_j^2 W_{ij} = 2\mathbf{y}^T L \mathbf{y}.$$

Therefore, the cost function in (1) can be replaced by quadratic form $\mathbf{y}^T L \mathbf{y}$ with L. A vector $\mathbf{y}$ that minimizes $\mathbf{y}^T L \mathbf{y}$ is the minimum eigenvalue solution to the generalized eigenvalue problem

$$L\mathbf{f} = \lambda D\mathbf{f}. \tag{5}$$

The following properties of matrix L ([8]) are useful for solving the eigenvalue problem.

(i) L is symmetric and positive semidefinite.
(ii) The smallest eigenvalue of L is 0, the corresponding eigenvector is the constant vector $\mathbf{1}$.
(iii) L has k non-negative, real-valued eigenvalues $0 = \lambda_1 \le \cdots \le \lambda_k$.

By (ii), the smallest eigenvalue solution to (5) is equal to 0, and the corresponding generalized eigenvector $\mathbf{y}$ is equal to $\mathbf{1}$ up to an arbitrary multiplier. This solution is trivial and we do not take it into account. To eliminate it, we change the minimization problem with the cost function (1) to the constraint minimization problem focused on

$$\operatorname*{argmin}_{\mathbf{f}^T D\mathbf{1}=0} \mathbf{f}^T L\mathbf{f}. \tag{6}$$

The solution of (6) is given by the generalized eigenvector with the smallest nonzero eigenvalue.

2.2 Standard Measures of Closeness

Let us emphasize that the selection of closeness among points in the initial dataset is essential for the dimensionality reduction result. This is because the reduction mapping respects this measure and produces new points such that they are close if and only if their prototypes are close as well. In the literature, the selection of closeness is connected with the evolution of a heat flow. Given the initial heat distribution $u(x, 0) = f(x)$ at time $t = 0$, its behavior in time $u(x, t)$ obey the *heat equation*

$$\left(\frac{\partial}{\partial t} + \mathcal{L} \right) u = 0.$$

The solution is given by $u(x, t) = \int_M H_t(x, y) f(y)$, where H_t is the heat kernel for this partial differential equation. If x and y are close and t is small, then

$$H_t(x, y) \approx (4\pi t)^{-\frac{m}{2}} \exp -\frac{\|x - y\|}{4t},$$

where manifold M has dimension m and is embedded in $\mathbb{R}^\ell$. From the heat equation we easily have

$$\mathcal{L} f(x) = -\left(\frac{\partial}{\partial t} \int_M H_t(x, y) f(y) dy \right)_{t=0},$$

and for small t we obtain

$$\mathcal{L} f(x) = \frac{1}{t} \left(f(x) - (4\pi t)^{-\frac{m}{2}} \int_M \exp -\frac{\|x - y\|}{4t} f(y) dy \right).$$

This representation of the Laplacian agrees with the given above informal description where we wrote that the Laplacian of a function gives the "difference between the average value of this function in the neighboring of a point, and its value at this point". Therefore, we obtain that closeness W_{ij} between points $\mathbf{x}_i$ and $\mathbf{x}_j$ can be estimated by

$$W_{ij} = \begin{cases} \exp -\frac{\|x-y\|}{4t}, & \text{if } \|x - y\| \leq \varepsilon, \\ 0, & \text{otherwise.} \end{cases}$$

Let us remind that function $f : M \to \mathbb{R}$ realizes the dimension reduction in the sense that

$$f(\mathbf{x}_i) = y_i, \ i = 1, \ldots, k,$$

if it preserves local distances between points M, where $\mathbf{x}_1, \ldots, \mathbf{x}_k$ are from manifold M, embedded in $\mathbb{R}^\ell$, and $\mathbf{y}_1, \ldots, \mathbf{y}_k$ are from the space $\mathbb{R}^m$, where $m \ll \ell$. The latter requirement reduces to the requirement that the most important projection (in absolute value) of f should be in the "direction" of the smallest eigenfunction.

2.3 Fuzzy Partition and Closeness

Since the very beginning, fuzzy models are known as compact and efficient tools of approximate representation of data. However, these models are not widely accepted, because the language they use is quite restrictive. Among the known exceptions, let us mention the theory of fuzzy (F-) transforms [4] that connects fuzzy and conventional approaches to mathematical modeling and numerical analysis. To extend the scope of F-transform applications, we show that this fuzzy theory can be explained in the language of dimensionality reduction. In short, the F-transform uses a fuzzy partition of a universe and is based on a linear transformation of considered objects to their simplified forms by weighted projections. Below, we remind the notion of a fuzzy partition and show how it relates to closeness. For simplicity, we consider one dimensional space $\mathbb{R}$, and remark that a similar definition can be made for any $\mathbb{R}^\ell$, $\ell \geq 1$.

Let $[a, b]$ be an interval on the real line $\mathbb{R}$. Fuzzy sets on $[a, b]$ are identified by their membership functions; i.e., they are mappings from $[a, b]$ into $[0, 1]$.

Definition 1 Let $[a, b]$ be an interval on $\mathbb{R}$, $n \geq 2$, and let $x_0, x_1, \ldots, x_n, x_{n+1}$ be nodes such that $a = x_0 \leq x_1 < \cdots < x_n \leq x_{n+1} = b$. We say that fuzzy sets $A_1, \ldots, A_n : [a, b] \to [0, 1]$, which are identified with their membership functions, constitute a *fuzzy partition* of $[a, b]$ if for $k = 1, \ldots, n$, if they fulfill the following conditions:

1. (locality)—$A_k(x) = 0$ if $x \in [a, x_{k-1}] \cup [x_{k+1}, b]$,
2. (continuity)—$A_k(x)$ is continuous,
3. (covering)—$A_k(x) > 0$ if $x \in (x_{k-1}, x_{k+1})$.

The membership functions $A_1, \ldots, A_n$ are called basic functions.

We say that the fuzzy partition $A_1, \ldots, A_n, n \geq 2$, is *h-uniform* if nodes $x_0, \ldots, x_{n+1}$ are *h-equidistant*; i.e., for all $k = 1, \ldots, n + 1$, $x_k = x_{k-1} + h$, where $h = (b - a)/(n + 1)$ and the following three additional properties are fulfilled:

4. for all $k = 1, \ldots, n$, $A_k(x)$ strictly increases on $[x_{k-1}, x_k]$ and strictly decreases on $[x_k, x_{k+1}]$,
5. for all $k = 1, \ldots, n$, and for all $x \in [0, h]$, $A_k(x_k - x) = A_k(x_k + x)$,
6. for all $k = 2, \ldots, n$, and for all $x \in [x_{k-1}, x_{k+1}]$, $A_k(x) = A_{k-1}(x - h)$.

It can be easily shown that for an *h*-uniform fuzzy partition $A_1, \ldots, A_n$, of $[a, b]$, there exists a continuous and even function $A_0 : [-1, 1] \to [0, 1]$ such that it vanishes on boundaries and for all $k = 1, \ldots, n$,

$$A_k(x) = A_0 \left(\frac{x - x_k}{h} \right), \quad x \in [x_{k-1}, x_{k+1}]. \tag{7}$$

We call A_0 a *generating function* of an *h*-uniform fuzzy partition.

2.4 F^m-Transforms

Let us fix $[a, b]$ and its fuzzy partition $A_1, \ldots, A_n$, with nodes $x_0, x_1, \ldots, x_n, x_{n+1}$, $n \geq 2$. Let k be a fixed integer from $\{1, \ldots, n\}$, and let $L_2(A_k)$ ($L_2([a, b])$) be a set of square-integrable functions on $[x_{k-1}, x_{k+1}]$ ($[a, b]$) with inner products

$$\langle f, g \rangle_k = \int\limits_{x_{k-1}}^{x_{k+1}} f(x)g(x)A_k(x)dx,$$

$$\langle f, g \rangle = \int\limits_{a}^{b} f(x)g(x)dx.$$

Spaces $L_2(A_k), k = 1, \ldots, n$, and $L_2([a, b])$ are Hilbert spaces on the corresponding domains. The functions $f, g \in L_2(A_k)$ are *orthogonal* in $L_2(A_k)$, if $\langle f, g \rangle_k = 0$. The function $f \in L_2(A_k)$ is orthogonal to a subspace B of $L_2(A_k)$ if $\langle f, g \rangle_k = 0$ for all $g \in B$.

Let us denote by $L_2^m(A_k)$ a linear subspace of $L_2(A_k)$ with the basis given by orthogonal polynomials $P_k^0, P_k^1, P_k^2 \ldots, P_k^m$. From [4], we know that

$$P_k^0 = 1|_{[x_{k-1}, x_{k+1}]}, \quad P_k^1 = x - x_k|_{[x_{k-1}, x_{k+1}]}. \tag{8}$$

Due to orthogonality of the basis polynomials, we have

$$\int\limits_{x_{k-1}}^{x_{k+1}} (x - x_k)A_k(x)dx = 0. \tag{9}$$

The following lemma gives analytic representation of the orthogonal projection on the subspace $L_2^m(A_k)$.

Lemma 1 ([4]) *Let function F_k^m be the orthogonal projection of $f \in L_2(A_k)$ on $L_2^m(A_k)$. Then,*

$$F_k^m = c_{k,0}P_k^0 + c_{k,1}P_k^1 + \cdots + c_{k,m}P_k^m, \tag{10}$$

where for all $i = 0, 1, \ldots, m$,

$$c_{k,i} = \frac{\langle f, P_k^i \rangle_k}{\langle P_k^i, P_k^i \rangle_k} = \frac{\int_{x_{k-1}}^{x_{k+1}} f(x)P_k^i(x)A_k(x)dx}{\int_{x_{k-1}}^{x_{k+1}} P_k^i(x)P_k^i(x)A_k(x)dx}. \tag{11}$$

The n-tuple $(F_1^m, \ldots, F_n^m)$ is an F^m-transform of f with respect to $A_1, \ldots, A_n$, or formally,

$$F^m[f] = (F_1^m, \ldots, F_n^m).$$

F_k^m is called the kth F^m-transform component of f.

In particular, let us consider the case where the basis of $L_2^m(A_k)$ is given by orthogonal polynomials $P_k^0, P_k^1, P_k^2 \ldots, P_k^m$ and P_k^0 is a constant function with the value 1. Then, the F^0-transform of f or simply, the F-transform of f with respect to the partition $A_1, \ldots, A_n$ is given by the n-tuple $(c_{1,0}, \ldots, c_{n,0})$ of constant functions (0-degree polynomials) where for $k = 1, \ldots, n$,

$$c_{k,0} = \frac{\langle f, 1 \rangle_k}{\langle 1, 1 \rangle_k} = \frac{\int_{x_{k-1}}^{x_{k+1}} f(x) A_k(x) dx}{\int_{x_{k-1}}^{x_{k+1}} A_k(x) dx}. \tag{12}$$

The F^1-transform of f with respect to $A_1, \ldots, A_n$ is given by the n-tuple $(c_{1,0} + c_{1,1}(x - x_1), \cdots, c_{n,0} + c_{n,1}(x - x_n))$ of linear functions (1-degree polynomials). The latter are fully represented by their 2D coefficients $((c_{1,0}, c_{1,1}), \ldots, (c_{n,0}, c_{n,1}))$, which in addition to (12), have the following particular representation:

$$c_{k,1} = \frac{\langle f, x - x_k \rangle_k}{\langle (x - x_k), (x - x_k) \rangle_k} = \frac{\int_{x_{k-1}}^{x_{k+1}} f(x)(x - x_k) A_k(x) dx}{\int_{x_{k-1}}^{x_{k+1}} (x - x_k)^2 A_k(x) dx}. \tag{13}$$

The *inverse F^m-transform* of function f with respect to the partition $A_1, \ldots, A_n$ is a function represented by the following *inversion formula*:

$$f_{F,n}^m(x) = \sum_{k=1}^{n} F_k^m A_k(x). \tag{14}$$

The following results demonstrate approximation properties of the direct and inverse F^m-transforms.

Lemma 2 ([4]) *Let $m \geq 0$, and let functions F_k^m and F_k^{m+1} be the k-th F^m- and F^{m+1}-transform components of f, respectively. Then,*

$$\| f - F_k^{m+1} \| \leq \| f - F_k^m \|.$$

Theorem 1 ([3, 4]) *Let $A_1, \ldots, A_n$, $n \geq 2$, be an h-uniform fuzzy partition of $[a, b]$, let functions f and A_k, $k = 1, \ldots, n$, be four times continuously differentiable on $[a, b]$, and let $F^1[f] = (c_{1,0} + c_{1,1}(x - x_1), \ldots, c_{n,0} + c_{n,1}(x - x_n))$ be the F^1-transform of f with respect to $A_1, \ldots, A_n$. Then, for every $k = 1, \ldots, n$, the following estimation holds true:*

$$c_{k,0} = f(x_k) + O(h^2),$$
$$c_{k,1} = f'(x_k) + O(h^2)$$

Theorem 2 ([3]) *Let f be a continuous function on $[a,b]$. For every $\varepsilon > 0$, there exist an integer n_ε and the related fuzzy partition $A_1, \ldots, A_{n_\varepsilon}$ of $[a, b]$ such that for all $x \in [a, b]$,*

$$|f(x) - f_{F,n_\varepsilon}(x)| < \varepsilon,$$

where f_{F,n_ε} is the inverse F-transform of f with respect to $A_1, \ldots, A_{n_\varepsilon}$.

Theorem 3 ([4]) *Let $A_1, \ldots, A_n$, $n \geq 2$, be an h-uniform fuzzy partition of $[a, b]$ that fulfills the Ruspini condition on $[a + h, b - h]$. Let functions f and A_k, $k = 1, \ldots, n$, be four times continuously differentiable on $[a, b]$, and let $f_{F,n}^m$ be inverse F^m-transform of f where $m \geq 1$. Then,*

$$\int_{a+h}^{b-h} |f_{F,n}^m(x) - f(x)| dx \leq O(h^2).$$

The discrete F-transforms were introduced in [3] and then further elaborated in a number of papers, see e.g., [9].

3 Dimensionality Reduction in the Space with a Fuzzy Partition

In this section, we discuss the problem of dimensionality reduction for the case where this technique assumed to be applied to datasets from a big family. As an example, we can consider a family of high resolution images on the same domain. According to the original idea, each dataset should be processed separately, so that the dimensionality reduction technique should be repeated as many times as many datasets we have in a family. In this case, we propose to decompose each data point into an argument and the dependent on it value and apply the dimensionality reduction technique to the common and only one dataset of arguments. Below, we give all necessary details.

We assume that dataset $\{\mathbf{x}_1, \ldots, \mathbf{x}_k\} \subset \mathbb{R}^\ell$ represents an ℓ-ary functional relation, i.e. every $\mathbf{x}_j$ is a couple $(\mathbf{p}_j, f_j)$, and there exists function $f : \{\mathbf{p}_1, \ldots, \mathbf{p}_k\} \to \mathbb{R}$, such that for all $i = 1, \ldots, k$, $f(\mathbf{p}_i) = f_i$. Function f can be a signal, time series, image, etc.

3.1 General Case of a Fuzzy Partition

The *dimensionality reduction* will be applied to the points in the domain $P = \{\mathbf{p}_1, \ldots, \mathbf{p}_k\}$ of f. For simplicity, we assume that $P \subseteq [a, b]$ and points in P are d-equidistant. Let us establish an h-uniform fuzzy partition of $[a, b]$ by the collection of fuzzy sets $A_1, \ldots, A_n$ and assume that $h = r.d$ and for every j, $1 \leq j \leq n$, there is a point $q_j \in P$ (*node*) such that $A_j(q_j) = 1$ and $A_i(q_j) = 0$, if $i \neq j$. We continue with each j, $j = 1, \ldots, n$, separately.

Fuzzy partition $A_1, \ldots, A_n$ determines geometry on the set P of points covered by it in the sense that two points p_i, p_s are close if and only if at least one of it coincides with a node of some fuzzy set from the partition, i.e.

$$W(p_i, p_s) = \begin{cases} A_j(p_i), & \text{if there exists } j, \; p_s = q_j, \\ A_j(p_s), & \text{if there exists } j, \; p_i = q_j, \\ 0, & \text{otherwise.} \end{cases}$$

It is easy to show that under the above given conditions, matrix W is symmetrical. Therefore, the weight assignment $W(p_i, p_s)$ determines the weighted graph $G = (P, E, W)$, where $(p_i, p_s) \in E$, if and only if $W(p_i, p_s) > 0$. Two other matrices D, L, that are required by the dimensionality reduction technique, are uniquely determined by W. Matrix L is a graph Laplacian.

We remind that the generalized eigenvector $\mathbf{y}_j^0$, corresponding to the minimal zero eigenvalue, is a constant vector $\mathbf{1}$. The generalized eigenvector $\mathbf{y}_j^1$ with the smallest nonzero eigenvalue fulfills (6).

Our next purpose is to give an explicit characterization of $\mathbf{y}_j^1$. This cannot be done in the general case of a fuzzy partition, because the distribution of basic functions $A_1, \ldots, A_n$ within $[a, b]$ is not limited. However, we can obtain the explicit characterization of $\mathbf{y}_j^1$ in one particular and important case.

3.2 Fuzzy Partition with One Basic Function

The constructed above weighted graph $G = (P, E, W)$ may have several connected components. In this case, each component is processed separately and independently. This means that the reduced representation will be obtained for each group of data points (vertices) that belong to one certain connected component. Then, the global reduction is a result of a unification of all local reductions. If graph $G = (P, E, W)$ arises after a fuzzy partition and assignment of closeness, then its connected components are associated with vertices covered by one particular basic function. This case will be considered below.

To simplify the denotation, we assume that all data points (vertices) are covered by one basic function, say A, so that $A : P \to (0, 1]$ where $P = \{p_1, \ldots, p_k\}$. Let k be an odd number, i.e., $k = 2q + 1$ and $A(p_{q+1}) = 1$. In this case, the weight matrix W is as follows:

$$W = \begin{bmatrix} 0 & 0 & \ldots & 0 & a_1 & 0 & \ldots & 0 \\ 0 & 0 & \ldots & 0 & a_2 & 0 & \ldots & 0 \\ \vdots & \vdots & & \vdots & \vdots & \vdots & & \vdots \\ 0 & 0 & \ldots & 0 & a_{q-1} & 0 & \ldots & 0 \\ a_1 & a_2 & \ldots & a_{q-1} & 1 & a_{q+1} & \ldots & a_k \\ 0 & 0 & \ldots & 0 & a_{q+1} & 0 & \ldots & 0 \\ \vdots & \vdots & & \vdots & \vdots & \vdots & & \vdots \\ 0 & 0 & \ldots & 0 & a_k & 0 & \ldots & 0 \end{bmatrix}, \tag{15}$$

where we use the denotation: $a_j = A(\mathbf{p}_j)$. The following two Theorems [2] characterize the spectrum of $L = D - W$ and the corresponding set of eigenvectors.

Theorem 4 *Let $k \times k$ matrix W (15), $k \geq 3$, be the weight matrix of the graph $G = (P, E, W)$ where all relevant sets are described above. Let $L = D - W$ be the graph Laplacian, where $D = (d_{ii})$ is the $k \times k$ diagonal matrix with $d_{ii} = \sum_{j=1}^{k} W_{ij}, i = 1, \ldots, k$. Then, the generalized eigenvalue problem*

$$L\mathbf{y} = \lambda D\mathbf{y},$$

has the three different eigenvalues $\lambda = 0, 1, 2 - \frac{1}{s}$, where the multiplicity of eigenvalue 1 is $k - 2$, and $s = \sum_{j=1}^{k} A(\mathbf{p}_j)$. Moreover, eigenvalue 1 is the second smallest.

Theorem 5 *Under the assumptions of Theorem 4, the eigenvector space EV of L is as follows: $EV \subset \mathbb{R}^k$, and*

$$EV = \{\mathbf{1}, \mathbf{y}_1^1, \ldots, \mathbf{y}_{k-2}^1, \mathbf{y}^2\},$$

where

(i) *vectors $\mathbf{y}_1^1, \mathbf{y}_{k-2}^1$ are orthogonal and each of them fulfills the condition*

$$\sum_{i=1}^{k} a_i y_{ji}^1 = 0 \quad \& \quad y_k = 0, \; j = 1, \ldots, k - 2; \tag{16}$$

(ii) *vector $\mathbf{y}^2$ is equal to $(1, \ldots, 1, \frac{1}{s} - 1, 1, \ldots, 1)^T$.*

3.3 Dimensionality Reduction Coincides with the F^1-Transform Component

We remind that we were focused on the problem of dimensionality reduction where datasets belong to a big family. We assumed that each dataset $\{\mathbf{x}_1, \ldots, \mathbf{x}_k\} \subset \mathbb{R}^\ell$ is further decomposed to $\{(\mathbf{p}_1, f_1), \ldots, (\mathbf{p}_k, f_k)\}$ and the domain $P = \{\mathbf{p}_1, \ldots, \mathbf{p}_k\}$ is partitioned by fuzzy sets $A_1, \ldots, A_n$. In the particular case $P \subset [a, b]$ and $n = 1$, we characterized closeness on P and constructed the corresponding graph Laplacian L and its eigenvectors.

The purpose of this section is to show that the reduced representation of the dataset that preserves local closeness agrees with the vector of F^1-transform components of $\mathbf{f}$ where $\mathbf{f} = (f_1, \ldots, f_k)$. This will be done for the particular case $n = 1$ considered in the preceding subsection. In other words, we assume that the set P is covered by one fuzzy set A.

By Sect. 2, the dimensionality reduction map preserves local distances (or closeness) between points in the dataset. In other words, the relevant projections of this map should be in the "direction" of the smallest eigenfunctions of the corresponding Laplacian. In the considered case, let us take the following two "smallest" Laplacian eigenvectors: $\mathbf{1}$ and $\mathbf{y}_1^1$, where $\mathbf{y}_1^1 = (p_1 - p_{q+1}, \ldots 0, \ldots, p_k - p_{q+1})$. It is easy to see that the coordinates of $\mathbf{y}_1^1$ lies on the line $x - p_{q+1}$. Therefore, this vector is in accordance with the two requirements: to be a basis polynomial for the F^1-transform (8) and to be an eigenvector with the second smallest eigenvalue (16). Moreover, we do not disregard the constant eigenvector $\mathbf{1}$ that corresponds to the smallest eigenvalue 0. This eigenvector corresponds to a constant basis polynomial for the F^1-transform (8).

Let us follow the technique of dimensionality reduction and project $\mathbf{f}$ onto two "smallest" Laplacian eigenvectors: $\mathbf{1}$ and $\mathbf{y}_1^1$. The sum of the obtained vector projections is the F^1-transform component of f. Therefore, the map $\mathbf{f} \mapsto F^1[\mathbf{f}]$ realizes dimensionality reduction and preserves local measure of closeness given by fuzzy set A.

It is worth noting that the important consequence of this result is the recommendation how to select the central node p_{q+1} of fuzzy set A and in general, of every basic function in a fuzzy partition. Due to the fact that the F^1-transform component of $\mathbf{f}$ realizes dimensionality reduction in the sense that the map $\mathcal{F}^1 : \mathbf{f} \mapsto F^1[\mathbf{f}]$ preserves local measure of closeness, the central node p_{q+1} of fuzzy set A should be close to all other points in P in the sense that the dependent value $f(p_{q+1})$ of an object from the given family is close to all other values $f(p_j)$ where $p_j \in P$.

4 Conclusions

In this contribution, we showed that the technique of dimensionality reduction based on the Laplacian eigenmaps can be used for the characterization of F^0 and F^1-transform components. Moreover, a fuzzy partition of a high-dimensional space determines its geometry and proposes principal factors (Laplacian eigenmaps) that can be used for efficient and reduced representation of high-dimensional data.

Acknowledgements This work was supported by the project LQ1602 IT4Innovations excellence in science. The additional support was also provided by *the project AI-Met4AI*, CZ.02.1.01/0.0/0.0/17-049/0008414.

References

1. M. Belkin, P. Niyogi, Laplacian eigenmaps for dimensionality reduction and data representation. Neural Comput. **15**(6), 1373–1396 (2003)
2. J. Janeček, I. Perfilieva, Dimensionality reduction and its F-transform representation, in *Proceedings of 11th Conference on EUSFLAT 2019*. Atlantis Studies in Uncertainty Modeling (2019)
3. I. Perfilieva, Fuzzy transform: theory and application. Fuzzy Sets Syst. **157**, 993–1023 (2006)

4. I. Perfilieva, M. Danková, B. Bede, Towards a higher degree F-transform. Fuzzy Sets Syst. **180**, 3–19 (2011)
5. C.O.S. Sorzano, J. Vargas, A. Pascual Montano, A survey of dimensionality reduction techniques (2014). arXiv:1403.2877v1 [stat.ML]
6. I. Perfilieva, Dimensionality reduction by fuzzy transforms with applications to mathematical finance, in *Econometrics for Financial Applications*, ed. by L. Anh, L. Dong, V. Kreinovich, N. Thach. ECONVN 2018. Studies in Computational Intelligence (Springer, Cham, 2018), p. 760
7. F.R.K. Chung, *Spectral Graph Theory* (American Mathematical Society, Providence, RI, 1997)
8. U. von Luxburg, A tutorial on spectral clustering. Stat. Comput. **17**, 395–416 (2007)
9. V. Novák, I. Perfilieva, A. Dvořák, *Insight into Fuzzy Modeling* (Wiley, Hoboken, New Jersey, 2016)

Perception and Reality

Nadipuram R. Prasad

Abstract The paper highlights a lifetime experience of knowledge gained through interactions with Professor Hung T. Nguyen and the profound impact he has made in advancing the state-of-the-art in modeling, simulating, and understanding the dynamics of uncertain systems. His seminal works on Random Fuzzy Sets have (a) significantly impacted our perceptions of seeking methods and approaches to acquire and assimilate data or information from uncertain sources, (b) enhanced our understanding to develop strategies for deep learning the interconnections, implications, and consequences, thereby allowing the processing of data to its logical conclusions and (c) has infinitely broadened the horizon towards conceiving machine-based decision-making architectures that optimize our perceived understanding to build systems that can mimic human decision-making capabilities. These three universal attributes fit the breadth and scope of Soft Computing. The paper focuses on three distinct topics wherein we explore human perceptions that merge the conscious attributes combining facts, myths, traditions, culture, beliefs, and a sense of reasoning towards a higher level of consciousness, and provides avenues for their adoption and physical realizations. The first topic discusses the possibilities of harvesting vast amounts of hydrokinetic energy from low-head flows in the Mekong Delta, Vietnam. The Mekong River has a perceived power potential of between 50 and 70 Gigawatts (GW). The reality is only one-quarter of the perceived power is plausible for harvesting. We discuss the possibility of transforming existing concrete structures across the arteries and tributaries of the Mekong River in the Delta Region to mini-hydropower plants. One example shows the possibilities of harvesting up to 250 KW at an existing structure that can serve the energy needs of local businesses involved in aquaculture, local commerce, and residential needs. Since recognizing conventional hydropower is ecologically damaging, Vietnam has focused its energy policy solely on Solar and Wind, which are 3–4 times higher in cost than Water. Low-head energy, which is currently unused, could be perceived as a significant

N. R. Prasad (✉)
Klipsch School of Electrical & Computer Engineering, New Mexico State University, Las Cruces, NM 88001, USA
e-mail: naprasad@nmsu.edu

V. Kreinovich (ed.), *Statistical and Fuzzy Approaches to Data Processing, with Applications to Econometrics and Other Areas*, Studies in Computational Intelligence 892,
https://doi.org/10.1007/978-3-030-45619-1_15

contributor to the renewable energy portfolio for Vietnam. The second topic discusses the possibility for Ayurvedic medicine, an ancient Indian practice of herbal medicine, as a complementary form of medication compared to modern medicine. The Tulsi plant is recognized for its life-saving properties that make Hindus believe it is a Holy plant. Scientific observations have shown the plant can produce substantial amounts of pure Oxygen by absorbing carbon dioxide from the ambient. Research has also shown that Tulsi can absorb fluorides from water. The discussions highlight the adoption of Ayurvedic medicine based on historical and traditional beliefs that Nature grown ingredients can induce natural body responses that aid in recovery from illnesses. Ayurvedic medicine exploits the power of the mind based on beliefs and traditions as a means for healing from illnesses. Finally, the third topic is a search for the meaning to the phrase "Pointing the finger at the Moon," and leading to the discovery of a three-dimensional holographic image on the Moon. The discovery sheds light on a long-standing question of what The Buddha was pointing to on the Moon. Historical background based on art is combined with basic Moon facts and a sense of geometry to uncover a three-dimensional holographic image that symbolizes Gautama Buddha's path to enlightenment. The form is identified as *Abhaya Mudra*, the Fearless Form. Among the many forms displayed in Buddhist sculptures of the meditation posture, the Abhaya Mudra is significant as it is a gesture for tolerance, fearlessness, and compassion. In Buddhist philosophy, the attributes of intolerance, fearfulness, and hatred are erased from the human mind through detachment.

Keywords Perception · Virtual reality · Electric power · Ayurveda · Buddhist philosophy

At the outset, it is an honor and a great privilege to write this paper honoring Professor Hung T. Nguyen for his immense contributions to the mathematical foundations of Random Fuzzy Set Theory, with applications to virtually anything imaginable that can be perceived and realized. His contributions are an ocean of knowledge that should motivate and inspire anyone in any discipline to explore and understand Life as a whole and the challenges towards meeting our professional goals and aspirations. If I could summarize what I learned from my deep engagements with Professor Hung Nguyen over three decades from a systems science perspective, it would be that "Everything in Life is fuzzy. Life is a continuum of actions and reactions. Reactions follow actions, and every reaction bears a consequence. There are reasons and motives for the consequences. Motives have aspirations. Aspiration is a perception of some future vision of the science, technology, and engineering needed to advance society. Virtual reality drives our perceptions. Perceptions of modelable and un-modelable uncertainty lead to realizations of robust systems."

In the age of high social interconnectivity and media interaction, there is a need to know relevant facts about everything past and present. It is essential to detect and disassociate false narratives. The future depends on the present. One might ask how facts can be preserved as pure facts and not as exaggerated facts, unless and until the root causes are uncovered. Our beliefs drive our perceptions that something could be true if it exists. The spectrum of possibilities ranges between mythological beliefs and the philosophy of science that overlaps with metaphysics exploring the nature of the human mind, ontology to study the existential possibilities, and epistemology in exploring the relationship between science and truth, and distinguishing between justified belief and opinion.

The publication of this paper ironically coincides with a momentous time in our history wherein our perceptions of normalcy has been suddenly disrupted. We fight the COVID-19 virus pandemic with a perception of discovering a vaccine soon enough to realize a sense of normalcy. The realities of finding a vaccine that is effective and will prevent acute infection is a temporal uncertainty. The discovery of an effective vaccine that can cure the disease is uncertain. The pandemic in its unpredictable ways has revealed a truth about the existential threat to Earth—an observable danger that is Physics-based. Shutdown and lockdown in Nations all across the globe to prevent the growth of infection have resulted in a significant drop in air pollution. Yet the perceptions of concerned scientists have little or no impact on those who believe otherwise and, therefore, Global Warming and Climate Change remains an issue.

1   Understanding Perception and Reality

As we look back in time, perceptions of the recent past are quite clear with little or no ambiguity, and which can be proven given that facts exist. Facts allow the backtracking of an event-sequence to the absolute truth. We can quote the exact words that someone spoke a hundred years ago in no uncertain terms with a high degree of specificity and certainty. Nearly a century ago, on January 27, 1921, Albert Einstein, in his address to the Prussian Academy of Sciences in Berlin on the topic of Geometry and Experience stated, *"So far as the mathematics refer to reality, they are not certain. And so far as they are certain, they do not refer to reality."* These statements reinforce the notions of impermanence and the theory of relativity and the degree to which mathematics can help humankind understand Life in its purest form.

Perceptions are mental formations based on the five senses for vision, hearing, touch, smell and taste, and a so-called sixth-sense that is a fusion of the five unique senses. Human perceptions are driven by facts, myths, traditions, customs, and beliefs, and a sense of reasoning that help to align the conscious attributes towards the desired vision. Recognizing that seeing is believing and that everything has a cause and an effect, one must see facts to assure the truth-value in decision making. Myths are unprovable and yet considered to be true in terms of cause and effect. We come to know about myths and practices during our upbringing and are conditioned to observe the conventions governing the myths. For instance, in Hinduism, each day of the week is associated with a specific Deity in the Hindu pantheon. The deep-rooted practice of Hinduism requires the observance of fasts and consumption of foods in a highly disciplined manner in commemoration of the deity worshipped each day. There are good days, and there are bad days for celebratory occasions. There is a good time of day and a bad time of day to conduct important daily chores. If you don't wear the sacred thread around your shoulder, then you are not a Brahmin. Scholars will argue the hidden powers behind these observances that can cause a metaphysical interaction between the soul and the higher being and allow one to achieve a transcendental state. It is the belief that one can be touched by a higher being and be granted one's wish in a paranormal state. Again, this is a belief, and one would dare not to refute this with a believer. The point is, myths exist, and superstitious belief has an overarching effect on everything. It is widely discussed and reported that political leaders of Nations consult with astrologers before making critical decisions. How and why planetary positions can influence the outcome of decision-making is in itself a deep-rooted belief that can neither be proven nor unproven, or disregarded.

Traditions transcend the sensation of touch, giving meaning to a feeling of closeness, a bond between entities, a mutual understanding of friendship, love, affection, gratitude, sympathy, and compassion. For instance, the culture to give flowers is a gesture of kindness and joy towards relatives and friends. The tradition is to offer a lotus flower arrangement as a heart-warming gift to someone you care and admire. Customs are analogous to the sweet smell of living within the confines of pleasure, comfort, and peace of mind. It is customary to place footwear at the front door-step to prevent outside dirt from entering the house. It is customary to bow in front of a deity with folded hands to receive the blessing from a higher entity. Customs induce moral values and respect for others. Belief lies in the understanding of the core values that speak the truth, and that whatever is written in the scriptures is true. Finally, the sixth-sense is a logical interpretation of all five sensory aggregates taken collectively. These perceptive attributes are clearly defined and described in canonical literature called the "Skandhas."

In Buddhist philosophy, the Sanskrit words "*Pancha Skandha*" translated as "five aggregates" sum up an individual's state of mind and physical existence." The five aggregates are (1) the environment, matter, body, shape, and form (rūpa, Avatār), which are the manifest form of the four elements—earth, air, fire, and water; it is a physical state that defines beauty and the individual persona and presentation. The Avatār epitomizes beauty, composition, and uniqueness of the shape and structure of everything in the Universe. It symbolizes the form of an object, which is living or inanimate. There is only one Planet Earth as we know and there is nothing like it anywhere we know so far, and NASA has named it the Blue Marble, a living and breathing object in Space. (2) Cognition, a state of thinking and building perceptions of things (samajñā), is a succession of thoughts that provide a clear sense for deductive reasoning, for grasping at the distinguishing features or characteristics of the object, and deep conceptual understanding. For instance, a river flowing from a higher elevation to a lower elevation with many waterfalls and fast-flowing waters has both potential and kinetic energy. It is a Physics-based conceptual understanding that can lead to the necessary mathematics and engineering. (3) Sensations, or feelings (vedanā), the urge to seek a greater understanding of the sensitivities and margins of pleasant, unpleasant, and neutral sensations that occur when our internal sense organs come into contact with external sense objects and the associated consciousness. What is good and acceptable for one may not be suitable and beneficial for the other. (4) Mental formations (saṃskāras), are planned and well thought out logical decision-making processes and connections, scenario-based decision-making clusters, and psychological forces that condition and spur mental activity. Mental formations are forces that shape moral and spiritual development through knowledge and insight. (5) Consciousness (vijñāna), the experience that brings together a greater understanding of perceived notions along with situational awareness and mindfulness towards those perceived notions. In short, the five aggregates, along with a sixth aggregate, a fusion of intuition, religious beliefs, and instincts, generate perceptions that lead to thinking, understanding, gaining knowledge, and raising the state of the conscious mind.

Increased awareness, according to the Skandhas, is an awakening. The practice of meditation is the type of training needed to achieve a higher state of the conscious mind. In principle, the human mind is a positive feedback system and is a characteristic that offers a glimpse into advances that can be made in deep learning and artificial intelligence. Whatever we wish to perceive exists somewhere, and in some shape and form. It is up to research, to uncover previously hidden facts, understand the effects and its possible implications, and create mental formations that allow us to perceive the realities in ways that reveal the truth and give rise to discovery.

With the Earth's precious environment in decline due to global warming and climate change, energy production and energy use must be moderated in ways that can sustain human civilization and bring economic growth and progress to the society. Methods to improve Earth's environment and processes for energy development must be explored simultaneously. Energy use no doubt will continue to rise as new cities and towns are born, communities grow, and small towns grow into large cities and explode amidst chaos and confusion due to overpopulation and overcrowded environment. With the year 2030 only a decade away, and Carbon emissions continuing to rise in India, China, and Vietnam, there is little chance that anything significant could be achieved within the next decade. Whether decarbonization will happen in the next two decades by the Year 2040 is doubtful despite efforts to reduce the use of fossil fuels by switching to Solar, Wind, and Water. Water is the cheapest resource among the three natural renewable energy sources. Large Solar and Wind power installations are 3–4 times more expensive and consume vast areas of land resources. The rate at which renewable energy resources are coming online is not fast enough to cause much change in the effects of global warming. Whether a noticeable shift in decarbonization will happen by 2050 remains pure speculation and a challenge for developers of renewable energy harvesting technology to conceive revolutionary ideas that are environmentally benign, and serve the energy needs efficiently.

While the perception that global use of renewable energy will peak by the Year 2050, the degree to which the perceived benefits will impact human society, in reality, remains uncertain. Both perception and reality are inseparable aggregates in our imaginative mind. One's perceptions and the realities in which the perceived attributes could materialize may be widely separated, and yet offer the possibilities for its realization. For instance, impressions of the Mekong River in terms of its total hydroelectric potential is over 50 GW, and some may say it is close to 70 GW. Size matters because it points to the immense amount of electric power that could uplift humanity in a giant leap. However, if all the perceived power were to be extracted, many more dams would be built across the Mekong River. Damming river flows would cause large reservoirs to drown out agricultural land, reduce the flows downstream, curtail fish migration, decimate mangrove forests, and destroy the ecosystem that supports aquatic and wildlife. Eventually, the entire ecosystem along the river would cease to exist. Impounding water upstream causes reduced flow downstream. Perceptions must, therefore, synchronize with the realities.

We choose to live within a realm of reality based on our imaginations and our own volition and understanding of what the environment offers. Deep thoughts, which probe the possibilities of existence, provide a level of fitness supporting the perception. Intuition is indeed a level of fitness. Something we had presumed to exist may be shown to exist if our hypothesis is correct, and hence, there exists a fitness function. For instance, gravitational waves, which Albert Einstein predicted must exist, were shown to exist. A massive Black hole at the center of the Milky Way galaxy that was known to exist was recently shown to exist. The LIGO-Virgo collaboration detected gravitational waves rippling out across space-time from the epic collision of two black holes from 2.4 billion light-years away. These are examples of how mathematics has influenced the development of technologies to discover Space and provide a better understanding of the structure of the Universe.

Human instincts to develop new technologies is motivated by the lack of existing resources to accomplish some desired objective adequately. It is uncertain if a technology that is commercially available today is suitable at a time when the conditions are not as previously anticipated. It may not work if the realities do not permit the use of the technology due to unpredictable uncertainties. A hydroelectric turbine developed for an irrigation canal where there are no fish cannot be used in a river where there is plenty of fish. Depth-of-perception could be a measure of the perceived reality. The more one focuses on the thought process, the more one learns and understands what the perception could mean and whether there is any truth in the perceived reality. The knowledge gained is a growth in our understanding of the physical laws that give the basic shape and form to the observed reality. It leads to a state of increased awareness of the perceived reality in the conscious mind. While the mathematics based on physical laws are exact, it excludes the realities that exist.

We need to explore the depths of human perception in ways that we may piece together our understanding of the reality in which our conscious attributes merge towards a higher level of consciousness. From an engineering viewpoint, perceptions of immensity and the potential of an immense body of water to serve human energy needs, for instance, must be realized in ways that appear not only reasonable at first glance, but lead to a location where exploration may follow. A bird's eye view motivates and inspires one to discover the ground-truth that exists and to find ways of exploring the ground-truth to align with a realistic vision.

During a road trip to the Mekong Delta, there was an urge to sense the immense amount of water in the Mekong River. It is the 12th largest river in the World. Short of taking an aerial view, the only option I had was to compare the size of bridges. From

Fig. 1 A large bridge across the main branch of Mekong River and a small bridge across its tributary provides a perception of the immensity of the river and the enormous amount of unused kinetic energy

an engineering viewpoint, perceiving something big or small depends on the context of how and in what ways our perceptions potentially benefit society. Whether it is a large bridge or a small bridge that connects two landmasses, as illustrated in Fig. 1, the purpose of the bridge remains the same, namely, to allow the transportation of humans and material goods. However, what lies under the bridge may be perceived differently. The figure illustrates a large bridge across the main branch of the Mekong River and a relatively smaller bridge across a tributary. The relative sizes are indicative of the shore-to-shore width of the river at the respective bridge locations of the main branch and its feeder as it flows out into the South China Sea.

Philosophical interpretation leads to transformation in the mindset towards seeking the ground-truth. In research, a shared ideology when all else fails is to philosophize and seek an understanding of the ideal truths. Our perceptions at first appear impossible and unreachable. However, a seemingly impossible task, when made possible, is a breakthrough. The following examples serve to identify how perceptions can lead to realizations.

2   Perceptions of Unused Low-Head Energy in the Mekong River

Historically rivers have been the birthplace of civilizations and have an inseparable link with humans. With food and water readily available, rivers have served the critical needs of societies for centuries. Humans, therefore, have a natural bond with waterways as a natural energy resource. Rivers and streams from the high mountainous regions of Cambodia and Laos flow across Vietnam from the West to the East, creating low-head energy as the flow discharges into the South Sea and the Gulf of Thailand. Since recognizing that conventional hydropower is harmful to the ecology and is not a viable option, Vietnam has made little effort to explore the vast

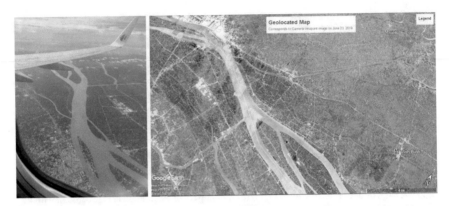

Fig. 2 Aerial view of the Mekong River on June 23, 2019, and co-located in Google Earth picture taken on January 1, 2019. Comparison points to perennial flows in the Mekong River

amount of unused low-head energy. Instead, the focus is on Solar and Wind energy. Low-head flows in the Mekong Delta offers immense amounts of hydropower to enable rapid growth in the Region's economy. The perception one can draw from a bird's eye view-point is that energy from the Mekong River alone could adequately serve the Delta Region while providing a path towards a Greener Vietnam. Yet many new coal-fired power plants are commissioned to meet the energy demands of the Delta Region. It is a dichotomy between our perceptions of clean air from Green energy and the realities driving economic growth. The cost to benefit ratio is reflected in one's perception to adopt the least cost technology that has the most benefit to society.

As luck would have it, on a flight from Kuala Lumpur, Malaysia, to Ho Chi Minh City, Vietnam, an aerial view of the Mekong River was awe-inspiring. Figure 2 illustrates the enormity of water flows, which would naturally increase during the monsoons and decrease during the off-season as it should.

The camera image shows river flows that are possibly enhanced by heavy monsoon rains upstream—a perception based on the knowledge of conditions that most likely exists during the monsoon season in the Lower and Upper Mekong River Basins of Laos and Cambodia. Co-located on a Google Earth image taken on January 1, 2019, one may infer from the relative size of the recent picture taken on June 23, 2019, that there are large quantities of perennial flows along the main arteries and tributaries.

A vast amount of electric power generation can be perceived along the entire length of the Mekong River. Flows in the Mekong Delta Region are along a low gradient due to the proximity of the South China Sea. As such, the flow velocity is low, but the discharge of water is large. Given that kinetic energy is one-half of the mass of water times the square of the flow velocity, it is easy to perceive the overall energy potential and the abundance of the natural renewable resource. If the flow-velocity is low, then an impeller placed along the axial length of the flow will also rotate at a low angular velocity. Coupled to gearboxes that convert low speed to higher speeds will enable effective power generation using commercially available generating equipment.

Novel approaches are needed to harness the vast amounts of unused low-head power, and we must seek out the methods through research and development. The following example shows in deterministic terms the amount of energy that could be harvested as a precursor to the engineering, design, and development of technology. The numbers reflect the magnitude and impact of producing large amounts of electric power from currently unused resources, and a basis for further exploration.

We take a look at one possibility in the Bach Lieu Province of the Mekong Delta, as an example, with the intent to transform an existing structure across the Bach Lieu River into a mini hydroelectric power plant. Figure 3 illustrates an existing concrete structure built across the river Song Bach Lieu, in the Bach Liew Province of the Mekong Delta. It serves as an example to perceive the possibilities for low-head hydropower generation at existing structures and to extend such perceptions at other similar structures along the river. The parameters are all based on visual observations.

The concrete structure allows water to flow through a gap approximately 20 m wide, 25 m deep, and 30 m long. By allowing 5 m depth for navigation purposes over the 30 m length, the available canal depth is 20 m. With the effective canal dimensions, 20 m wide and 20 m deep and 30 m long, the cross-sectional area A is 400 m^2.

Fig. 3 Existing concrete structure across the Mekong River tributary Song Bach Lieu, Bach Lieu Province, Vietnam

Fig. 4 Conceptual
illustration of five low-speed
impellers approximately 6 m
apart

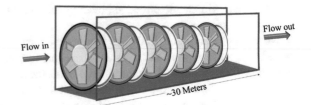

Flow in

Flow out

~30 Meters

Based on a flow velocity V of approximately 1 m/s, the discharge through the canal $Q = AV = 400\,\text{m}^3/\text{s}$. With the density of water $\rho \approx 1000\,\text{kg/m}^3$, the kinetic power from one impeller shaft is computed as $0.5\rho QV^2$ J/s, which is equal to $(1000)(400)(1^2) = 200{,}000$ W, or 200 kW per shaft.

If we assume five impeller shafts placed along the axial length of the canal, as illustrated in Fig. 4, we can perceive the hydropower potential as five times 200 kW or 1 MW.

The perception derived from a physical perspective points to the greater reality that a mass of water, about 400,000 kg in motion at a low velocity through the axial length of 30 m, must transfer a large amount of kinetic energy. The estimate of 1 MW is staggering. But this is precisely the outcome of laws that provide the perception for energy harvesting. The reality, however, is that the laws of Nature do not permit more than a quarter or perhaps one-third of the perceived estimate to extract. Therefore, one can believe that at a minimum, 250 kW can be harvested at this site to serve the local industries, businesses, and residential communities. The fact that there are many such concrete structures along Song Bach Lieu is highly encouraging towards perceiving the ground-truth for the potential low-head energy. Perception leads to innovation.

We turn now to something entirely different, where the belief in herbal medicine is gaining worldwide attention and transforming the mindset of societies towards health and wellness through herbal care.

3 Ayurveda as an Alternative for Modern Health Care

I recall as a kid learning to ride the bicycle, playing cricket, and playing soccer, and having frequent accidents on the playground, which caused bruises and, of course, pain. My mother would grind a special nut to make a paste and apply the paste to parts of the aching body and the injuries, and drink a little bit of the paste with warm milk to relieve the aching discomfort, and lo-and-behold the next day, I was good to go. The nut called Jaakai in India is none other than the nutmeg. Nutmeg is known to induce sleep and aid in the healing and recovery process. I recall using turmeric powder once to stop profuse bleeding from a deep cut. Turmeric is not only a disinfectant, but it is also a rapid coagulant. The same can be said about the treatment of colds and fever for which Ayurvedic medicine shows the ability to compete with modern medicine in terms of recovery from illness. There are many off-the-shelf products in modern pharmacy that offer the same or similar result but with the possibility for a reaction due to its synthesized chemical composition. It would be reasonable to perceive Ayurvedic medicine, therefore, as a complementary form of treatment to modern medical treatment based on evidence suggesting that there are indeed significant remedial benefits.

Unlike modern medicine that can have side effects due to chemical interaction, there are no side effects in Ayurvedic medicine because of natural ingredients and the absence of synthesized chemical compounds. The elements that comprise Ayurvedic treatment include the natural extracts of Nature grown herbs, tubers, roots, fruits, nuts, and plants that offer remedial comfort and relief to common illnesses. It is believed the combination of Yoga and Ayurveda can maintain the delicate balance between the mind, body, and spirit leading to health and wellness. Ayurvedic medicine is one of the World's oldest forms of holistic healing systems. Its primary purpose is to promote good health through healthy food and disciplined practice.

Among the many plants and herbs listed in the Ayurvedic apothecary, the Tulsi plant is considered a Holy plant by all Hindus because of its medicinal values, as described in Ayurveda, one of five Vedic scriptures in Hindu philosophy. An article in the Journal of Ayurveda and Integrative Medicine concludes *"Modern-day scientific research into Tulsi demonstrates the many psychological and physiological benefits from consuming Tulsi and provides a testament to the wisdom inherent in Hinduism and Ayurveda, which celebrates Tulsi as a plant that can be worshipped, ingested, made into a tea and used for medicinal and spiritual purposes within daily life."* [1]. Drawing a distinction between what we know today about the medicinal benefits of Tulsi, and how the human society may have first realized the potential for health and wellness leads one to perceive past reality and discover the truth. Modern science has shown that Tulsi plants produce Oxygen for 20 h every day while also producing Ozone for approximately 4 h every day. Maximum Oxygen production occurs at dawn and dusk when the humidity and dew point temperature are just right. Granted modern science with all the technology allows in-depth exploration that helps to understand the Chemistry of the Tulsi plant; there is no way of knowing exactly how people in the early days of civilization realized the benefit. It leaves one to speculate about the possibility that by observing the differences in health between those living close to the plants and those who were not, Tulsi was recognized to have remedial effects for those with breathing and pulmonary problems. Perceptions, therefore, that Tulsi could benefit those with pulmonary ailments can motivate and strengthen the willingness of people to adopt Ayurvedic medicine as a complementary form of treatment. Furthermore, the fact that the Tulsi plant produces large quantities of Oxygen by absorbing Carbon Dioxide from the air is particularly significant as it opens up the doors for widespread planting of Tulsi trees in parks, recreation centers, indoor terrariums, and greenhouses. Tulsi farming is already a big business venture that is primarily focused on marketing Tulsi tea. The Tulsi plant, therefore, appears to be an attractive solution for clean air and a clean environment.

Our beliefs are rooted in the customs and traditions of past realities that have withstood the test of time. The traditional practice in Hinduism requires one to accept "*Theertham*," water flavored with Tulsi leaves, as a blessing given to the devotee on behalf of the giver of knowledge, namely, the temple deity. No matter what, the core belief is that the *theertham* (water) is from the Holy River Ganges (although not true by any means), and is, therefore, readily accepted and consumed. The acceptance, based strictly on assuming the water is from the Ganges, is indeed blind faith. But having faith in religion is to have the willingness to put the power of the mind over matter towards healing and comfort, a belief that is engrained into the mindset. As such, the psychological response towards acceptance bears a direct effect on the physiological response as a consequence of one's belief. Scientific research shows Tulsi to drastically reduce the fluoride content in water due to its property of absorbing fluorides at a rapid rate and improving the taste of water. NASA intends to grow Tulsi plants in Space. The World Health Organization recommends a level of 0.5–1.0 part per million of fluorides in water that is safe for human consumption. Knowledge of all these facts tends to strengthen the belief and create a mindset for positive change. The perception, therefore, that Tulsi is indeed a miracle plant is believable. It is incredible to note that although ancient civilizations did not know the scientific facts as we know it today, they developed a deep bond with Nature in ways we in the present day are only left to imagine and speculate by perceiving the past and its realities as it existed then and establish pathways into the future.

There are no limits to one's imagination and creativity and no limits to what we can explore. Knowledge of facts is a driving force for the curious mind to study, understand, and investigate. If there is a question that begs an answer, we must find it. It does not matter where, when, or what. It is Why? We now turn to a discovery made in late 2007 and was reported in a blog under a pseudonym.[1] It is reported here for the first time in a formal presentation.

4 Pathways to Discovery

It is reasonable to expect a discovery of something every minute somewhere around the globe. NASA discoveries are bountiful in that constant observation of deep Space is bound to acquire new information about the structure and formation of the Universe. Seeing is believing. So how do we discover that which we cannot readily perceive? Something hidden behind the scenes and waiting to be discovered.

Perceptions of ancient historical facts are difficult to correlate. One has to accept the facts based on religious beliefs and unconditional acceptance of the truth as written in scriptures and ancient classical art. We can assume with a high level of confidence, the literary translations of ancient canonical texts are as accurate as can be. The circumstances under which these facts were generated form a basis to understand the present, and perceive the future. Our perceptions of the present are influenced by the effects and consequences of the past. While historical events dating

---

[1] http://sun-faced-buddha-moon-faced-buddha.blogspot.com/2011/08/discovery-of-buddha-form. html.

back over 2,600 years can be recounted with a level of certainty that the events did occur, the actual periods when the events occurred remain fuzzy and uncertain. The date and the period when events took place is based on scriptures and literary art, which may also be further supported by assertions embedded in the strictest customs and traditions that civilizations have observed, followed, and practiced throughout history. Religion and religious beliefs have withstood the test of time since their origins. Teachings from The Bhagwath Gita, The Bible, and The Koran are etched into the core beliefs of more than 7 billion humans on Earth.

The historical Buddha was a human who attained enlightenment approximately two thousand six hundred years ago. The UNESCO World Heritage Center has estimated the birth as 623 BCE, although there is a debate that he may have born sometime between 680 and 560 BCE. At the age of 29, he departed from his privileged life to seek answers to overcome human pain and suffering. His spiritual enlightenment paved a path to Nirvana, a transcendental state of mind devoid of pain and suffering, desires, and anger. He lived until the age of 80. These are facts.

Teachings of The Buddha appear in scriptures written in the Pali language, a derivative of Sanskrit. It is said that Gautama Buddha attained enlightenment following 49 days of intense meditation. Plate 1 depicts Buddha's first sermon to his five disciples in the presence of Gods and demi-Gods on the first full moon night following His enlightenment. Murals depicting the life and death of Gautam Buddha are displayed in every monastery, Pagoda, and Buddhist temple across Asia.

There is a vast body of literature that supports the belief that on His path to spiritual enlightenment in a deep meditative state, Gautama Buddha had to overcome all obstacles posed by Mara, the King of evil. Mara is an ideal personification of five deadly characters that could induce chaos, create deception, perform vile acts, exert cruelty, and cause death in one's personality and drive them into Hell[2] [2]. He attained a state of Nirvana, having detached himself from the clutches of attachment

Plate 1 Gautama Buddha's first sermon following His enlightenment, Bach Lieu Pagoda, Bach Lieu Province, Vietnam, July 14, 2019

---

[2]The notions of decision-making in the human mind can be described as a competition between a rational mind and an irrational mind. At no cost will the rational mind accept the irrational mind. There is only one winner and that's the rational mind. Mara is synonymous to irrational mind.

and desires. He had found a way of achieving a transcendental state of mind where there is no pain nor suffering, no desires or a sense of self, and the subject is released from the effects of prior sins (karma) and the cycle of death and rebirth. Nirvana is the final goal of Buddhism.

Legend has it that Mara's two daughters transformed into nymphs and unsuccessfully attempted to disrupt his focus and concentration through seduction and other vile acts. Mara himself transformed into five evil forces attacking Gautama's mind in every possible way, creating chaos, deception, engaging in despicable acts, demonstrating anger and cruelty and death. Plate 2 depicts The Buddha in a posture referred to as Abhaya Mudra (fearless form) when confronted by Mara. The large mural painting from Bach Lieu Pagoda in the Mekong Delta vividly portrays hostility, cruelty, violence, and death posed by Mara, and courage, love, compassion, peace, and harmony displayed by The Buddha.

The guiding principles of Buddhism are tolerance, mindfulness, and compassion attained through detachment. Tolerance induces patience and confidence, and a sense of fearlessness and courage to face odds and challenges. Mindfulness and compassion bring out one's sense of sharing, caring, giving, and empathy towards the well-being of others. These principles are firmly rooted in what is called *Mudras*.[3] *Mudras* are symbolic hand gestures that channel the internal energy towards desired outcomes in the state of mind. The five principle hand gestures that sculptures of the Buddha show in meditation postures represent the Wheel of Dharma, touching the Earth, charity, compassion, and boon-granting, and the fearless form. Statues of The Buddha in Abhaya Mudra posture, similar to Fig. 5, appear everywhere in Southeast Asia.

Historically, the Moon has been admired for its celestial beauty and is the center of attraction for all things considered beautiful in time immemorial. Literary pieces

Plate 2 Mural of The Buddha in a calm and peaceful state amidst the threat posed by Mara. Bach Lieu Pagoda, Bach lieu Province, Vietnam, July 14, 2019

---

[3]Mudras of the Great Buddha: Symbolic gestures and postures. https://web.stanford.edu/class/history11sc/pdfs/mudras.pdf.

Fig. 5 Buddha statue at the Truc Lam Monastery overlooking Da Nang Bay in Vietnam shows the Abhaya Mudra form

of art portray the Moon as a mirror for all humanity on Earth. The synergy in the Earth-Moon relationship is overwhelming because while the Earth is rotating, the Near Side of the Moon is continuously facing Earth. It is analogous for all Earthlings to looking into a mirror and asking who we are? The Full Moon night is celebrated in the Asian culture for its historical role honoring the Harvest Moon and family togetherness.

Attracted by gravitational force, Earth and Moon rotate along their axes in the anticlockwise direction as they orbit the Sun. The geometrical positions are such that only half the Moon's spherical surface, referred to as the near-side, is always facing Earth. Considering the libration of the Moon, which is a swaying phenomenon of the Moon as it orbits the Earth, nearly 59% of the Moon's surface is visible from Earth. The period of waxing towards a full Moon and then waning towards a new Moon is characterized by the Moon phases when the Moon exits from the shadow of Earth and then reenters, giving a cycle of motion that can be calculated precisely. Kepler's Laws of planetary motion accurately describe the positions of Earth and Moon around the Sun. There is no ambiguity in the Earth-Moon planetary motion around the Sun.

The Bhava-Chakra (*Life-Circle*), or the Circle of Life, is the epitome of Buddhist philosophy.[4] It is a symbolic representation of the three poisons, namely, ignorance, attachment, and aversion, that can lead one along a path of self-destruction. It is

---

[4] A concise summary of the various elements of the Bhavachakra appears in https://www.glensvensson.org/uploads/7/5/6/1/7561348/wheel_of_life.pdf.

Plate 3 The top portion of the Bhavechakra shows on the right Buddha pointing to the Moon. Mars is at the top left. With both hands, Mara, the embodiment of evil, holds the Circle of Life, the disc representing impermanence (symbolism of the Moon). The representation on the left upper corner is a state of Nirvana where one is freed from all sins

said that The Buddha himself created the first depiction and passed it on to King Rudrāyaṇa, who, after studying the drawing, was enlightened.

Plate 3 illustrates only the top segment of the Bhavachakra from a 5th Century Tibetan wall painting to highlight the key characteristics that are useful in seeking an answer to what Buddha was pointing to on the Moon. Scholars interpret the image of The Buddha pointing toward the moon as the path to liberation.

Like modern published works that have a title, an abstract, and the details of work, the layout of the Bhavachakra may be described similarly. The symbolic abstract of the Bhavachakra is indeed Buddha pointing his finger at the Moon.[5] Buddhist paintings depicting the root causes of human pain and suffering were recreated by word-of-mouth descriptions and through the words of Bodhidharma and the Bodhisattvas, followers, and messengers of The Buddha. They propagated Buddhist teaching throughout Southeast Asia.

When the Buddha gave his first sermon to his disciples on the first Full Moon night following his enlightenment, He explained his enduring struggles with Mara during his path to Nirvana. Legend has it that while pointing to the Moon, he said if one wishes to seek the truth, one must see where the truth is.

Plate 4 shows a Tibetan art icon from early 1st Century AD with Gautama Buddha pointing his finger at the Full Moon. The artist appears to have sketched an object on the Moon that is either a buffalo or an antelope. Growing up as kids, adults pointed

---

[5]The gesture of pointing to the Moon is very suggestive. The gesture does not proclaim anything except for one to look at the Moon and unravel the truth. What should one expect to see? It cannot be anything ordinary. Rather, it must be extraordinary. The expectation should be nothing short of a revelation. But what is it? We explore further.

Plate 4 A 1st century AD Tibetan wall painting depicts the famous gesture of "Pointing the finger at the Moon." The message is suggestive of the fact that there is something extraordinary on the Moon. What can it be?

to the Moon and asked if we see a Rabbit on the Moon. The Internet is abuzz with tracings of outlines that separate the dark and bright regions of the Moon.[6] Therefore, what appears on the Moon is unknown. So, what did Buddha point to on the Moon? What is the message to humanity in pointing to the Moon?

Since anything is possible, it is possible that in pointing to the Moon, Gautama Buddha described the existence of symbolic representations of life on Earth. Some image objects of life are in plain view. However, these are synonymous with trinkets one would find in a jewelry store while the most priced jewel remains hidden somewhere. One image that stands out is the notion of a family is quite striking as Fig. 6 illustrates the understanding of togetherness and the concept of family (*samsāra*).

In his seminal paper on the mathematical theory of communication theory [3], Claude Shannon states, "*The fundamental problem of communication is that of reproducing at one point either exactly or approximately a message selected at another point. Frequently the messages have meaning; that is, they refer to or are correlated according to some system with certain physical or conceptual entities.*" What did Shannon imply by decoding information either exactly or approximately? Every message has a meaning. So, what is the message? The question is really what The

---

[6]Caricature images http://wfmh.org.pl/thorgal/Moon/.

Fig. 6 Original image of the Moon (left), Perceived image *samsāra* (right)

Buddha was pointing to on the Moon in the context of His sermon, in which He describes his path to enlightenment.

There is no path to discoveries. It happens in unexpected ways. For brevity, details of the discovery are left out, and the reader is referred to a blog posting referenced in the Footnote.[7] The finding highlights the connectedness of all beings on Earth, and that everything we feel and experience in some form or the other gives us the strength and courage to sustain and strengthen the conditions of life on Planet Earth. Some excerpts from the blog are reproduced here for completeness.

Briefly, all NASA manned and robotic missions carried out thus far in Lunar exploration have used vertical landing as the means to land on the Lunar surface. Moon's gravity is one-sixth Earth gravity, and the lunar environment is in the vacuum of space. Because there is no environment on the Moon, aerodynamic drag does not exist to enable parachutes to slow down the descent. There is no other way for humans to land on the Moon safely other than to land vertically with rocket thrusters. Rocket propulsion is the only option. How about transporting cargo to the Moon and the possibilities for reducing the cost without rocket propulsion?

During the Summer of 2007, while engaged in a NASA fellowship at the Jet Propulsion Laboratory, there were many discussions of cargo transportation addressing the planned Lunar Outpost. Of course, the cost of transportation was a critical issue. Using inflatable objects to land safely on Mars had already been tested and proven. So, with a mission to deploy an Earth-observing telescope, an idea was conceived to land an inflatable ball-shaped robot on a curved rolling surface in much the same way as a plane lands on an airport runway. It would be a bowl-shaped rollway instead of a runway. Mathematically speaking, the bowl-shape had to satisfy the Lyapunov criteria for stability. All that had to be done immediately upon touchdown was

---

[7] http://sun-faced-buddha-moon-faced-buddha.blogspot.com/2011/08/discovery-of-buddha-form.html.

to deflate the ball in a controlled manner so it would slow down the rate of motion rapidly and allow the ball to deflate entirely upon reaching the target location and deploy the Earth-observing telescope. The question boiled down to finding a Crater with just the right characteristics that would enable such a robotic mission.

The Messier Crater, with its elliptical shape and the relatively smooth interior, offered the best possibility for landing a soccer-ball-shaped robotic object inside the crater. Located in Mare Fecunditatis, the Sea of Fertility, the Crater is known to have been formed over 3 Billion years ago. The idea was to let a ball-shaped robot to touchdown tangentially near the inside edge of the Messier Crater, roll towards the target location in a controlled manner, and deploy an Earth-observing telescope at the far end in the darkest spot on the Moon with a full view of Earth. The purpose of the telescope is to continuously observe Earth and beyond to detect near-Earth asteroids and serve as an early warning system against a threat from an asteroid collision. The mission would demonstrate the low cost of transporting cargo to the Moon. It would set the stage for commercial space cargo transportation, a necessary infrastructure towards the development of a Lunar habitat. It was just an idea then and is still a possibility for a future NASA mission. Figure 7 illustrates this concept.

The Craters reflect 99.9% of the incident sunlight and create rays of white light that point in the westerly direction (with Tycho as reference). The two beams of reflected UV radiation are mesmerizing. Figure 8 shows pictures taken during Apollo 15 and 17 missions and highlights the reflectivity of the crater interiors and the high energy beams that clearly show potential for a wide range of Lunar-based applications.

It is theorized that the impact of a comet or a meteor or an asteroid occurred at a low incidence angle ($<2^0$), causing the formation of Messier A and then Messier— analogous to how a rock would skip the surface of the water if it were thrown at a low incidence angle. The high reflection coefficient of the interior of the Craters causes the cone of reflected light to shine brightly like a searchlight from the East to the West. Because there is no environment on the Moon, the rays of reflected light travel linearly over many kilometers of the lunar surface with no edge dispersion. The reflected light sources provide a means to navigate precisely to the location.

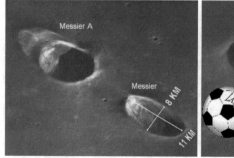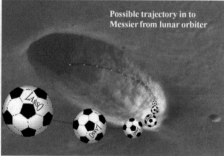

Fig. 7 Messier Crater pair (left). Deploying an Earth-observing telescope in Messier (right)

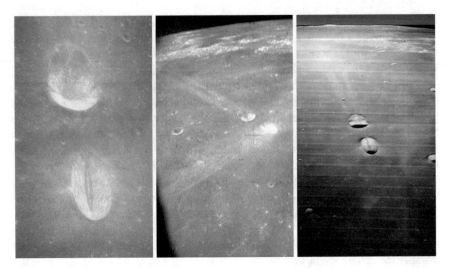

Fig. 8 Photographs from Apollo 15 and Apollo 17 missions: Messier A and Messier Lunar Craters show high reflectivity of the crater interior. The interior has a higher albedo than the surrounding Mare

Simple geometry using a straight edge and a protractor shows the regions from where the Messier rays may be visible to the naked eye. One could surmise that the cone of radiated light energy in the vacuum of space could cover all the Marias that are West of Mare Fecunditatis. The point here is that an observer located anywhere above the region covered by Marias Nubium, Cognitum, Insularum, Vaporum, Tranquillitatis, and Nectaris can see the point sources of light from Messier A and Messier. The convenience of using Matlab® image-processing functions to analyze the key features of the light patterns of high-resolution images of the Moon provided some degree of confirmation. There was a degree of truth to conjectures and speculative thinking. The premise that rays of light could give directional guidance for orbiting Lunar modules to eject cargo payload that will reach the Messier Crater and land safely appeared feasible. However, the cone of light formed by the reflected light sources from the Messier Craters and the countless number of craters causes an interference pattern above the Lunar surface.

In the course of examining ways to isolate the rays from Messier Craters from the interference pattern, out of the blues, a 3D holographic image of the Buddha form was discovered above Tycho, the South Pole of the Moon. My immediate thought was: Could this be the answer to what The Buddha pointed to the Moon? It is indeed ironic that one could stumble into something so different, unique, and relevant and in an unexpected way.

The image is composed of all wavelengths in the electromagnetic spectrum. The holographic image is located within the region enveloped by Marias Nubium, Cognitum, and Insularum on the left, Mare Vaporum at the top, and Marias Tranquillitatis and Nectaris on the right. The face of a human form is located just below Maria labeled "Mare Vaporum" in Fig. 9.

The holographic image is perceived to be formed by directed radiation crisscrossing the lunar surface from multitudes of lunar craters that span the regions enveloped by Marias Nubrium, Cognitum, and Insularum on the left, Mare Vaporum on top, and Marias Tranquillitatis and Nectaris on the right. These are point sources of reflected UV radiation in addition to the radiated beams from the Messier Craters

Figure 10 illustrates the Full moon in visible light and a false-color image taken by the Galileo spacecraft from approximately 425,000 km from the Moon. The highlighted segment shows a discernible object.

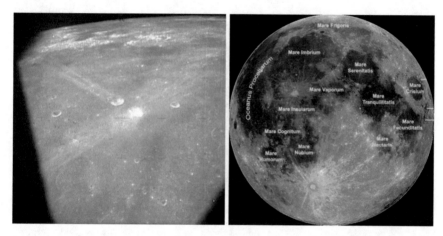

Fig. 9 Messier Craters, Mare Fecunditatis, from Apollo 15 mission. Location of Lunar Marias showing an image representative of the Abhaya Mudra

Fig. 10 Location of the Buddha form at full-scale shown above Tycho

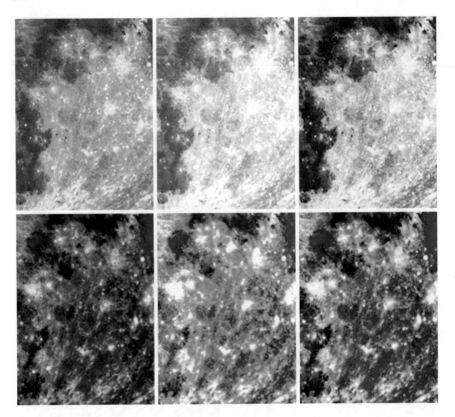

Fig. 11 View of the image at different contrast shows the posture of the Abhaya Mudra form

A closeup of the image formed in visible light and a false-color image is illustrated in Fig. 11 at different contrast levels to aid in identifying the 3D image.

Figure 12 is a further close-up of the 3D holographic image near the facial area. It shows how legendary knowledge of the horrifying character described as Mara may have appeared within the mental formations of The Buddha.

Figures 10, 11, and 12 points to the location and distinct features of the holographic image on the Moon. Images in visible light and those in false colors are shown with slightly different contrast levels to enhance the viewing experience and aid in feature identification. Figure 12 shows the facial characteristics in what appears to be a composite image of a human face with a mask that partially covers the top half of the human face. By deduction, one may conclude the foreground is the face of Mara, the evil King with his crown and intimidating mustache. Mara's face masks part of the human face. We must contend with the pixel resolution of the human face behind the mask. It is interesting that the shade of grey-green clearly distinguishes the hologram of what appears as the form of a human with the right ear fully visible. Part of the right side of the head, the chin and mouth, are also easily distinguishable

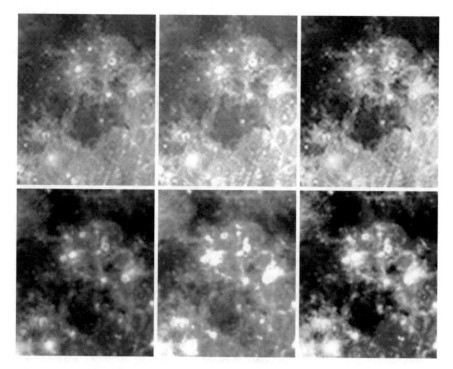

Fig. 12 Close-up of the face at different contrast shows the right ear, mouth, and chin of a human-like face, which appears Green in the false-color image. A demonic face-mask covers the eyes and the forehead in Red, showing a crown, a prominent nose, and grotesque mustache of Mara, the evil

in grey-green. The right shoulder is recognized by tracing upwards from the right palm.[8]

What is the significance of this finding? If The Buddha was indeed pointing to the image, then what is the message that He wished to convey to humanity? How could He have perceived or seen such an object with his naked eyes? Observing the finger provides no information while what is on the Moon is many more than one can imagine. His battle of wits with Mara, as described in the literature, is consistent with details in the Bhavachakra. The three-dimensional holographic image, which is an epitome of His enlightenment, exemplifies the battle of wits in a singular and eternal shape and form on the Moon.

The rhythm of the Moon's phases has for time immemorial guided humanity in developing the pillars of religions and the ethical and moral guidelines that foster the beliefs in the existence of Gods and demi-Gods in their abstract forms, and Saints who have lived in flesh and blood to provide the inner strength and wisdom. Gautama Buddha, as a human, understood human intolerance as an outgrowth of ignorance and desire, which is the root cause of all the pain and suffering.

---

[8]The images of the Moon are obtained from the NASA website: https://solarsystem.nasa.gov/resources/2460/lunar-near-side/?category=moons_earths-moon.

In closing, one could surmise that the intent in pointing to the Moon is to unlock the doors of the human mind and arouse the consciousness of humanity at its roots to build a mindset towards tolerance, confidence, fearlessness, mindfulness, and compassion. We may perceive the message to mean that a human is always under the grip of a force that, if unrecognized, can lead to a path of self-destruction. The symbolic representations in the Bhavachakra epitomize the paths one can lead in their life. The findings are at the tip of an iceberg leaving much more to be explored and understood. Of course, as humans explore the Moon, it is likely the shapes and forms we know now will not remain the same, raising issues of how humans contribute to the impermanence. Exploration requires one to leave things behind the way you find it so the future generations can experience, understand, and develop a mindset for positive change.

On Earth, while one can perceive a zero-Carbon environment with Solar, Wind, and Water alone and nothing else, the realities are that humans will continue the use of fossil fuels unless and until the cost of electric power consumption is a fraction of a cent per kilowatt-hour. Although we know it will happen sometime in the future, it leaves one to perceive the future and the reality that may exist then.

As health and wellness are a significant aspect of our daily life, Ayurveda is a discipline that I have adopted to practice. Our perceptions of healing come from a belief that herbs of all types had to have been the only remedial elements of the ancient apothecary. We may, therefore, perceive the adoption of Ayurvedic medicine as a complementary form to modern medicine that helps to moderate the mind and body to stay healthy and fit.

5   Epilogue

Professor Hung T. Nguyen is among the greatest mathematicians of our times who has made an everlasting impact on my life and others close to me, and on the lives of many of his students who I know very well. Enjoying life in the beautiful surroundings of Chiang Mai, Thailand, his presence is felt across the globe. Hung is truly an epitome of what General MacArthur said in his departure address to the US Congress more than a half-century ago in 1951, stating, *"old soldiers never die, they just fade away."* He is indeed a great inspiration to anyone considering retirement. Retire, but don't disappear. It is a challenge that shows commitment to advancing knowledge to the highest degree regardless of where you are or what you are doing.

Hung Nguyen's seminal contributions to the Theory of Random Fuzzy Sets embodies everything one can imagine. It applies to all disciplines, including engineering, economics, environmental science, natural sciences, physical science, social sciences, and medicine. With all things considered fuzzy, our interactions were driven from an engineering perspective to integrate fuzzy information to the control of real-World physical processes and systems. Hung's contribution to the mathematical foundations of statistical and fuzzy inferencing, data fusion, large data analysis, sampling, filtering, clustering, image analysis, anomaly detection, and for estimation, prediction, and control, greatly influenced our explorations in Control Theory. Modeling ambiguity and uncertainty, present in virtually every aspect of human decision-making, had become a primary focus of our research in the area of threat assessment, analysis, and threat mitigation.

During a trip to Bhubaneswar, India, in 1998, we visited the City of Puri, a historically famous city and known for the Jagannath Temple, a temple that every Hindu believes must visit at least once in their lifetime. Before arriving in India, Hung had asked me if we could go to a place where he could see Indian Buddhist monks practicing Buddhism, and I was searching for places to visit. Although there are many Hindu temples all across India, I was not sure where to find one that was in our travel plans. For instance, many monasteries and numerous Buddhist temples and pagodas dot the landscape in every city and town in Thailand and Vietnam, and I was looking to visit a Buddhist monastery in India. I had never been to one in India. As we approached the entrance of the Jagannath Temple, a temple administrator without saying a word, and chewing a mouthful of beetle-nut and leaves, pointed his finger at a sign on the wall that read "Hindus Only." I knew Hung is a Buddhist, and I am a Hindu, and I thought, does it mean I can go in and Hung cannot? Looking at me, the temple administrator said in Hindi that he would permit me inside but not Hung. Confused at first, I told the administrator that Hung was a Buddhist. But the administrator insisted in Hindi and pointing to the sign that said Hindus Only! Surprised and still not satisfied with his answer, I asked him why he would not let a Buddhist in, the administrator's reply was stunning. He said, "Even India Gandhi, the Prime Minister, was not allowed in because her husband Feroze Gandhi was a Muslim." But here is the irony.

Buddhism is an offspring of Hinduism. At the entrance to any Hindu temple, there is a pillar called Ashoka Pillar, named after King Ashoka the Great, who adopted and practiced Buddhism during the time of his reign between 273 and 236 BCE. King Ashoka, a devout follower of Buddhism, called it the Buddha Stupa, only about 300 years following the historical Buddha. The Buddha Stupa symbolizes the metaphysical transfer of knowledge and wisdom from all higher-level deities (Gods and Goddesses in the Hindu mythology) through The Buddha, giving meaning to human contact with all other human beings entering the deity's abode. In other words, the notion of transfer of energy from higher-beings to ordinary human beings is through The Buddha. Knowing that every Hindu temple has a representation of The Buddha, the refusal to let Hung enter the temple seemed conflicting. The administrator had perhaps failed to recognize that the Buddha Stupa stands as a bridge between all humans and God. Or maybe, he was obeying orders as dictated by the signposted message on the wall. It was an overwhelming feeling to realize that knowledge exists, and one must find it. I even asked the temple administrator how he would know that I was a Hindu just by looking at my appearance. Of course, his reply in Hindi was, "You look Hindu." Although his guess was right, my physical looks at that time could also have misled their perceptions that I was not a Hindu! At our next stop in Delhi during the same trip, a pious-looking individual sitting on the steps of Jama Masjid, a famous 16th Century mosque, while staring at my grey and white beard asked in Hindi "Are you Muslim?". I was intrigued by the manner and the circumstances under which it happened. There appeared to be some form of a fitness function in identifying religious affiliation based upon a perception of an individual's persona, the color of the skin, the shape of the eyes, height and other stigmas that give the means for biometric identification. Many such thoughts deeply inspired me towards engaging in perception modeling, a topic that shaped my research with Hung following the September 11, 2001, terrorist attack. Of course, the realities then and the realities now are far different. The consequences of technology must be perceived in light of a shift in the human mindset.

As a colleague and a close friend, Hung motivated me, inspired me, and advised me on life-changing matters. During his tenure at New Mexico State University, we engaged in research and teaching for over three decades [4, 5]. Hung brought to light many mathematical tools required in decision-making in ways that an engineer could understand. He has had a transformational effect on my life and many others. It is indeed an honor knowing him for all these years.

References

1. M.M. Cohen, Tulsi—*Ocimum sanctum*: a herb for all reasons. J. Ayurveda. Integr. Med. **5**(4), 251–259 (2014), https://www.ncbi.nlm.nih.gov/pmc/articles/PMC4296439/
2. A. Guruge, The Buddha's Encounters with Mara the Tempter: Their Representation in Literature and Art, https://www.accesstoinsight.org/lib/authors/guruge/wheel419.html
3. C. Shannon, A mathematical theory of communication. Bell Syst. Tech. J. **27**, 379–423, 623–656 (1948)
4. H.T. Nguyen, N.R. Prasad (eds.), *Fuzzy Modeling and Control: Selected Works of M. Sugeno* (CRC Press, Boca Raton, 1999)
5. H.T. Nguyen, N.R. Prasad, C.L. Walker, A.E. Walker, *A First Course in Fuzzy and Neural Control* (CRC Press 2002)

The Falsificationist Foundation for Null Hypothesis Significance Testing

David Trafimow

Abstract There have been many criticisms of null hypothesis significance testing (NHST), including important ones made by Dr. Hung Nguyen. The goal of this chapter is to honor Dr. Nguyen for his important scientific contributions by introducing a new argument against NHST that complements the arguments he already has made. Specifically, NHST aficionados sometimes justify NHST by arguing that it is based on Popper's falsificationist philosophy of science, which has a history of being very popular among scientists in the 20th century, and much of this popularity continues into the 21st century. In contrast, I argue that NHST is inconsistent, rather than consistent, with Popper. I further argue that there are problems with Popper's philosophy, at least in the simple form used to justify NHST.

1 Introduction

It is a great honor for me to have been invited to write a chapter for the present volume, in tribute of Dr. Hung Nguyen. Dr. Nguyen is a top-level researcher who has devoted his life not only to his own area, but more generally to the betterment of science. I got to know Dr. Nguyen through my ban on p-values in the journal, *Basic and Applied Social Psychology* [16], and a chapter he wrote about it [7]. I was lucky to meet Dr. Hung when he gave a colloquium on the topic at the math department at New Mexico State University, and we had a wonderful dinner afterwards and conversed more about the topic.[1] It is our conversation over dinner, in the context of his colloquium and chapter, that provides the starting point for the present chapter in Dr. Nguyen's honor. The present chapter can be considered as providing a short philosophical supplement to our discussion on p-values.

---

[1] In addition, we wrote an article together that included authors from many countries [14].

D. Trafimow (✉)
Department of Psychology, MSC 3452, New Mexico State University, Box 30001, Las Cruces, NM 88001-8003, USA
e-mail: dtrafimo@nmsu.edu

V. Kreinovich (ed.), *Statistical and Fuzzy Approaches to Data Processing, with Applications to Econometrics and Other Areas*, Studies in Computational Intelligence 892, https://doi.org/10.1007/978-3-030-45619-1_16

219

The most prevalent statistical approach to hypothesis testing is the null hypothesis significance testing (NHST) procedure. As NHST is usually performed, the researcher sets up a test hypothesis. Usually, the test hypothesis specifies that there is no difference between the population parameters of two populations. For example, the test hypothesis might specify that the means in populations from which an experimental group and a control group are derived, are the same. Although the test hypothesis does not have to specify that there is no difference, it almost always does. In addition, the researcher almost never intends to support the test hypothesis. Rather, the test hypothesis is a straw person to be disconfirmed in favor of an alternative hypothesis that the researcher favors. For example, if the researcher wishes to support the effectiveness of a blood pressure medicine, the test hypothesis would be that the blood pressure of the populations from which the experimental and control conditions came are the same. If the sample means from the experiment differ in the desired direction, and $p < .05$, the researcher rejects the test hypothesis of no effect in favor of the alternative (and desired) hypothesis that the blood pressure medicine really does lower blood pressure. Stated more abstractly, the goal of proposing the test hypothesis is to "nullify" it in favor of the preferred alternative hypothesis. Thus, the test hypothesis is typically called the "null" hypothesis, and the goal of NHST is to reject that hypothesis to fulfill the goal of favoring the alternative hypothesis preferred by the researcher.

NHST bears a surface similarity to the popular philosophy of science proposed by Karl Popper [8]. Popper considered the problem of proving theories in science by data collection. According to what might be considered the "received view" at the time, scientists are supposed to derive predictions from their theories and then test those predictions against empirical findings. If the empirical findings are in line with predictions, the researcher can declare an empirical victory. But if empirical findings are not in line with predictions, the researcher is saddled with an empirical defeat. Contrary to the predilections of researchers, Popper felt that researchers should make less out of empirical victories, and more out of empirical defeats. Here is why.

Commencing with empirical victories, Popper argued that there is the logical fallacy of affirming the consequent to be considered. To see that this is so, read the following invalid syllogism, Syllogism I.

1. If my theory is true, I should obtain an empirical victory.
2. I obtain an empirical victory.
3. Therefore, my theory is true.

The problem, of course, is that one might obtain an empirical victory for a reason other than the truth of the theory. The history of science is littered with empirical victories obtained from wrong theories.

In contrast, Popper argued that researchers would be better off attempting to falsify their theories, rather than attempting to confirm them. The logical structure of falsification can be seen in the following logically valid syllogism, Syllogism II.

1. If my theory is true, I should obtain an empirical victory.
2. I do not obtain an empirical victory.

3. Therefore, my theory is not true.

Note that in this latter syllogism, the researcher obtains an empirical defeat rather than an empirical victory and disconfirms rather than confirms the theory. Based strictly on logic, the latter syllogism works just fine and so Popper made the elegant point that researchers should use falsification, which is based on a logically valid syllogism; as opposed to verification, which is based on a logically invalid syllogism.

But if researchers are to use empirical defeats to reject theories, rather than empirical victories to prove them, how is science to progress? Popper's answer is that science progresses by researchers falsifying theories through empirical defeats, replacing them with better theories which are themselves subject to falsification, and so on into the indefinite future. Theories that withstand falsification attempts are "corroborated" but not proved. And strongly corroborated theories—such as Einstein's theory of relativity—are valuable in science.[2] Although matters are not nearly as clean as been described thus far, Popper's philosophy has attractive features, not the least of which is the wonderful spirit of bold conjectures and surprising predictions, refutations, and even bolder future conjectures and surprising predictions. Popper's philosophy was extremely popular for much of the 20th century and continues to be popular in the 21st century, despite much strong criticism.

To return to NHST, it is easy to see the parallels between it and Popper's falsificationist philosophy of science. Consider that in NHST, one proposes a null hypothesis to be rejected, and in falsificationist philosophy one proposes a theory to be rejected. Thus, it seems that NHST goes with falsificationist thinking as bacon goes with eggs. There is a seeming strong underlying philosophical basis for NHST that is often touted by NHST aficionados.

2 Appearances Are Deceiving

But despite appearances, there are issues. As will be discussed later, there are problems with Popper's falsificationist philosophy of science, or at least the simple version used to justify NHST. And as will be discussed right now, NHST is not faithful to Popper anyway, despite appearances.

To commence, consider the null hypothesis, which is extremely unlikely right from the start. Because there is an infinitude of possibilities, the probability that any one possibility is true approaches zero. Thus, the null hypothesis is likely wrong even without data. But to make matters a good deal worse, consider that there are assumptions that go with statistical tests [7]. For example, Berk and Freedman [2] emphasized that almost all statistical tests assume random and independent sampling from a defined population and this is almost never true. So even in the unlikely event that the null hypothesis itself might be true, the whole model being tested—that includes assumptions that are practically never true along with the null hypothesis—is just about never true or simply never true. That the model is completely

---

[2]The interested reader can consult Einstein [4] for an accessible account, in Einstein's own words.

implausible, even prior to data collection, is inconsistent rather than consistent with Popper's falsificationist philosophy of science. For Popper, the idea was to test plausible theories, falsify them (or corroborate them), and then replace them with better theories. If one begins with complete implausibility, falsification provides no strong stimulus towards something better. Put another way, Popper wanted researchers to start with the best theories possible, so that falsification really means something, and not to pose completely implausible theories that are trivially easy to falsify. Thus, we see that falsifying the null hypothesis is certainly not in the Popperian spirit.

The problems worsen when one considers the alternative hypothesis that the researcher really wishes to support. Recall that the null hypothesis generally specifies that there is no effect so that the alternative hypothesis, merely being "not the null," includes any effect that is not the null. The researcher is certainly right as the null hypothesis is not true! Consider again a main attraction of Popper's falsificationist philosophy of science, which is the emphasis on bold conjectures accompanied by surprising predictions that are falsified and replaced with even bolder conjectures and more surprising predictions. To choose a famous physics example, Galileo was replaced with Newton who, in turn, was replaced with Einstein. And it is difficult, indeed, to become bolder than Einstein! (Nevertheless, the hope is that Einstein too will be replaced with something even better and bolder.) Well, then, with the considerations of boldness and surprising predictions salient, how bold and surprising is the researcher's preferred hypothesis?

You know the answer to that one! The hypothesis that the null is wrong is almost certainly right, which renders it not bold at all. The researcher is taking very little risk.[3] Therefore, between the twin crimes of a null hypothesis that is almost certainly untrue embedded in a model that is certainly untrue, and an alternative hypothesis that is almost certainly true embedded in a model that is certainly true (that is, the explicit model including the null is certainly wrong), NHST is in strong opposition to Popper's bold falsificationist spirit. Consequently, one cannot claim Popper's falsificationism as providing a philosophical foundation for NHST. But even if it did, there are problems with the philosophy itself.

3 Problems with Popper's Falsificationist Philosophy

Although we have seen that Syllogism II is logically valid, that does not mean it is without problems. And the main problem is that the major premise is false. To see the falsity, consider a famous case in the history of science involving Haley's prediction of the timing of the return of the comet that now bears his name. Although he used Newton's theory, he also needed additional assumptions such as about the current position of the comet, its velocity, and the presence or absence of other astronomical

---

[3] Some hypotheses are directional, in which case the alternative hypothesis has a 50% chance of being correct, rather than a close to 100% chance of being correct. But even 50% cannot be considered to approach anything close to what Popper would have said to be bold.

bodies. These assumptions, not in the theory but nevertheless necessary for deriving predictions, are often termed auxiliary assumptions (e.g., [11, 13]). Thus, we see that it is theory and auxiliary assumptions that eventually lead to the prediction. Because of this, Syllogism II needs to be revised as is shown by Syllogism III below. Syllogism III uses the same minor premise as Syllogism II, but the change in the major premise also necessitates a change in the conclusion. I present Syllogism III below and then discuss the philosophical implications.

1. If my theory is true and my auxiliary assumptions are true, I should obtain an empirical victory.
2. I do not obtain an empirical victory.
3. Therefore, either my theory is not true or at least one auxiliary assumption is not true.

Note how the necessity to include auxiliary assumptions in the major premise (first assumption) forces a watering down of the conclusion to maintain the logical validity of the syllogism. Specifically, because the empirical defeat can be blamed on either the theory or at least one auxiliary assumption, the researcher is not led inevitably to theory falsification. The point has been stressed by philosophers ranging from Duhem [3] to Lakatos [5], and Popper was well-aware of the difficulty. Because theory falsification is not nearly as clear as it seems in what might be termed the naïve falsificationist scheme (Syllogism II), a major attraction of Popper's falsificationist philosophy of science—clarity and philosophical elegance—is compromised.[4]

Nor is this the only problem for falsification. There are theories that are not particularly falsifiable that are nevertheless useful. An example from my own area of psychology is what has been termed classical test theory or classical true score theory (CTT). To see the issue, consider the following quotation from Trafimow [12]:

CTT is based on the notion of a "true score" which is defined as the expected value of observed scores across an infinite number of test-taking occasions under the ideal condition that test-taking occasions are independent. Also, an "error score" is defined as the remainder after the true score is subtracted from the observed score. From these two definitions, it follows that true scores and error scores are uncorrelated [9, 17]. Finally, CTT defines parallel tests as tests with the same true scores and error variances. Based on these definitions, it is possible to derive the attenuation formula that describes the extent to which measurement error decreases the magnitudes of correlations from what would be obtained in the absence of random measurement error. It is possible to add assumptions, which could be true or false, to the two main definitions described above, and thereby generate falsifiable CTT models. However, at base, CTT clearly is not falsifiable because it depends on definitions rather than on assumptions that could be true or false. Does that mean that CTT is not a good theory?

---

[4]To be fair, although auxiliary assumptions are generally brought up to criticize Popper or neo-Popperian falsificationist philosophies, Trafimow [13] showed that they are problem for verification too. That is, empirical victories can be credited to the theory or to auxiliary assumptions. The ambiguity creates problems for theory verification and not just theory falsification.

Trafimow [12] answered his rhetorical question by pointing out that CTT makes important and surprising points, such as the derivations of the attenuation and dis-attenuation formulas. Using CTT, it is possible to calculate good estimates of the extent to which effect sizes using dependent variables subject to random measurement error decrease obtained effect sizes. It also is possible to use CTT equations to estimate what empirically determined effect sizes likely would be if it were possible to eliminate all random measurement error. The importance of CTT in the history of psychology would be difficult to overestimate. Thus, despite the falsifiability issues that pertain to CTT, it nevertheless is valuable, thereby indicating that something is wrong with Popper's falsificationist philosophy of science.

Lest we be too critical of Popper, CTT adheres to Popper's spirit in the sense of being a bold conjecture from which surprising things follow. But the letter of Pop-per's falsificationist philosophy of science clearly does not work. Thus, a reasonable person might agree with Popper in spirit while recognizing that there nevertheless are important problems.

4   Conclusion

Although there are many critiques of NHST, including critiques by Dr. Hung Nguyen (e.g., [7] and myself (e.g., [10, 14–16]), the philosophical foundations of NHST have been relatively ignored (but see [6]) for an exception). The present chapter in honor of Dr. Hung Nguyen remedies the lack by focusing exclusively on this topic and providing yet another reason for researchers to eschew NHST. The arguments are summarized easily.

Because Popper's falsificationist philosophy provides the philosophical founda-tion for NHST, there are two issues. First, is NHST consistent with Popper's teach-ings? Second, are there problems with the philosophy itself? The answer to both questions is in the affirmative.

Most important, NHST is very inconsistent with Popper's emphasis on bold con-jectures and surprising predictions. Because the model containing the null hypothe-sis is guaranteed wrong, and so the alternative and preferred model (not the model containing the null hypothesis) is guaranteed right, there is very little risk for the researcher, and very little boldness. If the researcher collects a sufficiently large sample size, he or she is almost guaranteed to obtain statistical significance and reject the null hypothesis in favor of the preferred alternative hypothesis. This is con-trary to what Popper taught and so his falsificationist philosophy of science scarcely can be used to justify NHST.

In addition, however, there are problems with Popper's philosophy, or at least the naïve version used to justify NHST.[5] The necessity to include auxiliary assumptions in the major premise of Syllogism III, with the associated inability to unambiguously

---

[5]There are more sophisticated falsificationist positions, such as that suggested by Andersson [1]. However, to my knowledge, nobody has used Andersson's more sophisticated approach to justify

falsify theories, compromises the clarity and elegance of the falsificationist philosophy. Moreover, there are useful theories in science, of which CTT is merely an example, that are not easily falsified because they depend on definitions rather than assumptions. Although it is possible to argue that theories such as CTT are at least in the spirit of CTT, the letter of Popper's falsificationist philosophy is blatantly violated. Worse yet, at least from the standpoint of NHST, an aficionado might defend Popper from the implications of the CTT example by making an "in the spirit" argument. But NHST fails even here because it blatantly is not in the Popperian spirit, as we already have seen.

Therefore, NHST fails on both counts. It does not follow either the letter or the spirit of Popper's falsificationist philosophy of science. And the philosophy itself—at least the naïve version typically used—is difficult to defend (though I still like it in spirit). Therefore, there is no way to avoid the conclusion that Popper's falsificationist philosophy fails to justify NHST. With this philosophical foundation removed from the NHST edifice, NHST is rendered more vulnerable to previous criticisms, including those having been made by Dr. Hung Nguyen.

References

1. G. Andersson, *Criticism and the history of science: Kuhn's, Lakatos's, and Feyerabend's criticisms of critical rationalism* (E. J. Brill, New York, 1994)
2. R.A. Berk, D.A. Freedman, Statistical assumptions as empirical commitments, in *Law, Punishment, and Social Control: Essays in Honor of Sheldon Messinger*, ed. by T.G. Blomberg, S. Cohen, 2nd edn. (Aldine de Gruyter, 2003), pp. 235–254
3. P. Duhem, *The Aim and Structure of Physical Theory* (P.P. Wiener, Trans). (Princeton University Press, Princeton, NJ, 1954). (Original work published 1906)
4. A. Einstein, *Relativity: The Special and the General Theory* (Robert W. Lawson, Trans.) (Crown Publishers, New York, 1961)
5. I. Lakatos, *The Methodology of Scientific Research Programmes: Philosophical Papers*, vol. 1 (J. Worrall, G. Currie, Eds.). (Cambridge University Press, Cambridge, United Kingdom, UK, 1978)
6. P.E. Meehl, Theoretical risks and tabular asterisks: Sir Karl, Sir Ronald, and the slow progress of soft psychology. J. Consult. Clin. Psychol. **46**, 806–834 (1978)
7. H.T. Nguyen, On evidential measures of support for reasoning with integrated uncertainty: a lesson from the ban of p-values in Statistical Inference, in *Integrated Uncertainty in Knowledge Modelling and Decision Making. IUKM 2016*, ed. by V.N. Huynh, M. Inuiguchi, B. Le, B. Le, T. Denoeux. Lecture Notes in Computer Science, vol. 9978 (Springer, Cham, 2016), pp. 3–15
8. K.R. Popper, *The Logic of Scientific Discovery* (Basic Books, New York, NY, 1959)
9. T. Raykov, G.A. Marcoulides, *Introduction to psychometric theory*. (Routledge, London, UK, 2011)
10. D. Trafimow, Hypothesis testing and theory evaluation at the boundaries: surprising insights from Bayes's theorem. Psychol. Rev. **110**, 526–535 (2003)
11. D. Trafimow, The theory of reasoned action: a case study of falsification in psychology. Theory Psychol. **19**, 501–518 (2009)

NHST. In my judgment, even more sophisticated falsificationist positions are incompatible, rather than compatible, with NHST.

12. D. Trafimow, Are measurement theories falsifiable and should we care? Theory Psychol. **23**(3), 397–400 (2013)
13. D. Trafimow, Implications of an initial empirical victory for the truth of the theory and additional empirical victories. Philos. Psychol. **30**, 411–433 (2017)
14. D. Trafimow, V. Amrhein, C.N. Areshenkoff, C.J. Barrera-Causil, E.J. Beh, Y.K. Bilgiç, R. Bono, M.T. Bradley, W.M. Briggs, H.A. Cepeda-Freyre, S.E. Chaigneau, D.R. Ciocca, J.C. Correa, D. Cousineau, M.R. de Boer, S.S. Dhar, I. Dolgov, J. Gómez-Benito, M. Grendar, J.W. Grice, M.E. Guerrero-Gimenez, A. Gutiérrez, T.B. Huedo-Medina, K. Jaffe, A. Janyan, A. Karimnezhad, F. Korner-Nievergelt, K. Kosugi, M. Lachmair, R.D. Ledesma, R. Limongi, M.T. Liuzza, R. Lombardo, M.J. Marks, G. Meinlschmidt, L. Nalborczyk, H.T. Nguyen, R. Ospina, J.D. Perezgonzalez, R. Pfister, J.J. Rahona, D.A. Rodríguez-Medina, X. Romão, S. Ruiz-Fernández, I. Suarez, M. Tegethoff, M. Tejo, R. van de Schoot, I.I. Vankov, S. Velasco-Forero, T. Wang, Y. Yamada, F.C.M. Zoppino, F. Marmolejo-Ramos, Manipulating the alpha level cannot cure significance testing. Front. Psychol. **9**, 699 (2018)
15. D. Trafimow, B.D. Earp, Null hypothesis significance testing and the use of P values to control the Type I error rate: the domain problem. New Ideas Psychol. **45**, 19–27 (2017)
16. D. Trafimow, M. Marks, Editorial. Basic Appl. Soc. Psychol. **37**, 1–2 (2015); D. Trafimow, M. Marks, Editorial. Basic Appl. Soc. Psychol. **38**, 1–2 (2016)
17. D.W. Zimmerman, Probability spaces, Hilbert spaces, and the axioms of test theory. Psychometrika. **40**, 395–412 (1975)

A Wavelet Strategy to Buy Low and Sell High

Lanh Tran

This paper is dedicated to Professor Hung Nguyen on the occasion of his 75th birthday

Abstract Suppose the price of a commodity or stock fluctuates indefinitely, is there any explicit strategy for a trader to "ride the price waves" by buying low and selling high to eventually win even if price does not increase? In this paper, we explain what "high" and "low" mean and how much to buy and sell to make a profit in the long run. The strategy is contained in my website https://AgateTrading.com.

1 Introduction

The buy low and sell high problem can be summarized like this: suppose the price of a commodity or stock fluctuates indefinitely, is there any explicit strategy for a trader to "ride the price waves" by buying low and selling high to eventually win even if price does not increase? Buy low and sell high is the basic tenet of most forms of trading. However, people who engage in this kind of trading usually lose since it is not clear what"high" and "low"mean and how much to sell when price is high and how much to buy when price is low. Overall, trading is a very difficult business [1]. There is a large amount of literature on the buy low and sell high problem. See [2] and the references therein for an account of this information.

L. Tran (✉)

Department of Statistics, Indiana University, Bloomington, IN 47408, USA

e-mail: LanhTran14@gmail.com

© The Editor(s) (if applicable) and The Author(s), under exclusive license to Springer Nature Switzerland AG 2021

V. Kreinovich (ed.), *Statistical and Fuzzy Approaches to Data Processing, with Applications to Econometrics and Other Areas*, Studies in Computational Intelligence 892, https://doi.org/10.1007/978-3-030-45619-1_17

This paper shows how these problems can be dealt with by employing my web-page "Portfolio Checker" on my website https://AgateTrading.com. The website showcases a strategy that can be used as a guide for the trading of a commodity or stock. A trader using it always makes a profit in the long run. The only condition assumed is that the standard deviation of the stock price traded stays bounded away from zero. The strategy is designed mainly to trade stocks or commodities with no clear upward or downward trends. See [3] for a guide to commodity trading,

The trading strategies presented are based on information obtained from the movements of wavelets created by fluctuations of market prices. They do not involve any forecasting or prediction of future prices.

2   The Portfolio Checker Program

This section explains how to operate the portfolio checker at the website AgateTrading.com. The user starts by entering the "Magic Number" on the bottom left side of the screen right below the "Generate Table" button. The Magic Number is currently set to be 8141946.

Remark 2.1 It is important to enter the Magic Number correctly on the screen.

She needs to come up with three numbers before she makes a trade: the current market price of a share of stock being considered for selling or buying, the number of shares to buy (a positive integer) or sell (a negative integer) and the commission for the trade.

She then asks PC if such a trade is feasible. This is done by entering the three numbers on the screen with a comma or a space as a separator and then activating the computer program by clicking on the button with the words "Generate Table".

On the screen, don't separate digits in a number by using commas. For example: write 1250 instead of 1,250. PC analyzes these numbers and answers the question "Is my Portfolio OK?" with a simple "YES" or "NO". The user needs to make sure that she gets a "YES" each time she makes a trade.

The following rules are set by PC:

Rule 1. The first trade must be a buy of at least 20 shares. In addition, the cost of the first trade must be at least $1,000.00.

Rule 2. Don't sell if the current price of a share is less than the price at its last trade.

Rule 3. Don't buy if the current price of a share is higher than the price at its last trade.

Rule 4. Don't buy or sell an amount less than 50 times the commission.

Remark 2.2 If you buy a stock of $2.50 a share then to satisfy Rule 1, you need to start with at least 400 shares, while if you buy a stock of $250.00 a share, then you need to start with at least 20 shares, which cost $5,000.00.

The number of shares you buy in your first transaction depends on the total amount of money you have allocated for that particular stock. Some funds should be available for additional purchases if necessary.

Rules 2 and 3 are essential elements of the buy low and sell high scheme. Rule 4 is to ensure that you don't buy or sell an amount that is too small to justify the commission that you pay.

PC does not recommend a number of shares to buy or sell. This gives more flexibility to the trader in deciding as to how much to buy or sell. The computer program is designed to assure the user that she will eventually make a profit if she gets a "YES" answer to the portfolio question before each trade that she makes.

The trader controls her account balance. If her account balance gets too low, she should buy more often than sell to raise it up.

A "NO" answer occurs due to various reasons. The most common ones are:

(1) The price of the stock traded has not increased enough or decreased enough from the price at its last trade to warrant, respectively, a "sell" or a "buy". A regular trader with 300 shares of SPY and a bank of about $100,000.00 should wait for the price to change about 2% or more from the price at its last trade before she attempts to trade.

(2) The number of shares entered is too high or too low.

Example 2.1 Below is an example of 5 trades of the stock SPY, made respectively by a trader, on January 2nd, 6th, 8th, 14th and February 3rd in the year 2015. Assume that the trader uses Interactive Brokers as a brokerage. The brokerage charges commissions of $1.00 per share if the number of shares per trade is 200 or less or $0.005 per share if a trade involves more than 200 shares.

Before each trade, the trader used the Portfolio Checker to make sure that she got "YES" for an answer. The trader's 5 transactions are summarized in the table below:

Date	Price	Sha	Com	CCom	Cost	CumCost	CSh	MV	Profit
1/2/15	197.05	300	1.50	1.50	59115.00	59115.00	300	59115.00	−1.50
1/6/15	191.66	50	1.00	2.50	9583.00	68698.00	350	67081.00	−1619.50
1/8/15	197.50	−36	1.00	3.50	−7110.00	61588.00	314	62015.00	423.50
1/14/15	192.66	36	1.00	4.50	6935.76	68523.76	350	67431.00	−1097.26
2/3/15	196.48	−28	1.00	5.50	−5501.44	63022.32	322	63266.56	238.74

The first four columns list respectively the trading dates, share prices at which trades are made, number of shares traded, and the corresponding commissions. A positive number for the shares means a "buy" and a negative one means a "sell".

The fifth column (CCom) lists cumulative commissions, which are total commissions paid up to the trading day. For example, on 1/8/15, after she sold 36 shares at the price of 197.50 a share, she has paid a commission of $1.50 + $1.00 + $1.00, which equals $3.50.

The sixth column (Cost) lists the cost of each trade which equals the number of shares traded multiplied by the price at which it is traded. You have to pay money if it is a positive number and you get back cash if it is a negative number. A positive number for cost indicates a buy and a negative number indicates a sell.

The seventh column (CumCost) lists the cumulative cost paid by the trader. For example, after the fourth transaction is made on 1/14/15, the trader's cumulative cost is obtained by adding $59,115.00, $9,583.00, −$7,110.00 and $6,935.76, which equals $68,523.76.

The eighth column (CSh) lists the total number of shares held by the trader after each trade. For example, after the fifth trade was carried out on 2/3/15, the trader has a total of shares equal to 322, which is the sum of all numbers in Column 3.

The ninth column lists the market value of the trader on each trading day. The market value (MV) is computed by multiplying the cumulative number of shares of the trader by the corresponding price. For example, on 1/14/15, the market value of the trader is $192.66 × 350, which equals $67,431.00.

Finally, the tenth column lists the profit after each trade. The profit made by the trader after each trade is given by:

$$\text{Profit} = \text{MV} - \text{CumCost} - \text{CumCom}.$$

The profit after the last trade is

$$\$63,266.56 - \$63,022.32 - \$5.50 = \$238.74.$$

Note the interesting fact that the trader has made a profit of $238.74 while the market price drops from $197.05 a share to $196.48 a share. This example is simple, but it indicates that the trader can win by riding the "price waves".

The trader was able to utilize fluctuations of the price of the stock to make profits. Let us see what happens if we arbitrarily replace 191.66 by 189.66, 197.50 by 200.50 and 192.66 by 187.66. The table above would change to:

Date	Price	Sha	Com	CCom	Cost	CumCost	CSh	MV	Profit
1/2/15	197.05	300	1.50	1.50	59115.00	59115.00	300	59115.00	−1.50
1/6/15	189.66	50	1.00	2.50	9483.00	68598.00	350	66381.00	−2219.50
1/8/15	200.50	−36	1.00	3.50	−7218.00	61380.00	314	62957.00	1573.50
1/14/15	187.66	36	1.00	4.50	6755.76	68135.76	350	65681.00	−2459.26
2/3/15	196.48	−28	1.00	5.50	−5501.44	62634.32	322	63266.56	626.74

She makes a profit of $626.74 which is much more than $238.74 since the price of the stock fluctuates more.

It is a good idea for you to copy and paste the data into a text or Excel file and keep it as your record. If you mistype and then activate the program, the whole data that you have entered may disappear. In case this happens, you still have a record of what you have typed, you can copy and paste your data back onto the computer screen, correct the mistype, and then reactivate the program. The website does not hold any data for you after you logout.

Exercise 2.1 As an exercise, you may try to enter the prices, shares, commissions in the table above on the screen as follows: 197.05, 300, 1.50, 189.66, 50, 1.00, 200.50, −36, 1.00, 187.66, 36, 1.00, 196.48, −28, 1.00. The numbers need to be separated by a comma or a space. The answer to the portfolio question is a "YES" after you activate the program.

Remark 2.3 Following Example 2.1, what should your next move be? There are 322 shares left. How about selling ten percent of the shares (32 shares) if your stock price goes up 2% or buying 32 additional shares if your stock price decreases 2%. The stock price becomes .98 × $196.48 = $192.55 if it decreases 2% and 1.02 × $196.48 = $200.41 if it increases 2%. You find out both of the trades get a "YES" answer by asking PC. Then you can make an advance order with your brokerage to perform whichever transaction that comes first.

There are many ways to enter the data on the "DATA TO CALCULATE" area. You may find it more to your liking to enter the data above as follows:
    197.05, 300, 1.50
    189.66, 50, 1.00
    200.50, −36, 1.00
    187.66, 36, 1.00
    196.48, −28, 1.00

Exercise 2.2 Assume that Scottrade is your brokerage and that your bank is $100,000.00. Note that Scottrade charges a commission fee of 7 dollars for each transaction no matter how many shares are involved. Using the computer program and the five prices of the stock SPY above, but starting with a buy of 350 shares on 1/2/2105, find 4 additional feasible trades on, respectively, the 6th, 8th, 14th and February 3rd of 2015. To get a feasible trade, you have to get a "YES" answer to the portfolio question before each trade is made. Do you make a profit after you finish with the last trade?

Example 2.1 shows that it is possible for you to make a profit even if the share price of the traded stock decreases if you trade properly. You can also lose if the share price of the traded stock increases if trading is not done right. An example is given below.

Example 2.2 Consider the following list of three transactions made by a trader using Interactive Brokers as the brokerage. After the last transaction is made, she has a total of 1100 shares. If the share price drops down to $75.60, the market value of these shares would be 1100 × $75.60 = $83, 160.00, which is less than the cumulative cost of $83,481.00. She ends up with a loss of $83, 481.00 − $83, 160.00 + $7.00, which equals, $328.00, not counting the commission she has to pay if she cashes out. She loses money while the price of a share goes up from $75.49 to $75.60.

Date	Price	Sha	Com	CCom	Cost	CumCost	CSh	MV	Profit
1/8/19	75.49	1000	5.00	5.00	75490.00	75490.00	1000	75490.00	−5.00
1/9/19	86.03	−100	1.00	6.00	−8603.00	66887.00	900	77427.00	10534.00
1/10/19	82.97	200	1.00	7.00	16594.00	83481.00	1100	91267.00	7779.00

Let us ask the computer if these transactions are OK by entering

75.49, 1000, 5.00

86.03, −100, 1.00

82.97,   200, 1.00

on the screen and activate the program. The answer is a "NO". The reason is because the amount of shares she bought on the last transaction is too high.

How much should she buy then? She should just try a smaller number that she is comfortable with and then ask the computer if buying the latter is ok before making such a trade. How about asking the portfolio checker if buying 60 at $82.97 is ok? This is done by entering

75.49, 1000, 5.00

86.03, −100, 1.00

82.97,   60, 1.00

on the screen. The answer is "YES". Suppose she buys 60 shares at $82.97 per share, and then 40 shares if the price drops back to the starting price of $75.49 on 1/11/19, she would make a profit of $597.20 as shown in the table below.

Date	Price	Sha	Com	CCom	Cost	CumCost	CSh	MV	Profit
1/8//19	75.49	1000	5.00	5.00	75490.00	75490.00	1000	75490.00	−5.00
1/9/19	86.03	−100	1.00	6.00	−8603.00	66887.00	900	77427.00	10534.00
1/10/19	82.97	60	1.00	7.00	4978.20	71865.20	960	79651.20	7779.00
1/11/19	75.49	40	1.00	8.00	3019.60	74884.80	1000	75490.00	597.20

She again holds 1,000 shares which is the same number of shares she started with. The choice of buying 60 shares and then 40 is purely arbitrary. Buying 40 shares at $82.97 a share and then 60 shares if price falls to $75.49 would also get a "YES" answer from the Portfolio Checker. The answer to the question as to how many shares to buy or how many shares to sell depends on many variables: the commission you have to pay, cost of a share, among others. The best way to learn how to use the website is for you to play with it.

Exercise 2.3 In Example 2.2 above, buying 200 shares at $82.97 a share is not an acceptable trade. What is the maximum allowable number of shares that she can buy at $82.97 to get a "YES" answer?

Hint: *Use portfolio checker and get the answer by trial and error. Decrease the number 200 slowly until you get "YES" for an answer.*

It is fairly easy for you to get a "YES" answer from the portfolio checker after some practice. Before you quit using the website, make sure you save the historical data of your trades in some file. Click on the "Select Table" button on the upper right hand corner of the screen and use copy and paste to save your data.

You need to upload all the data back to the screen in case you want to ask the portfolio checker some time later if your "next" trade is ok or not. Whether your "next" trade is feasible or not depends not only on your "last" move but on the entire historical data of your trades.

3 When Does PC Work Well?

The table below shows the trading of SPY from 4/6/98 to 9/8/03 when its share price moved from $79.60 on 4/6/98 to $79.03 on 9/8/03, which is a period of over 5 years. During this period, PC makes a profit of $14,721.15 while share price comes down slightly.

PC does best for the trading of commodities or stocks with fluctuating prices and with no upward or downward trends.

Date	Price	Sha	Com	CCom	Cost	CumCost	CSh	MV	Profit
4/6/98	79.60	1000	5.00	5.00	79600.00	79600.00	1000	79600.00	−5.00
4/27/98	77.49	100	1.00	6.00	7749.00	87349.00	1100	85239.00	−2116.00
5/20/98	80.11	−100	1.00	7.00	−8011.00	79338.00	1000	80110.00	765.00
9/2/98	71.03	200	1.00	8.00	14206.00	93544.00	1200	85236.00	−8316.00
11/4/98	80.54	−150	1.00	9.00	−12081.00	81463.00	1050	84567.00	3095.00
11/23/98	85.65	−200	1.00	10.00	−17130.00	64333.00	850	72802.50	8459.50
1/7/99	91.29	−150	1.00	11.00	−13693.50	50639.50	700	63903.00	13252.50
4/22/99	98.26	−100	1.00	12.00	−9826.00	40813.50	600	58956.00	18130.50
6/2/99	93.72	150	1.00	13.00	14058.00	54871.50	750	70290.00	15405.50
7/13/99	100.88	−150	1.00	14.00	−15132.00	39739.50	600	60528.00	20774.50
8/4/99	94.55	150	1.00	15.00	14182.50	53922.00	750	70912.50	16975.50
8/25/99	100.16	−100	1.00	16.00	−10016.00	43906.00	650	65104.00	21182.00
9/22/99	94.80	120	1.00	17.00	11376.00	55282.00	770	72996.00	17697.00
11/5/99	100.08	−120	1.00	18.00	−12009.60	43272.40	650	65052.00	21761.60
3/17/00	107.19	−150	1.00	19.00	−16078.50	27193.90	500	53595.00	26382.10
12/1/00	96.92	200	1.00	20.00	19384.00	46577.90	700	67844.00	21246.10
1/30/01	101.32	−120	1.00	21.00	−12158.40	34419.50	580	58765.60	24325.10
3/1/01	91.62	150	1.00	22.00	13743.00	48162.50	730	66882.60	18698.10
4/2/01	84.20	150	1.00	23.00	12630.00	60792.50	880	74096.00	13280.50
4/18/01	91.42	−100	1.00	24.00	−9142.00	51650.50	780	71307.60	19633.10
8/30/01	83.78	100	1.00	25.00	8378.00	60028.50	880	73726.40	13672.90
9/25/01	75.51	125	1.00	26.00	9438.75	69467.25	1005	75887.55	6394.30
10/23/01	80.83	−100	1.00	27.00	−8083.00	61384.25	905	73151.15	11739.90
7/9/02	71.65	150	1.00	28.00	10747.50	72131.75	1055	75590.75	3431.00
7/19/02	63.48	150	1.00	29.00	9522.00	81653.75	1205	76493.40	−5189.35
8/14/02	69.11	−150	1.00	30.00	−10366.50	71287.25	1055	72911.05	1593.80
9/23/02	62.98	150	1.00	31.00	9447.00	80734.25	1205	75890.90	−4874.35
11/6/02	70.04	−100	1.00	32.00	−7004.00	73730.25	1105	77394.20	3631.95
1/27/03	64.45	150	1.00	33.00	9667.50	83397.75	1255	80884.75	−2546.00
5/2/03	70.80	−125	1.00	34.00	−8850.00	74547.75	1130	80004.00	5422.25
9/8/03	79.03	−100	1.00	35.00	−7903.00	66644.75	1030	81400.90	14721.15

Note that the price of a share dropped from $79.60 on 4/6/98 to $79.03 on 9/8/03 but a profit of $14,721.15 was made by riding the price waves.

4 Commodity Trading

You can see monthly prices of various commodities in [4]. Below is a table of yearly Aluminium price from July 1989 to July 2019. The price is in US dollars per metric ton.

Date	Price
Jul-89	1,755.93
Jul-90	1,570.96
Jul-91	1,296.78
Jul-92	1,313.05
Jul-93	1,202.13
Jul-94	1,492.42
Jul-95	1,860.10
Jul-96	1,458.70
Jul-97	1,591.99
Jul-98	1,309.21
Jul-99	1,403.76
Jul-00	1,563.50
Jul-01	1,416.39
Jul-02	1,338.09
Jul-03	1,436.09
Jul-04	1,709.27
Jul-05	1,778.79
Jul-06	2,512.71
Jul-07	2,732.44
Jul-08	3,071.24
Jul-09	1,667.96
Jul-10	1,988.27
Jul-11	2,525.43
Jul-12	1,876.25
Jul-13	1,769.61
Jul-14	1,948.30
Jul-15	1,639.50
Jul-16	1,629.05
Jul-17	1,902.96
Jul-18	2,082.24
Jul-19	1,796.99

The price of aluminium fluctuates significantly without any upward or downward trends. This is a common behavior of many commodities, for example, crude oil and sugar.

The table below shows that all trades got an "YES" from the Portfolio Checker.

Date	Price	Sha	Com	CCom	Cost	CumCost	CSh	MV	Profit
Jul-89	1,755.93	100	1	1	175,593.00	175,593.00	100	175,593.00	−1
Jul-91	1,296.78	80	1	2	103,742.41	279,335.41	180	233,420.41	−45,917.00
Jul-94	1,492.42	−40	1	3	−59,696.80	219,638.61	140	208,938.81	−10,702.80
Jul-95	1,860.10	−40	1	4	−74,404.00	145,234.61	100	186,010.00	40,771.39
Jul-96	1,458.70	75	1	5	109,402.50	254,637.11	175	255,272.48	630.38
Jul-97	1,591.99	−50	1	6	−79,599.50	175,037.61	125	198,998.75	23,955.14
Jul-98	1,309.21	100	1	7	130,921.00	305,958.63	225	294,572.25	−11,393.38
Jul-00	1,563.50	−50	1	8	−78,175.00	227,783.63	175	273,612.50	45,820.88
Jul-01	1,416.39	100	1	9	141,639.00	369,422.63	275	389,507.25	20,075.63
Jul-04	1,709.27	−100	1	10	−170,927.00	198,495.63	175	299,122.25	100,616.63
Jul-06	2,512.71	−40	1	11	−100,508.40	97,987.23	135	339,215.84	241,217.63
Jul-08	3,071.24	−60	1	12	−184,274.41	−86,287.18	75	230,343.00	316,618.19
Jul-09	1,667.96	75	1	13	125,097.00	38,809.82	150	250,194.00	211,371.19
Jul-10	1,988.27	−40	1	14	−79,530.80	−40,720.98	110	218,709.70	259,416.69
Jul-11	2,525.43	−50	1	15	−126,271.50	−166,992.47	60	151,525.80	318,503.25
Jul-12	1,876.25	50	1	16	93,812.50	−73,179.97	110	206,387.50	279,551.47
Jul-15	1,639.50	50	1	17	81,975.00	8,795.03	160	262,320.00	253,507.97
Jul-17	1,902.96	−50	1	18	−95,148.00	−86,352.97	110	209,325.59	295,660.56
Jul-19	1,796.99	50	1	19	89,849.50	3,496.53	160	287,518.41	284,002.88

The price of aluminium hardly changes from 1989 to 2019, but a trader using PC makes a profit of $284,002.88. The table gives another example that a trader can make profit by using PC to buy low and sell high. Real life trading involves more frequent trading than shown in the table above. Prices of commodities often move up and down without an upward or downward trend for very long periods of time. A buy and hold investor of such commodities does not make much profit. The reader is referred to [4] for more information on prices of many commodities.

References

1. B.M. Barber, T. Odean, Trading is hazardous to your wealth: the common stock investment performance of individual investors. J. Financ. **2**, 773–806 (2000)
2. Commodity Prices. http://www.investorwords.com/11534/buy_low_sell_high.html
3. How to Trade Commodities. https://optionsmethods.com/how-to-trade-commodities-online/
4. Aluminum Futures End of Day Settlement Price. https://www.indexmundi.com/commodities/? commodity=aluminum&months=360

A Note on the Relation Between Liu's Uncertain Measure and Choquet Capacity

Guo Wei, Zhiming Li, and Mei Cai

Abstract In this note, it is proved that Liu's uncertain measure and Choquet capacity differ from each other. Specifically, over any locally compact separable metric space such as $\mathbb{R}^n$, a Liu's uncertain measure is generally not a Choquet capacity. Further, Liu's uncertain measure is not an ordinary probability measure in the sense of Kolmogorov. Notice that Choquet theory is consistent with Kolmogorov's probability theory, as a direct generalization of the latter by allowing the random elements to take closed sets of the underlying space as the values.

Keywords Choquet capacity · Liu's uncertain measure

1 Choquet Capacities and Choquet Theorem

Let E be a locally compact separable (LCS) metric space with metric d, $\mathcal{P}$ the class of all subsets of E, $\mathcal{F}$ the hyperspace of all closed subsets of E equipped with the Fell topology (Fell [3], [9]), and $\mathcal{K}$ the class of all compact subsets of E.

A capacity is a map $T : \mathcal{P} \to [0, 1]$ satisfying the following three conditions (Matheron [7]):

G. Wei
Department of Mathematics & Computer Science, University of North Carolina at Pembroke, Pembroke, NC 28372, USA
e-mail: guo.wei@uncp.edu

Z. Li
School of Mathematics, Northwest University, Xi'an 710127, Shaanxi, China
e-mail: china-lizhiming@163.com

M. Cai (✉)
School of Management Science and Engineering, Nanjing University of Information Science and Technology, Nanjing 210044, Jiangsu, China

V. Kreinovich (ed.), *Statistical and Fuzzy Approaches to Data Processing, with Applications to Econometrics and Other Areas*, Studies in Computational Intelligence 892, https://doi.org/10.1007/978-3-030-45619-1_18

(i) If $A, B \in \mathcal{P}$ and $A \subseteq B$, then $T(A) \leq T(B)$;

(ii) For $A, A_n \in \mathcal{P}$ $(n \in \mathbb{N})$, if $A_n \uparrow A$ in $\mathcal{P}$ (i.e., $A_n \subseteq A_{n+1}$ for $n \in \mathbb{N}$ and $\bigcup_{n=1}^{\infty} A_n = A$), then $T(A_n) \uparrow T(A)$;

(iii) For $K, K_n \in \mathcal{K}$ $(n \in \mathbb{N})$, if $K_n \downarrow K$ in $\mathcal{K}$ (i.e., $K_{n+1} \subseteq K_n$ for $n \in \mathbb{N}$ and $\bigcap_{n=1}^{\infty} K_n = K$), then $T(K_n) \downarrow T(K)$.

A Choquet capacity T requires an additional probability condition, alternating of infinite order which is described as follows.

Let $K, K_1, K_2, \ldots$ be compact sets of E, and let $\Delta_n(K; K_1, \ldots, K_n)$ be the probability for the RACS X to hit $K_1, \ldots, K_n$ but miss K. Then all functions Δ_n defined below must be non-negative (see Matheron [7]): $\Delta_1(K; K_1) = T(K \cup K_1) - T(K)$

$$\Delta_2(K; K_1, K_2) = \Delta_1(K; K_1) - \Delta_1(K \cup K_2; K_1) \tag{1}$$

$$\Delta_n(K; K_1, \ldots, K_n) = \Delta_{n-1}(K; K_1, \ldots, K_{n-1}) - \Delta_{n-1}(K \cup K_n; K_1, \ldots, K_{n-1})$$

It can be proved that

$$\Delta_n(K; K_1, \ldots, K_n) = \sum_{\emptyset \neq I \subseteq \{1,2,\cdots,n\}} (-1)^{|I|+1} T\left(K \cup \bigcup_{i \in I} K_i\right) - T(K), \tag{2}$$

and consequently, (1) is equivalent to the following condition: T is monotonically increasing on $\mathcal{K}$ and the inequality

$$T\left(\bigcap_{i=1}^{n} K_i\right) \leq \sum_{\emptyset \neq I \subseteq \{1,2,\cdots,n\}} (-1)^{|I|+1} T\left(\bigcup_{i \in I} K_i\right) \tag{3}$$

holds for all $n \geq 2$ (Nguyen [12], p. 117). If T is a probability measure, the inequality (3) becomes Poincaré's equality (see Nguyen [12], p. 12). Moreover, T can be extended to $\mathcal{P}$ by first to any open subset G and then to arbitrary subset A of E (see Matheron [7] and Wei et al. [17]).

Remark 1.1 A belief function F is monotone of infinite order (Nguyen [12], p. 72), which is also referred as complete monotonocity (Molchanov [11], p. 7 and p. 127), and such a condition is characterized by the following:

$$F\left(\bigcup_{i=1}^{n} K_i\right) \geq \sum_{\emptyset \neq I \subseteq \{1,2,\cdots,n\}} (-1)^{|I|+1} F\left(\bigcap_{i \in I} K_i\right) \tag{4}$$

If F is a probability measure, this inequality becomes Poincaré's equality too (see Nguyen [12], p. 12).

Nguyen interpreted the belief function as a coarsening of an ordinary random variable α, where the latter is not observable. In such a situation, we observe a random closed set S that contains α with probability 1. So we have $F(A) = P(S \subseteq A) = 1 - P(S \cap A^c \neq \emptyset)$, i.e., $F(A) = 1 - T(A^c)$, where T is the Choquet capacity of S. This relation is referred as the conjugation between Choquet capacities and belief functions. □

A RACS X is an $\mathcal{A}$-$\mathcal{B}(\mathcal{F})$ measurable function from a probability space $(\Omega, \mathcal{A}, \mathrm{P})$ to the measurable space $(\mathcal{F}, \mathcal{B}(\mathcal{F}))$. Regular letter P will represent a probability function, italic P will represent the induced probability measure ($P = \mathrm{P}X^{-1}$), and $\mathbb{N}$ will denote the set of all non-negative integers.

The Choquet capacity T of X characterizes the probability law P of X through the Choquet Theorem below. The Choquet Theorem has been extensively investigated in the context of probability, e.g. [1, 4, 7, 11–17].

Choquet Theorem (see Matheron [7]) Let E be a LCS metric space and $\mathcal{F}$ the space of all closed sets of E equipped with the Fell topology. Then there exists a (necessarily unique) probability measure P on the Borel σ-field $\mathcal{B}(\mathcal{F})$ generated by the topology of $\mathcal{F}$ satisfying $T(K) = \mathrm{P}(X \cap K \neq \emptyset) = P(\mathcal{F}_K)$ for $K \in \mathcal{K}$ if and only if T is a Choquet capacity on $\mathcal{K}$ with $0 \leq T \leq 1$ and $T(\emptyset) = 0$. □

Choquet Theorem is the foundation of random set theory. It reveals the relationship between the Choquet capacity T of X and the probability measure P of X. In this theorem, the probability law P describes the random evolution of X; the Choquet capacity T plays the role of the distribution functions of ordinary random vectors; the Fell topology governs the convergence of closed sets, which is consistent with the standard convergence of closed sets (see Matheron [7]; Wei and Wang [16, 17]). The theory of RACS originates from the hit-probability (miss-probability).

In particular, any ordinary random variable taking point values in a metric space E can be viewed as a special RACS, a random singleton set, and hence it is completely characterized by its associated Choquet capacity. See e.g., Molchanov [11], Wei and Wang [16, 17].

Specifically, if $E = \mathbb{R}^n$, any ordinary random vector $\alpha = (\alpha_1, \alpha_2, \ldots, \alpha_n)$: $(\Omega, \mathcal{A}, \mathrm{P}) \to (\mathbb{R}^n, \mathcal{B}(\mathbb{R}^n))$ induces a RACS $X : (\Omega, \mathcal{A}, \mathrm{P}) \to (\mathcal{F}(\mathbb{R}^n), \mathcal{B}(\mathcal{F}(\mathbb{R}^n)))$ by $X(\omega) = \prod_{j=1}^n [\alpha_j(\omega), \infty)$. The relation between the distribution function $\mathbf{F}$ of α and the Choquet capacity T of X is given by

$$T(\prod_{j=1}^n [a_j, b_j]) = \mathrm{P}(X \cap \prod_{j=1}^n [a_j, b_j] \neq \emptyset) = \mathrm{P}(\alpha_1 \leq b_1, \alpha_2 \leq b_2, \cdots, \alpha_n \leq b_n) = \mathbf{F}(b_1, b_2, \ldots, b_n).$$

An ordinary finite measure, which is additive, characterizes the probability law of an ordinary random variable (a random point) into an Euclidean space or a metric space. The notion of additive measures has been generalized to non-additive measures for different purposes. For example, the Choquet capacity and Liu's uncertain measure.

The Choquet capacity, which characterizes the probability law of random closed sets, plays a fundamental role in random sets theory, where the latter provides a natural framework for representing in an elegant way the imprecision of the data available to a statistician (Nguyen [12]), and has been widely studied by many authors, e.g., Choquet [1], Meyer [8], Matheron [7], Nguyen [12, 13] and Molchanov [11] among others. Choquet capacity of a random set generalizes the notion of probability distribution of ordinary random variables and is of completely alternating lower probability (also called belief function in the finite case), which is an imprecise probability model (Miranda and Nguyen [10]; Nguyen [12]). For the definition of Choquet capacity and its properties, we refer to the above references.

2 Liu's Uncertain Measure

Uncertainty theory was introduced by Professor Baoding Liu in 2007 (Liu [5, 6]), and it is developed based on normality, monotonicity, self-duality, countable subadditivity, and product measure axioms. The key concepts in the theory include the uncertain measure as well as uncertain variables defined in terms of the uncertain measure. Let $(\Gamma, \mathcal{L})$ be a measurable space, where $\mathcal{L}$ is a σ-algebra. According to Professor Liu, an uncertain measure satisfies three axioms, Normality Axiom, Duality Axiom and countable subadditivity Axiom: [5, 6]:

Axiom 1. (Normality Axiom) $\mathcal{M}(\Gamma) = 1$ for the universal set Γ.

Axiom 2. (Duality Axiom) $\mathcal{M}(\Lambda) + \mathcal{M}(\Lambda^c) = 1$ for any event $\Lambda \in \mathcal{L}$.

Axiom 3. (Subadditivity Axiom) For every countable sequence of events $\Lambda_1, \Lambda_2, \ldots$, it holds that

$$\mathcal{M}(\bigcup_{k=1}^{\infty} \Lambda_k) \leq \sum_{i=1}^{\infty} \mathcal{M}(\Lambda_k).$$

The triple $(\Gamma, \mathcal{L}, \mathcal{M})$ is an uncertainty space. The fourth axiom is the product measure axiom:

Axiom 4. (Product Measure Axiom) Let $(\Gamma_k, \mathcal{L}_k, \mathcal{M}_k)$ be uncertainty spaces for $k \in \mathbb{N}$. Then the product measure $\mathcal{M}$ is an uncertain measure defined on the product σ-algebra satisfying

$$\mathcal{M}(\prod_{k=1}^{\infty} \Lambda_k) = \bigwedge_{k=1}^{\infty} \mathcal{M}_k(\Lambda_k).$$

Definition of Independent Events: Let $(\Gamma, \mathcal{L})$ be a measurable space. Events $\Lambda_1, \Lambda_2, \ldots, \Lambda_n$ in $\mathcal{L}$ are said to be independent if

$$\mathcal{M}(\bigcap_{k=1}^{n} \Lambda_k^*) = \bigwedge_{k=1}^{n} \mathcal{M}(\Lambda_k^*)$$

where Λ_k^* are arbitrarily chosen from $\{\Lambda_k, \Lambda_k^c, \Gamma\}$.

Moreover, there is a Maximum Uncertainty Principle: For any event, if there are multiple reasonable values that an uncertain measure may take, then the value as close to 0.5 as possible is assigned to the event.

According to Professor Liu, "an event has no uncertainty if its uncertain measure is 1 or 0 because we in the former case the event is believed to happen and in the latter case not to happen. An event is the most uncertain if its uncertain measure is 0.5 because the event and its complement may be regarded as equally likely. Uncertainty theory is concerned with to what extent something can be known."

The uncertain variables and their distributions are defined similar to the ordinary random variables and their distributions.

3 Examples

In this section, we present uncertain measures that are not Choquet capacities, and vice versa. Liu gave the following example of uncertain measure [5, 6].

Example 3.1 Assume Γ is the usual space of real numbers. Let α be a number with $0 < \alpha \leq 0.5$. Define an uncertain measure as follows:

$$\mathcal{M}(\Lambda) = \begin{cases} 0, & \text{if } \Lambda = \emptyset \\ \alpha, & \text{if } \Lambda \text{ is upper bounded and } \Lambda \neq \emptyset \\ 0.5, & \text{if both } \Lambda \text{ and } \Lambda^c \text{ are upper unbounded} \\ 1 - \alpha, & \text{if } \Lambda^c \text{ is upper bounded and } \Lambda \neq \Gamma \\ 1, & \text{if } \Lambda = \Gamma. \end{cases}$$

(i) Let $\Lambda_i = (-\infty, i]$ for $i = 1, 2, \cdots$ Then $\Lambda_i \uparrow \Gamma$ and $\lim_{i \to \infty} \mathcal{M}(\Lambda_i) = \alpha$.
(ii) Let $\Lambda_i = [i, +\infty)$ for $i = 1, 2, \cdots$ Then $\Lambda_i \downarrow \emptyset$ and $\lim_{i \to \infty} \mathcal{M}(\Lambda_i) = 1 - \alpha$.

This uncertain measure is neither a Choquet capacity, nor a belief function, as elaborated below:

(i)' In above (i), it does not hold $\mathcal{M}(\Lambda_i) \uparrow \mathcal{M}(\Gamma)(= 1)$. This is different from Choquet capacities which are continuous from below.

(iii) For any non-empty compact sets Λ_i, $i = 1, 2, \ldots$, satisfying $\Lambda_i \downarrow \Lambda$, it holds that $\mathcal{M}(\Lambda_i) \downarrow \mathcal{M}(\Lambda)(= \alpha)$. This is similar to Choquet capacities.

(iii)' When, like in above (ii), Λ_i, $i = 1, 2, \ldots$ are closed sets satisfying $\Lambda_i \downarrow \Lambda$, it does not guarantee $\mathcal{M}(\Lambda_i) \downarrow \mathcal{M}(\Lambda)$. This is seen in above (ii), due to $\mathcal{M}(\Lambda_i) = 1 - \alpha$ but $\mathcal{M}(\emptyset) = 0$. This is similar to Choquet capacities, i.e. generally not continuous from above.

(iv) For any non-empty compact set K, it holds that $\mathcal{M}(K) = \alpha$. This uncertain measure is alternating of infinite order, which can be verified using the equivalence condition (3) and the first equality in the following lemma.

Lemma 3.2 *For any $n, m \in \mathbb{N}$, it holds that*

$$\sum_{\emptyset \neq I \subseteq \{1,2,\dots,n\}} (-1)^{|I|+1} = 1$$

and

$$\sum_{I \subseteq \{1,2,\dots,n\}, J \subseteq \{1,2,\dots,m\}, I \cup J \neq \emptyset} (-1)^{|I|+|J|+1} = 1. \quad \square$$

The first equality in Lemma 3.2 can be obtained through the Binomial expansion of $[(-1) + 1]^n$. The second equality is implied by the first.

(v) This uncertain measure is not monotone of infinite order, which can be verified using the equivalence condition (4) as follows: Take $K_1 = [0, 1]$ and $K_2 = [2, 3]$. Then the left hand side of (4) gives $\mathcal{M}(K_1 \cup K_2) = \alpha$, but the right hand side yields $\mathcal{M}(K_1) + \mathcal{M}(K_2) - \mathcal{M}(K_1 \cap K_2) = 2\alpha$. Hence, this uncertain measure is neither a belief function.

The following example is taken from our previous paper (Li et al. [4]).

Example 3.3 Let $\{W_t \mid t \geq 0\}$ be a stationary Poisson process with intensity rate λ, and $(\Omega, \mathcal{A}, P)$ the probability space for each of the random variables $W_t, t \geq 0$. For $t \geq 0$, the expected value of W_t is λt. Also, let α denote the uniform random variable on the torus $\mathbb{T}^2$, with $(\Omega', \mathcal{A}', P')$ the corresponding probability space.

Given $t > 0$, let $\alpha_1, \alpha_2, \dots, \alpha_{W_t}$ be a collection of W_t many uniform random variables on the torus $\mathbb{T}^2$, identically distributed and independent each other. Then, we define a random finite set $X_t = X_{\alpha_1, \alpha_2, \dots, \alpha_{W_t}}$, whose probability space $(\Omega'', \mathcal{A}'', P'')$ is obtained as the product probability space of $(\Omega, \mathcal{A}, P)$ and $(\Omega', \mathcal{A}', P')$, by letting $X_t(\omega, \omega')$ contain $\alpha_1(\omega'), \alpha_2(\omega'), \dots, \alpha_{W_t(\omega)}(\omega')$. If $W_t(\omega) = 0$, define $X_t = \emptyset$ (thus $P(X_t = \emptyset) = P(W_t = 0) = e^{-\lambda t}$).

Now, for each $t > 0$, we will determine the Choquet capacity T_t of X_t (strictly, the notation is T_{X_t}), i.e., for any compact subset K of $\mathbb{T}^2$, we will calculate the capacity $T_t(K)$, the probability of X_t hitting K, as follows:

$$T_t(K) = P(X_t \cap K \neq \emptyset) = \sum_{n=1}^{\infty} P(W_t = n \text{ and } X_t \cap K \neq \emptyset) \tag{5}$$
$$= 1 - e^{-\lambda t \mu(K)},$$

where μ is the Haar measure on $\mathbb{T}^2$ (Lebesgue measure up to normalization).

By the Choquet Theorem, the probability measure P_t of X_t satisfies

$$P_t(\mathcal{F}_K) = T_t(K) = 1 - e^{-\lambda t \mu(K)}.$$

Clearly, $T_t(\emptyset) = 0$ and $T_t(\mathbb{T}^2) = 1 - e^{-\lambda t} < 1$, due to $P(X = \emptyset) = e^{-\lambda t}$. To normalize T_t, let $C_t(\cdot) = \frac{T_t(\cdot)}{1 - e^{-\lambda t}}$.

This Choquet capacity C_t is not an uncertain measure since

$$T_t(K) + T_t(K^c) = \frac{1}{1 - e^{-\lambda t}}(2 - e^{-\lambda t \mu(K)} - e^{-\lambda t \mu(K^c)}) \tag{6}$$

$$= \frac{1}{1 - e^{-\lambda t}}(2 - e^{-\lambda t \mu(K)} - e^{-\lambda t(1 - \mu(K))}),$$

which is equal to 1 when $K = \emptyset$ or $K = \mathbb{T}^2$, but not equal to 1 otherwise.

By the way, this capacity is ergodic under Arnold's CAT map, a hyperbolic toral automorphism defined on the torus defined as follows (Li et al. [4]):

$$f : \mathbb{T}^2 \to \mathbb{T}^2 \text{ defined by } f(x, y) = (2x + y \mod 1, x + y \mod 1). \tag{7}$$

4 Relation Between Liu's Uncertain Measure and Choquet Capacity

Uncertain measure describes a belief degree. The duality is a critical condition in the definition. A question arises: Does there exist a Choquet capacity that satisfies this duality?

In this note, we are going to answer this question assuming the underlying space is a LCS metric space, which is a standard requirement for probability settings.

Recall that E stands for a LCS metric space. Let $\mathcal{O} = \{O_n \mid n \in \mathbb{N}\}$ be a countable topological base of E, and $\mathcal{B}$ the σ-algebra generated by the open sets of E. Recall that $\mathcal{F}$ denotes the set of all closed subsets of E, equipped with the Fell topology. Put $\mathcal{F}_1 = \{\{x\} \mid x \in E\}$, the subset of $\mathcal{F}$ that contains only the singleton sets of E, and $\mathcal{F}_{>1} = \{F \in \mathcal{F} \mid |F| > 1\}$, the subset of $\mathcal{F}$ that contains only the subsets of E that have more than one point. It is clear that $\mathcal{F}_{>1} = \mathcal{F} \setminus (\mathcal{F}_1 \cup \{\emptyset\})$.

By Choquet Theorem, a Choquet capacity T characterizes the probability law of a random closed set X through an (additive) probability measure P defined on the hyperspace $\mathcal{F}$ where the Fell topology is equipped: For any capacitable subset A in E, it holds that $T(A) = P(X \cap A \neq \emptyset) = P(\mathcal{F}_A)$ where $\mathcal{F}_A = \{F \in \mathcal{F} \mid F \cap A \neq \emptyset\}$.

With this setting, our question becomes: Does there exist a RACS X whose Choquet capacity T satisfies the following: For any T-capacitable set $A \subseteq E$,

$$T(A) + T(A^c) = 1? \tag{8}$$

A Borel set is capacitable, but not the vice versa. We will prove that, unless T degenerates to an ordinary additive measure, there exists no a Choquet capacity satisfying (8) for all Borel sets A.

Theorem 4.1 *Let A be Borel in E. If $T(A) + T(A^c) = 1$, then $P(\mathcal{F}_A \cap \mathcal{F}_{A^c}) = 0$.*

Proof By Choquet Theorem, $T(A) + T(A^c) = 1$ in (8) becomes

$$P(\mathcal{F}_A) + P(\mathcal{F}_{A^c}) = 1, \tag{9}$$

which can be decomposed into the following using the additivity of P:

$$P(\mathcal{F}_A^{A^c} \cup (\mathcal{F}_A \cap \mathcal{F}_{A^c})) + P(\mathcal{F}_{A^c}^A \cup (\mathcal{F}_{A^c} \cap \mathcal{F}_A)) = 1,$$

i.e.,

$$P(\mathcal{F}_A^{A^c}) + P(\mathcal{F}_{A^c}^A) + 2P(\mathcal{F}_A \cap \mathcal{F}_{A^c}) = 1. \tag{10}$$

On the other hand, the probability measure P satisfies $P(\mathcal{F}) = 1$, which can be decomposed into

$$P(\mathcal{F}_A^{A^c}) + P(\mathcal{F}_{A^c}^A) + P(\mathcal{F}_A \cap \mathcal{F}_{A^c}) = 1. \tag{11}$$

To guarantee (8), from (10) and (11), the following must hold: For any Boreal set A in E,

$$P(\mathcal{F}_A \cap \mathcal{F}_{A^c}) = 0. \tag{12}$$

$\square$

If (8) holds for every Borel set A in E, then T is necessarily an ordinary additive measure, as to be proved in the next theorem.

Theorem 4.2 *If for every Borel set A in E, $T(A) + T(A^c) = 1$, then the support of P is $\mathcal{F}_1$. Hence, P is additive and X is actually a random singleton set.*

Proof It suffices to show $P(\mathcal{F}_{>1}) = 0$. Take a countable topological base of E (see [2]), $\mathcal{O} = \{O_n \mid n \in \mathbb{N}\}$. By the assumption and Theorem 4.1, we have $P(\mathcal{F}_{O_n} \cap \mathcal{F}_{(O_n)^c}) = 0$ for all $n \in \mathbb{N}$.

Claim: For every $F \in \mathcal{F}_{>1}$, there exists $n_F \in \mathbb{N}$ such that $F \in \mathcal{F}_{O_{n_F}} \cap \mathcal{F}_{(O_{n_F})^c}$.

In fact, pick up any two different points x and y from F. By the Hausdorff separability of E, there exist a pair of open sets from the base $\mathcal{O}$, say O_{n_x} and O_{n_y}, that separate x and y, i.e., $x \in O_{n_x}$, $y \in O_{n_y}$ and $O_{n_x} \cap O_{n_y} = \emptyset$. Clearly, $O_{n_y} \subseteq (O_{n_x})^c$. Hence, we have $F \cap O_{n_x} \neq \emptyset$ and $F \cap (O_{n_x})^c \neq \emptyset$, i.e., $F \in \mathcal{F}_{O_{n_x}} \cap \mathcal{F}_{(O_{n_x})^c}$.

It follows the above claim that $\mathcal{F}_{>1}$ is a union of countably many sets, each of measure 0, i.e.,

$$\mathcal{F}_{>1} = \bigcup_{n \in \mathbb{N}} \mathcal{F}_{O_n} \cap \mathcal{F}_{(O_n)^c}. \tag{13}$$

Therefore, we have, by applying the subadditivity of P and Theorem 4.1 subsequently, $P(\mathcal{F}_{>1}) \leq \sum_{n \in \mathbb{N}} P(\mathcal{F}_{O_n} \cap \mathcal{F}_{(O_n)^c}) = 0$.

Moreover, $P(\mathcal{F}_{>1}) = 0$ implies $P(\mathcal{F}_1) = 1$ assuming $P(\{\emptyset\}) = 0$, i.e., the support of P is $\mathcal{F}_1$ and the corresponding RACS X is a random singleton set. $\square$

Hence we conclude that Liu's uncertain measure and Choquet capacity differ from each other. Also, Liu's uncertain measure may not possess the complete monotonocity, monotone of infinite order.

Moreover, Liu's uncertain measure is different from Kolmogorov's probability measure due to not only its weakened Duality Axiom, but substantially the modifications in the definitions of product measures and independent events.

Ending note: In analogy to the concept of random closed sets, we can introduce the concept of uncertain closed sets where the probability space is replaced by a uncertain space. Subsequently, an uncertain variable taking values in a space, say $\mathbb{R}$, may be viewed as a special uncertain closed set, a singleton set or an half-closed and half-unbounded open interval.

Let $E = \mathbb{R}^n$. Any uncertain vector $\alpha = (\alpha_1, \alpha_2, \ldots, \alpha_n) : (\Gamma, \mathcal{L}, \mathcal{M}) \to (\mathbb{R}^n, \mathcal{B}(\mathbb{R}^n))$ induces an uncertain closed set $X : (\Gamma, \mathcal{L}, \mathcal{M}) \to (\mathcal{F}(\mathbb{R}^n), \mathcal{B}(\mathcal{F}(\mathbb{R}^n)))$ by letting $X(\gamma) = \prod_{j=1}^{n}[\alpha_j(\gamma), \infty)$. The relation between the distribution function Φ of α and a sort of "capacity T" of X is given by

$$
\begin{aligned}
T(\prod_{j=1}^{n}[a_j, b_j]) &= \mathcal{M}(X \cap \prod_{j=1}^{n}[a_j, b_j] \neq \emptyset) \qquad (14) \\
&= \mathcal{M}(\alpha_1 \leq b_1, \alpha_2 \leq b_2, \cdots, \alpha_n \leq b_n) \\
&= \bigwedge_{i=1}^{n} \mathcal{M}_i([-\infty, b_i]).
\end{aligned}
$$

We need a Choquet theorem for uncertain closed sets.

Finally, let us also point out that these two non-additive measures are useful for generalizing the traditional ergodic theory over additive measures to the ergodic theory over non-additive measures.

Acknowledgements This work was supported by National Natural Science Foundation of China (NSFC) (11871394, 71871121), and HRSA, US Department of Health & Human Services (No. H49MC0068).

References

1. G. Choquet, Theory of capacities. Ann. Inst. Fourier. **V**, 131–295 (1953–54)
2. R. Engelking, General topology, PWN-Polish Scientific Publishers, Warszawa, 1977 (or Heldermann, Berlin, 1989)
3. J.M.G. Fell, A Hausdorff topology for the closed subsets of a locally compact non-Hausdorff space. Proc. Amer. Math. Soc. **13**, 472–476 (1962)
4. R. Li, G. Wei, T. Dooling, S. Bourquin, Fell topology, Choquet capacity and capacity-ergodic systems. Pure Appl. Math. **32**, 416–432 (2016)
5. B. Liu, *Uncertainty Theory*, 2nd edn. (Springer, Berlin, 2007) (3rd ed, 2010; 4th ed, 2015)

6. B. Liu, *Uncertainty Theory*, 5th edn. (Uncertainty Theory Laboratory, Beijing, 2018). http://orsc.edu.cn/liu/ut.pdf
7. G. Matheron, *Random Sets and Integral Geometry* (Wiley, New York, New York, 1975), pp. 1–35
8. P.A. Meyer, *Probabilitiés et Potentiel* (Hermann, Paris, 1966)
9. E. Michael, Topologies on spaces of subsets. Trans. Amer. Math. Soc. **71**, 152–182 (1951)
10. E. Miranda, H.T. Nguyen, Random sets and imprecise probabilities. Int. J. Approx. Reason. **46**, 1–2 (2007)
11. I. Molchanov, *Theory of Random Sets* (Springer, London, 2005), pp. 90, 92, 402–403
12. H.T. Nguyen, *An Introduction to Random Sets* (Chapman and Hall/CRC, New York, 2006)
13. H.T. Nguyen, Y. Wang, G. Wei, On Choquet theorem for random upper semicontinuous functions. Int. J. Approx. Reason. **46**, 3–16 (2007)
14. D. Stoyan, Models and statistics. Int. Stat. Rev. **66**, 1–27 (1998)
15. D. Stoyan, W.S. Kendall, J. Mecke, *Stochastic Geometry and Its Applications*, 2nd edn. (Wiley, Chichester, 1995), pp. 385–420
16. G. Wei, Y. Wang, Formulating stochastic convergence of random closed sets on locally compact separable metrizable spaces using metrics of the hit-or-miss topology. Int. J. Intell. Tech. Appl. Stat. **1**, 33–57 (2008)
17. G. Wei, Y. Wang, H.T. Nguyen, D.E. Beken, On the upper semicontinuity of Choquet capacities. Int. J. Approx. Reason. **51**, 429–440 (2010)

The Joint Distribution of the Discrete Random Set Vector and Bivariate Coarsening at Random Models

Zheng Wei, Baokun Li, and Tonghui Wang

Abstract In this paper, the characterization of the joint distribution of random set vector by the belief function is investigated. A method for constructing the joint distribution of discrete bivariate random sets through copula is given, and a routine of calculating the corresponding bivariate coarsening at random model of finite random sets is obtained. Several examples are given to illustrate our results.

Keywords Random set · Copula · Coarsening at random model · Joint belief function · Jointly monotone of infinite order

1 Introduction

Random sets can be used to model imprecise observations of random variables where the outcomes are assigned as set valued instead of real valued. The theory of random sets is viewed as a natural generalization of multivariate statistical analysis. Random set data can also be viewed as imprecise or incomplete observations which are frequent in today's technological societies. The distribution of the univariate random set and its properties can be found in Nguyen [12], Shafer [18], Dempster [2] and Nguyen and Wang [13], and Molchanov [10]. Recently, the characterization of joint

Z. Wei · T. Wang (✉)
Department of Mathematical Sciences, New Mexico State University, Las Cruces, NM, USA
e-mail: twang@nmsu.edu

Z. Wei
e-mail: zheng.wei@maine.edu

B. Li
School of Statistics, Southwestern University of Finance and Economy, Chengdu, China
e-mail: bali@swufe.edu.cn

Z. Wei
Department of Mathematics and Statistics, University of Maine, Orono, ME, USA

© The Editor(s) (if applicable) and The Author(s), under exclusive license
to Springer Nature Switzerland AG 2021
V. Kreinovich (ed.), *Statistical and Fuzzy Approaches to Data Processing, with Applications to Econometrics and Other Areas*, Studies in Computational Intelligence 892,
https://doi.org/10.1007/978-3-030-45619-1_19

distributions of random sets on co-product spaces was discussed by Schmelzer [15, 17] and Nguyen [14]. In this paper, this characterization is modified for discrete random set vector.

Copulas are used to model multivariate data as they account for the dependence structure and provide a flexible representation of the multivariate distribution. Copulas are multivariate distributions with $[0, 1]$-uniform marginals, which contain the most of the multivariate dependence structure properties and do not depend on the marginals. It is known that copulas connect with marginals to obtain possible joint distributions. In order to investigate the dependence relationship between two random sets, it is necessary to built a bridge for connecting the joint belief functions of random set vector and copulas. For references, see Schmelzer [16, 17], Nguyen [14], Joe [6], Nelsen [11], and Wei et al. [22–24]. In this paper, a method for constructing the joint distribution of the discrete bivariate random set vector through copula is given.

In the univariate discrete application, we usually partition the set E into finitely subsets $A_i \in \mathcal{A}$, the σ-field of subsets of E, $i = 1, 2, \ldots, N$. Consider the random set $\mathcal{S}$ that takes values from the class $\{A_1, A_2, \ldots, A_N\}$ with probability $f_{\mathcal{S}}(A_i) = P(\mathcal{S} = A_i)$, and the random variable X with probability $P_X(A_i) = P(X \in A_i)$, $i = 1, 2, \ldots, N$. Heitjan and Rubin [7] provide a definition of *coarsening at random* (CAR) model: for all $A \in \mathcal{A}$,

$$P(\mathcal{S} = A | X = x) = \text{constant}, \qquad \text{for all } x \in A.$$

The importance of the univariate CAR model has been discussed by Li and Wang [9], Grünwald and Halpern [5] and Jaeger [8]. The examples and applications of coarsening at random are discussed in Heitjan and Rubin [7], Gill et al. [3], Tsiatis [20], Nguyen [12], Tsiatis et al. [21]. However, the bivariate CAR model has not been discussed in literature. In this paper, the bivariate CAR model is introduced based on the joint distribution of random set vector. Also, a routine for calculating the bivariate CAR model solutions is provided.

This paper is organized as follows. The characterization of the joint distribution of random set vector by its joint belief functions is obtained in Sect. 2. A method of connecting the joint belief function of random set vector with given marginals and copula(subcopula) is given in Sect. 3. As an application of random set vector, the bivariate CAR model and its properties are investigated in Sect. 4. Also, the computational aspects of CAR model are provided. To illustrate our main results, several examples are given.

2   Joint Belief Functions

Throughout this paper, let $(\Omega, \mathcal{A}, P)$ be a probability space and let E_1 and E_2 be finite sets, where Ω is sample space, $\mathcal{A}$ is a σ-algebra on subsets of Ω and P is a probability measure. Recall that a finite random set $\mathcal{S}$ with values in powerset of a finite E is a map $S : \Omega \rightarrow 2^E$ such that $S^{-1}(\{A\}) = \{\omega \in \Omega : S(\omega) = A\} \in \mathcal{A}$ for any $A \subseteq E$. Let $f : 2^E \rightarrow [0, 1]$ be $f(A) = P(S = A)$, then f is a probability density function of S on 2^E. In the following, we will extend this definition to the cases of the random set vector.

Definition 2.1 A random set vector (S_1, S_2) with values in $2^{E_1} \times 2^{E_2}$ is a map $(S_1, S_2) : \Omega \rightarrow 2^{E_1} \times 2^{E_2}$ such that $\{\omega \in \Omega : S_1(\omega) = A, S_2(\omega) = B\} \in \mathcal{A}$, for any $A \subseteq E_1$ and $B \subseteq E_2$.

The function $h : 2^{E_1} \times 2^{E_2} \rightarrow [0, 1]$ is said to be a **joint probability density function** of (S_1, S_2) if $h \geq 0$ and $\sum_{A \subseteq E_1} \sum_{B \subseteq E_2} h(A, B) = 1$, where $h(A, B) = P(S_1(\omega) = A, S_2(\omega) = B)$, $A \subseteq E_1$ and $B \subseteq E_2$.

The function $H : 2^{E_1} \times 2^{E_2} \rightarrow [0, 1]$ given by

$$H(A, B) = P(S_1 \subseteq A, S_2 \subseteq B) = \sum_{C \subseteq A} \sum_{D \subseteq B} h(C, D), \quad A \in 2^{E_1}, \quad B \in 2^{E_2}.$$

(2.1)

is said to be the **joint belief function** of (S_1, S_2).

It can be shown H in (2.1) satisfies the following properties:

(i) $H(\emptyset, \emptyset) = H(\emptyset, B) = H(A, \emptyset) = 0$, and $H(E_1, E_2) = 1$;

(ii) H is **monotone of infinite order on each component**, i.e., for any B in 2^{E_2} and any distinct sets $A_1, A_2, \ldots, A_k$ in 2^{E_1}, $k \geq 1$,

$$H\left(\bigcup_{i=1}^{k} A_i, B\right) \geq \sum_{\emptyset \neq I \subseteq \{1,2,\ldots,k\}} (-1)^{|I|+1} H\left(\bigcap_{i \in I} A_i, B\right),$$

(2.2)

and for any $A \in 2^{E_1}$ and any distinct sets $B_1, B_2, \ldots, B_\ell$ in 2^{E_2}, $\ell \geq 1$,

$$H\left(A, \bigcup_{j=1}^{\ell} B_j\right) \geq \sum_{\emptyset \neq J \subseteq \{1,2,\ldots,\ell\}} (-1)^{|J|+1} H\left(A, \bigcap_{j \in J} B_j\right);$$

(2.3)

and

(iii) H is **jointly monotone of infinite order**, i.e., for distinct sets $A_1, A_2, \ldots, A_k$ in 2^{E_1} and distinct $B_1, B_2, \ldots, B_\ell$ in 2^{E_2}, where k, ℓ are positive integers,

$$
\begin{aligned}
H\left(\bigcup_{i=1}^{k} A_i, \bigcup_{j=1}^{\ell} B_j\right) \geq & - \sum_{\emptyset \neq I \subseteq \{1,2,\ldots,k\}} \sum_{\emptyset \neq J \subseteq \{1,2,\ldots,\ell\}} (-1)^{|I|+|J|} H\left(\bigcap_{i \in I} A_i, \bigcap_{j \in J} B_j\right) \\
& + \sum_{\emptyset \neq I \subseteq \{1,2,\ldots,k\}} (-1)^{|I|+1} H\left(\bigcap_{i \in I} A_i, \bigcup_{j=1}^{\ell} B_j\right) \\
& + \sum_{\emptyset \neq J \subseteq \{1,2,\ldots,\ell\}} (-1)^{|J|+1} H\left(\bigcup_{i=1}^{k} A_i, \bigcap_{j \in J} B_j\right).
\end{aligned}
$$

(2.4)

The relationship between the joint belief function H and the probability density function h is given as follows.

Given any joint belief function H of $(\mathcal{S}_1, \mathcal{S}_2)$, there exists a probability density function $h : 2^{E_1} \times 2^{E_2} \to [0, 1]$ corresponding to H. In fact, let $H : 2^{E_1} \times 2^{E_2} \to [0, 1]$ be such that

(i) $H(\emptyset, \emptyset) = H(\emptyset, B) = H(A, \emptyset) = 0$, and $H(E_1, E_2) = 1$,

(ii) H is monotone of infinite order on each component, and

(iii) H is joint monotone of infinite order. Then for any $(A, B) \in 2^{E_1} \times 2^{E_2}$, there exists a nonnegative set function $h : 2^{E_1} \times 2^{E_2} \to [0, 1]$, called the **Möbius inverse** of H, such that

$$H(A, B) = \sum_{C \subseteq A} \sum_{D \subseteq B} h(C, D) \tag{2.5}$$

and

$$\sum_{C \subseteq E_1} \sum_{D \subseteq E_2} h(C, D) = 1. \tag{2.6}$$

The function $h : 2^{E_1} \times 2^{E_2} \to [0, 1]$ is of the form

$$h(A, B) = \sum_{C \subseteq A} \sum_{D \subseteq B} (-1)^{|A \setminus C| + |B \setminus D|} H(C, D), \tag{2.7}$$

where $A \setminus C = A \cap C^c$ and C^c is the complement of C.

Remark 2.1 Several remarks on H and h given in Definition 2.1 are listed below.

(a) Consider the set function $F_1(A) = H(A, E_2)$, $A \in 2^{E_1}$. It is easy to show that $F_1(A)$ is a belief function of random set $\mathcal{S}_1$ over E_1, which is called the **marginal belief function** of random set $\mathcal{S}_1$. Similarly, $F_2(B) = H(E_1, B)$, $B \in 2^{E_2}$ is the marginal belief function of random set $\mathcal{S}_2$ over E_2. More details on belief functions of univariate random sets are given in Nguyen [12].

(b) For any given $B \subseteq E_2$, let $f_2(B)$ be the Möbius inverse of $F_2(B)$. Then

$$P(\mathcal{S}_1 \subseteq A, \mathcal{S}_2 = B) = \sum_{C \subseteq A} h(C|B) f_2(B) = H_{\mathcal{S}_1|\mathcal{S}_2}(A|B) f_2(B),$$

where $h(C|B) = P(\mathcal{S}_1 = C|\mathcal{S}_2 = B)$ is the conditional probability of $\mathcal{S}_1 = C$ given $\mathcal{S}_2 = B$. We call $H_{\mathcal{S}_1|\mathcal{S}_2}(A|B)$ be the **conditional belief function** of $\mathcal{S}_1$ given $\mathcal{S}_2 = B$. Similarly, we can obtain $H_{\mathcal{S}_2|\mathcal{S}_1}(B|A)$ the **conditional belief function** of $\mathcal{S}_2$ given $\mathcal{S}_1 = A$. For a given joint belief function $H(A, B)$ of random set vector $(\mathcal{S}_1, \mathcal{S}_2)$, we say $\mathcal{S}_1$ and $\mathcal{S}_2$ are **independent** if and only if $H(A, B) = F_1(A)F_2(B)$, for all $A \in 2^{E_1}$ and $B \in 2^{E_2}$.

(c) If $B_1 = B_2 = \cdots = B_\ell \equiv B$, then the Property (iii) of H is reduced to an equality, so Property (ii) is needed for characterizing the marginal belief functions of $\mathcal{S}_1$ and $\mathcal{S}_2$, respectively.

(d) In view of the direct product $(2^{E_1} \times 2^{E_2}, \leq)$ of two locally finite posets $(2^{E_1}, \subseteq)$ and $(2^{E_1}, \subseteq)$, where $(C, D) \leq (A, B)$ means $C \subseteq A$ and $D \subseteq B$, with its Möbius function

$$\mu : (2^{E_1} \times 2^{E_2}) \times (2^{E_1} \times 2^{E_2}) \to \mathbb{Z} \text{ with } \mu((C, D), (A, B)) = (-1)^{|A \setminus C| + |B \setminus D|},$$

we have

$$H(A, B) = \sum_{(C,D) \leq (A,B)} h(C, D),$$

where $h(A, B)$ is the **Möbius inverse** of H,

$$h(A, B) = \sum_{(C,D) \leq (A,B)} (-1)^{|A \setminus C| + |B \setminus D|} H(C, D). \tag{2.8}$$

$h(., .)$ is also called the probability assignment of random set vector $(\mathcal{S}_1, \mathcal{S}_2)$. Note that there is a bijection between the joint belief function H and its the corresponding joint density h (See e.g. Nguyen [14]).

(e) Similar to Property (iii), there is a property called completely monotone, given by Schmelzer [15, 16] and Nguyen [14] as follows.

A set function $H_1 : 2^{E_1} \times 2^{E_2} \to [0, 1]$ is said to be **completely monotone**, if for any $k \geq 2$ and $(A_i, B_i) \in 2^{E_1} \times 2^{E_2}, i = 1, 2, \ldots, k,$

$$H_1 \left(\bigcup_{i=1}^{k} A_i, \bigcup_{i=1}^{k} B_i \right) \geq \sum_{\emptyset \neq I \subseteq \{1,2,\ldots,k\}} (-1)^{|I|+1} H_1 \left(\bigcap_{i \in I} A_i, \bigcap_{i \in I} B_i \right). \tag{2.9}$$

Note that the difference between (2.2)–(2.4) and (2.9) is that (A_i, B_j)'s in (2.2)–(2.4) are distinct sets while (A_i, B_i)'s in (2.9) are not necessarily distinct sets and can be duplicated many times if needed. In the following, we will show that (2.9) is equivalent to (2.2)–(2.4).

Proposition 2.1 *If* $H : 2^{E_1} \times 2^{E_2} \to [0, 1]$ *is such that* $H(\emptyset, \emptyset) = H(\emptyset, B) = H(A, \emptyset) = 0$, *and* $H(E_1, E_2) = 1$, *then* H *is completely monotone in each component given* (2.9) *if and only if* H *is monotone of infinite order on each component given in* (2.2), *and* (2.3) *and* H *is joint monotone of infinite order given in* (2.4).

Proof For "only if" part, assume that (2.9) holds. Let $A_1, A_2 \ldots, A_k \in 2^{E_1}$ be distinct, $B \in 2^{E_2}$, if we set $B_1 = B_2 = \cdots = B_k = B$, then (2.9) is reduced to

$$H \left(\bigcup_{i=1}^{k} A_i, B \right) = H \left(\bigcup_{i=1}^{k} A_i, \bigcup_{i=1}^{k} B_i \right) \geq \sum_{\emptyset \neq I \subseteq \{1,2,\ldots,k\}} (-1)^{|I|+1} H_1 \left(\bigcap_{i \in I} A_i, \bigcap_{i \in I} B_i \right)$$

$$= \sum_{\emptyset \neq I \subseteq \{1,2,\ldots,k\}} (-1)^{|I|+1} H_1 \left(\bigcap_{i \in I} A_i, B \right),$$

so that (2.2) holds. Similarly, if $A \in 2^{E_1}$, we set $A_1 = A_2 = \cdots = A_k = A$, and $B_1, B_2 \ldots, B_\ell \in 2^{E_2}$ be distinct, then (2.9) implies (2.3). Now let $A_1, \ldots, A_k \in 2^{E_1}$ and $B_1, \ldots, B_\ell \in 2^{E_2}$ be distinct. Define sets C_t and D_t, $1 \leq t \leq k + \ell$ by

$$
C_t = \begin{cases} A_t & \text{if } 1 \leq t \leq k \\ \bigcup_{i=1}^{k} A_i & \text{if } k+1 \leq t \leq k+\ell, \end{cases}
\qquad
D_t = \begin{cases} \bigcup_{j=1}^{\ell} B_j & \text{if } 1 \leq t \leq k \\ B_{t-k} & \text{if } k+1 \leq t \leq k+\ell. \end{cases}
$$

Then, (2.9) can be written to:

$$
H\left(\bigcup_{i=1}^{k} A_i, \bigcup_{j=1}^{\ell} B_j\right) = H\left(\bigcup_{t=1}^{k+\ell} C_t, \bigcup_{t=1}^{k+\ell} D_t\right) \geq \sum_{\emptyset \neq I \subseteq \{1,\ldots,k+\ell\}} (-1)^{|I|+1} H\left(\bigcap_{t\in I} C_t, \bigcap_{t\in I} D_t\right)
$$

$$
= \sum_{\emptyset \neq I \subseteq \{1,\ldots,k\}} (-1)^{|I|+1} H\left(\bigcap_{t\in I} C_t, \bigcap_{t\in I} D_t\right) + \sum_{\emptyset \neq I \subseteq \{k+1,\ldots,k+\ell\}} (-1)^{|I|+1} H\left(\bigcap_{t\in I} C_t, \bigcap_{t\in I} D_t\right)
$$

$$
+ \sum_{\substack{I \cap \{1,\ldots,k\} \neq \emptyset \\ I \cap \{k+1,\ldots,k+\ell\} \neq \emptyset}} (-1)^{|I|+1} H\left(\bigcap_{t\in I} C_t, \bigcap_{t\in I} D_t\right)
$$

$$
= \sum_{\emptyset \neq I \subseteq \{1,\ldots,k\}} (-1)^{|I|+1} H\left(\bigcap_{i\in I} A_i, \bigcup_{j=1}^{\ell} B_j\right) + \sum_{\emptyset \neq J \subseteq \{1,\ldots,\ell\}} (-1)^{|J|+1} H\left(\bigcup_{i=1}^{k} A_i, \bigcap_{j\in J} D_j\right)
$$

$$
- \sum_{\emptyset \neq I \subseteq \{1,\ldots,k\}} \sum_{\emptyset \neq J \subseteq \{1,\ldots,\ell\}} (-1)^{|I|+|J|} H\left(\bigcap_{i\in I} A_i, \bigcap_{j\in J} B_j\right),
$$

so that (2.4) holds.

For "if" part, assume (2.2)–(2.4) holds. There exists a nonnegative set function $h : 2^{E_1} \times 2^{E_2} \to [0, 1]$, such that

$$
H(A, B) = \sum_{C \subseteq A} \sum_{D \subseteq B} h(C, D)
$$

and

$$
\sum_{C \subseteq E_1} \sum_{D \subseteq E_2} h(C, D) = 1.
$$

Now, for any $k \geq 2$ and $(A_i, B_i) \in 2^{E_1} \times 2^{E_2}$, $i = 1, 2, \ldots, k$. For any $C \subseteq E_1$ and $D \subseteq E_2$, let $\mathcal{J}(C, D) = \{i = 1, 2, \ldots, k$ such that $C \subseteq A_i$, and $D \subseteq B_i\}$. Then $C \subseteq \bigcap_{i \in \mathcal{J}(C,D)} A_i$ and $D \subseteq \bigcap_{i \in \mathcal{J}(C,D)} B_i$ if $\mathcal{J}(C, D) \neq \emptyset$. Clearly,

$$H\left(\bigcup_{i=1}^{k} A_i, \bigcup_{i=1}^{k} B_i\right) = \sum_{\substack{C \subseteq \bigcup\limits_{i=1}^{k} A_i \ D \subseteq \bigcup\limits_{i=1}^{k} B_i}} h(C, D) \geq \sum_{\substack{C,D \\ \mathcal{J}(C,D) \neq \emptyset}} h(C, D)$$

$$= \sum_{\substack{C,D \\ \mathcal{J}(C,D) \neq \emptyset}} \left[\sum_{\emptyset \neq I \subseteq \mathcal{J}(C,D)} (-1)^{|I|+1} \right] h(C, D) = \sum_{\emptyset \neq I \subseteq \{1,...,k\}} (-1)^{|I|+1} \left[\sum_{\substack{C,D \\ \mathcal{J}(C,D) \supseteq I}} h(C, D) \right]$$

$$= \sum_{\emptyset \neq I \subseteq \{1,...,k\}} (-1)^{|I|+1} H\left(\bigcap_{i \in I} A_i, \bigcap_{i \in I} B_i\right),$$

hence (2.9) holds. □

Given a set function $H : 2^{E_1} \times 2^{E_2} \to [0, 1]$, it is natural to ask whether if it is a well-defined joint belief function. By Properties (i)–(iii) of H, we only need to check if (2.2)–(2.4) holds for all distinct sets A_i's and B_j's.

Example 2.1 Let $E_1 = \{1, 2\}$ and $E_2 = \{3, 4\}$. The set function H is given in Table 1.

It is easy to verify from (2.2) to (2.4) that H is not a joint belief function. Indeed, if we choose $A_1 = \{1\}$, $A_2 = \{2\}$, and $B_1 = \{3\}$, $B_2 = \{4\}$, from (2.4) we obtain,

$$1 = H(A_1 \cup A_2, B_1 \cup B_2) \geq$$
$$H(A_1, B_1 \cup B_2) + H(A_2, B_1 \cup B_2) + H(A_1 \cup A_2, B_1) + H(A_1 \cup A_2, B_2)$$
$$-H(A_1, B_1) - H(A_2, B_1) - H(A_1, B_2) - H(A_2, B_2)$$
$$= 0.6 + 0.6 + 0.5 + 0.6 - 0.1 - 0.2 - 0.3 - 0.3 = 1.4.$$

Note that in (2.9), we need to choose $A_1 = \{1\}$, $A_2 = \{2\}$, $A_3 = \{1, 2\} = A_4$, and $B_1 = B_2 = \{3, 4\}$, $B_3 = \{3\}$ $B_4 = \{4\}$, so that

$$1 = H\left(\bigcup_{i=1}^{4} A_i, \bigcup_{i=1}^{4} B_i\right) \geq \sum_{\emptyset \neq I \subseteq \{1,2,3,4\}} (-1)^{|I|+1} H_1\left(\bigcap_{i \in I} A_i, \bigcap_{i \in I} B_i\right) = 1.4.$$

On the other hand, it is easy to see that $h(A, B)$ given in (2.7) is not nonnegative. Indeed, from (2.7) we obtain from

Table 1 Set function H

H	{3}	{4}	{3, 4}
{1}	0.2	0.3	0.6
{2}	0.1	0.3	0.6
{1, 2}	0.5	0.6	1

$$h(\{1, 2\}, \{3, 4\}) = H(\{1, 2\}, \{3, 4\}) - H(\{1, 2\}, \{3\}) - (\{1, 2\}, \{4\}) - H(\{1\}, \{3, 4\})$$
$$- (\{2\}, \{3, 4\}) + H(\{1\}, \{3\}) + H(\{1\}, \{4\}) + H(\{2\}, \{3\}) + H(\{2\}, \{4\})$$
$$= -0.4 < 0.$$

Therefore, the Property (ii) and the Property (iii) of H are necessary for characterizing the joint distribution of the random set vector. $\qquad\square$

3 Connections Between Joint Belief Functions and Subcopulas

From the joint belief function of random set vector(S_1, S_2), it is not easy to check the dependence relationship between S_1 and S_2. Copula is a useful tool for modeling dependence of random variables as they account for the dependence structure and provide a flexible representation, see Nelson [11], Sklar [19], Hung [14], and Wei et al. [22, 24]. Copulas connect marginals to obtain possible joint distributions. Therefore, it is necessary to built a bridge for connecting the joint belief functions of random set vector and copulas.

Definition 3.1 [11] A *two-dimensional* **subcopula** is a function C' with the following properties:

(a) Domain of C' is $I_1 \times I_2$, where I_1 and I_2 are subsets of $[0, 1]$ containing 0 and 1;

(b) C' is 2-increasing, i.e., $C(u, v) - C(u', v) - C(u, v') + C(u', v') \geq 0$ for $u' \leq u$, $v' \leq v$ and $C'(u, 0) = C'(0, v) = 0$;

(c) For every u in I_1 and every v in I_2,

$$C'(u, 1) = u \quad \text{and} \quad C'(1, v) = v.$$

A **copula** $C : [0, 1]^n \to [0, 1]$ is a subcopula C' with $I_1 = I_2 = [0, 1]$.

Let H be the joint distribution function of a random vector (X, Y) with marginals F and G, then there exists a copula C such that $H(x, y) = C(F(x), G(y))$. Furthermore, if the marginals are continuous, then the copula C is unique. An initial approach of using copulas for random sets was discussed by Alvarez [1]. Schmelzer [17] showed that combining univariate belief functions by a single copula does not necessarily yield a joint belief function. In this section, we are going to make connections between joint belief functions of discrete random set vector and copulas(subcopulas) by equiping the order on focal sets.

3.1 An Algorithm for Constructing Joint Belief Functions Through Copulas

Given two univariate belief functions $F_1(A) = P(S_1 \subseteq A)$, $A \in 2^{E_1}$ and $F_2(B) = P(S_2 \subseteq B)$, $B \in 2^{E_2}$ on finite domains E_1 and E_2, respectively, what are all possible joint belief functions, $H(A, B)$, with these given marginals? In the following, we will introduce a method for constructing joint belief functions from given marginal belief functions and copula.

Let $f_1(A) = P(S_1 = A)$ and $f_2(B) = P(S_2 = B)$ be the densities of S_1 and S_2, respectively. Since E_i's are finite sets, we can order the elements of $\mathcal{F}(f_1) = \{A \subseteq E_1 : f_1(A) > 0\}$, by the Lexicographical order (also known as lexical order, dictionary order) as

$$\mathcal{F}(f_1) = \{A_1, A_2, \ldots, A_m\},$$

Here $\mathcal{F}(f_1)$ is called the **focal sets**, see [12]. Similarly, we can obtain

$$\mathcal{F}(f_2) = \{B \subseteq E_2 : f_2(B) > 0\} = \{B_1, B_2, \ldots, B_n\}.$$

Let $\mathcal{B}_1$ is the collection of all Borel subsets of $[0, 1]$ and $\lambda(dx)$ is the Lebesgue measure on $\mathcal{B}_1$. Consider the probability space $([0, 1], \mathcal{B}_1, \lambda(dx))$. For $F_1(A)$, partition $[0, 1]$ into m intervals $I_1, I_2, \ldots, I_m$ with length $f_1(A_i)$, $A_i \in \mathcal{F}(f_1)$. Similarly, for $F_2(B) = P(S_2 \subseteq B)$, partition $[0, 1]$ into n intervals $J_1, J_2, \ldots, J_n$ with length $f_2(B_j)$, $B_j \in \mathcal{F}(f_2)$. Define

$$S_1' : [0, 1] \to \mathcal{F}(f_1) \quad S_1'(x) = A_i \quad \text{for } x \in I_i, i = 1, \ldots, m, \quad (3.1)$$

and

$$S_2' : [0, 1] \to \mathcal{F}(f_2) \quad S_2'(y) = B_j \quad \text{for } y \in J_j, j = 1, \ldots, n. \quad (3.2)$$

Observe that the Lebesgue measure $\lambda(dx)$ on $[0, 1]$ corresponds, by Lebesgue-Stieltjes theorem, to the distribution function $x \to x$, on $[0, 1]$, of the uniform random variable on it. As such, a joint distribution on $[0, 1]^2$ with uniform marginals is precisely some copula $\mathcal{C}$.

Note that each copula $\mathcal{C}$, as a bivariate distribution, generates a probability measure on $\mathcal{B}_1 \times \mathcal{B}_1$ of $[0, 1]^2$, denoted as $d\mathcal{C}$, by

$$d\mathcal{C}\{[u_1, u_2] \times [v_1, v_2]\} = \mathcal{C}(u_1, v_1) - \mathcal{C}(u_1, v_2) - \mathcal{C}(u_2, v_1) + \mathcal{C}(u_2, v_2). \quad (3.3)$$

Now we can consider the random set vector $(S_1', S_2') : [0, 1]^2 \to 2^{E_1} \times 2^{E_2}$ which has marginal densities f_1, f_2. Let $\mathcal{C}$ be a copula. If we equip the measurable space $([0, 1]^2, \mathcal{B}_1 \times \mathcal{B}_1)$ with the probability measure $d\mathcal{C}$, then the function $H_{\mathcal{C}} : 2^{E_1} \times 2^{E_2} \to [0, 1]$, defined by

$$H_{\mathcal{C}}(A, B) = d\mathcal{C}\{(x, y) \in [0, 1]^2 : S'_1(x) \subseteq A, S'_2(y) \subseteq B\}$$
$$= d\mathcal{C}\{(x, y) \in [0, 1]^2 : S'_1(x) \times S'_2(y) \subseteq A \times B\}. \tag{3.4}$$

In summary, we have the following result.

Theorem 3.1 *For any given univariate belief functions F_1 and F_2 of random sets S_1 and S_2, a Lexico order on focal sets, and a copula C, the joint belief function of random set vector (S_1, S_2) can be constructed by (3.4). Furthermore, the joint density of random set vector (S_1, S_2) can be obtained by its Möbius inverse of H_C given in (2.8).*

Note that it is easy to verify that $H_C(., .)$ has marginal belief functions F_1 and F_2. The following example is an illustration of our construction method.

Example 3.1 Let $E_1 = \{1, 2\}$ and $E_2 = \{3, 4, 5\}$. Suppose the densities of random sets S_1 and S_2 are given by

$$f_1(\{1\}) = f_1(\{2\}) = 0.25, f_1(\{1, 2\}) = 0.5,$$

and

$$f_2(\{3\}) = 0.2, f_2(\{4\}) = f_2(\{3, 5\}) = 0.3, f_2(\{3, 4, 5\}) = 0.2,$$

respectively. Now if we equip $\mathcal{F}(f_i)$ with Lexicographical order, we will obtain a unique joint belief function of S_1 and S_2 for any given copula C.

Consider the orders given by

$$\mathcal{F}(f_1) = \{\{1\}, \{2\}, \{1, 2\}\} \quad \text{and} \quad \mathcal{F}(f_2) = \{\{3\}, \{4\}, \{3, 5\}, \{3, 4, 5\}\}.$$

Define $S'_1 : [0, 1] \to \mathcal{F}(f_1)$ and $S'_2 : [0, 1] \to \mathcal{F}(f_2)$ respectively by by

$$S'_1(x) = \begin{cases} \{1\}, & \text{if } x \in [0, 0.25], \\ \{2\}, & \text{if } x \in (0.25, 0.5], \\ \{1, 2\}, & \text{if } x \in (0.5, 1] \end{cases} \qquad S'_2(y) = \begin{cases} \{3\}, & \text{if } y \in [0, 0.2], \\ \{4\}, & \text{if } y \in (0.2, 0.5], \\ \{3, 5\}, & \text{if } y \in (0.5, 0.8], \\ \{3, 4, 5\}, & \text{if } y \in (0.8, 1]. \end{cases}$$

If we apply Farlie-Gumbel-Morgenstern copula $\mathcal{C}(u, v) = uv(1 + \rho(1 - u)(1 - v))$ with $\rho = \frac{1}{2}$, then from (3.4), we obtain the joint distribution given bellow (Table 2),

Then, from (2.8), we can calculate the joint density, h_C, of (S_1, S_2) given below (Table 3).

Remark 3.1 Note that the construction method given above shows that the joint distribution H_C depends not only on copula C but also on the order of $\mathcal{F}(f_1)$ and $\mathcal{F}(f_2)$. Nguyen [14] suggested us to use the principle of maximum entropy for

Table 2 Joint distribution of $(\mathcal{S}_1, \mathcal{S}_2)$

H_C	{3}	{4}	{5}	{3, 4}	{3, 5}	{4, 5}	{3,4,5}
{1}	13/200	267/3200	0	19/128	421/3200	367/3200	1/4
{2}	11/200	249/3200	0	17/128	407/3200	249/3200	1/4
{1, 2}	1/5	3/10	0	1/2	1/2	3/10	1

Table 3 Joint density of $(\mathcal{S}_1, \mathcal{S}_2)$

h_C	{3}	{4}	{5}	{3, 4}	{3, 5}	{4, 5}	{3, 4, 5}
{1}	13/200	267/3200	0	0	213/3200	0	7/200
{2}	11/200	249/3200	0	0	231/3200	0	9/200
{1, 2}	2/25	111/800	0	0	129/800	0	3/25

selecting the orders of A_i's and B_j's so that the definition of joint distribution is determined. However, this selection of A_i's and B_j's is not unique so that same maximum entropy can result different joint distribution of $(\mathcal{S}_1, \mathcal{S}_2)$. For example, let $E_1 = \{1, 2\}$ and $E_2 = \{3, 4\}$, $f_1(\{1\}) = 1/3$, $f_1(\{1, 2\}) = 2/3$ and $f_2(\{4\}) = 3/4$, $f_2(\{3, 4\}) = 1/4$, all four different orders on $\mathcal{F}(f_1) = \{A_1 = \{1\}, A_2 = \{1, 2\}\}$ and $\mathcal{F}(f_2) = \{B_1 = \{4\}, B_2 = \{3, 4\}\}$ give different distributions. However, all four different orders give the same entropy,

$$\mathcal{E}_{nt}(h_C) = - \sum_{A \in 2^{E_1}, B \in 2^{E_2}} h_C(A, B) \log_2 h_C(A, B) = 0.5183131.$$

3.2 Constructing a Subcopula from the Joint Belief Function

The previous subsection shows that given marginal belief functions and a copula, we can construct a joint belief function. Conversely, given joint belief function, we can obtain a subcopula. Note that any subcopula can be extended to a copula, but its extension is not generally unique.

Now given a joint belief function $H : 2^{E_1} \times 2^{E_2} \to [0, 1]$ of discrete random set vector $(\mathcal{S}_1, \mathcal{S}_2)$, we can find its marginal belief functions $F_1(A) = H(A, E_2)$, $F_2(B) = H(E_1, B)$ and the corresponding marginal densities $f_1 : 2^{E_1} \to [0, 1]$, $f_2 : 2^{E_2} \to [0, 1]$. By the construction method given in (3.1)–(3.4), we can solve for a subcopula $\mathcal{C}'$.

Theorem 3.2 *Given a joint belief function $H : 2^{E_1} \times 2^{E_2} \to [0, 1]$ of discrete random set vector $(\mathcal{S}_1, \mathcal{S}_2)$ and a lexical order on focal sets, there is a unique subcopula $\mathcal{C}'$, such that $H(A, B) = \mathcal{C}'(F_1(A), F_2(B))$, for any $A \subseteq E_1$, $B \subseteq E_2$, where F_1 and F_2 are marginal belief functionals.*

Proof Given $H : 2^{E_1} \times 2^{E_2} \to [0, 1]$, we can find two marginal belief functions $F_1 : 2^{E_1} \to [0, 1]$ and $F_2 : 2^{E_2} \to [0, 1]$ by $F_1(A) = H(A, E_2)$ and $F_2(B) = H(E_1, B)$ for any $A \subseteq E_1$ and $B \subseteq E_2$. Furthermore, we can find two marginal densities f_1 and f_2. Since E_i's are finite sets, we can order the elements of $\mathcal{F}(f_1) = \{A \subseteq E_1 : f_1(A) > 0\}$, by the Lexicographical order as

$$\mathcal{F}(f_1) = \{A_1, A_2, \ldots, A_m\}.$$

Similarly, we can obtain

$$\mathcal{F}(f_2) = \{B \subseteq E_2 : f_2(B) > 0\} = \{B_1, B_2, \ldots, B_n\}.$$

For $F_1(A)$, partition $[0, 1]$ into m intervals $I_1, I_2, \ldots, I_m$ with length $f_1(A_i)$, $A_i \in \mathcal{F}(f_1)$, assume the partition is $0 = i_0 < i_1 < \cdots < i_{m-1} < i_m = 1$. Similarly, for $F_2(B) = P(\mathcal{S}_2 \subseteq B)$, partition $[0, 1]$ into n intervals $J_1, J_2, \ldots, J_n$ with length $f_2(B_j)$, $B_j \in \mathcal{F}(f_2)$, assume the partition is $0 = j_0 < j_1 < \cdots < j_{n-1} < j_n = 1$.

Define

$$\mathcal{S}_1' : [0, 1] \to \mathcal{F}(f_1) \qquad \mathcal{S}_1'(x) = A_i \text{ for } x \in I_i, i = 1, \ldots, m, \qquad (3.5)$$

and

$$\mathcal{S}_2' : [0, 1] \to \mathcal{F}(f_2) \qquad \mathcal{S}_2'(y) = B_j \text{ for } y \in J_j, j = 1, \ldots, n. \qquad (3.6)$$

Then, $\mathcal{S}_i'$s are two marginal random sets with densities f_i. Let $I_1 = \{i_0, \ldots, i_m\}$ and $I_2 = \{j_0, \ldots, j_n\}$, then $I_i \subseteq [0, 1]$ which contains 0 and 1. Then we can define $C' : I_1 \times I_2 \to [0, 1]$ by (3.4): Define $C'(u, 0) = 0 = C'(0, v)$, and $C'(i_1, j_1) = C'(i_1, j_1) - C'(i_1, 0) - C'(0, j_1) + C'(0, 0) = H(A_1, B_1)$, following this pattern, we can define C' for all $i \in I_1$ and $j \in I_2$ recursively and it is easy to verify C' is indeed a subcopula by the definition. $\qquad \square$

The following example is an illustration of the above construction algorithm.

Example 3.2 Given a joint belief function H in Table 4.

From the last row and the last column of the joint belief function, we can calculate the marginal densities, f_1 and f_2, which are given below (Table 5).

Define $\mathcal{S}_1 : [0, 1] \to 2^{E_1}$ and $\mathcal{S}_2 : [0, 1] \to 2^{E_2}$ respectively by

$$\mathcal{S}_1(x) = \begin{cases} \{1\}, & \text{if } x \in [0, \frac{1}{3}], \\ \{2\}, & \text{if } x \in (\frac{1}{3}, \frac{2}{3}], \\ \{1, 2\}, & \text{if } x \in (\frac{2}{3}, 1]. \end{cases} \qquad \mathcal{S}_2(y) = \begin{cases} \{3\}, & \text{if } y \in [0, \frac{1}{6}], \\ \{4\}, & \text{if } y \in (\frac{1}{6}, \frac{2}{6}], \\ \{5\}, & \text{if } y \in (\frac{2}{6}, \frac{3}{6}], \\ \{3, 4, 5\}, & \text{if } y \in (\frac{3}{6}, 1]. \end{cases}$$

Solving $C'(u, v)$ in (3.4), we obtain a subcopula C' given in Table 6.

For example, for calculating $C'(\frac{2}{3}, \frac{2}{6})$, which belongs to the interval $(\frac{1}{3}, \frac{2}{3}] \times (\frac{1}{6}, \frac{2}{6}]$, we have

Table 4 The joint belief function H

H	{3}	{4}	{5}	{3, 4}	{3, 5}	{4, 5}	{3, 4, 5}
{1}	1/12	1/12	1/12	1/6	1/6	1/6	1/3
{2}	1/12	1/12	1/12	1/6	1/6	1/6	1/3
{1,2}	1/6	1/6	1/6	1/3	1/3	1/3	1

Table 5 Marginal densities $f_1(A)$ and $f_2(B)$

A	{1}	{2}	{1, 2}	B	{3}	{4}	{5}	{3, 4}	{3, 5}	{4, 5}	{3, 4, 5}
f_1	1/3	1/3	1/3	f_2	1/6	1/6	1/6	0	0	0	1/2

Table 6 The subcopula C'

C'	1/6	2/6	3/6	1
1/3	1/12	1/6	1/4	1/3
2/3	1/6	2/6	2/4	2/3
1	1/6	2/6	3/6	1

$$dC' \left\{ \left(\frac{1}{3}, \frac{2}{3} \right] \times \left(\frac{1}{6}, \frac{2}{6} \right] \right\} = C' \left(\frac{2}{3}, \frac{2}{6} \right) + C' \left(\frac{1}{3}, \frac{1}{6} \right) - C' \left(\frac{1}{3}, \frac{2}{6} \right) - C' \left(\frac{2}{3}, \frac{1}{6} \right)$$

$$= H(\{2\}, \{4\}) = \frac{1}{12}.$$

Thus,

$$C' \left(\frac{2}{3}, \frac{2}{6} \right) = H(\{2\}, \{4\}) - C' \left(\frac{1}{3}, \frac{1}{6} \right) + C' \left(\frac{1}{3}, \frac{2}{6} \right) + C' \left(\frac{2}{3}, \frac{1}{6} \right)$$

$$= \frac{1}{12} - \frac{1}{12} + \frac{1}{6} + \frac{1}{6} = \frac{1}{3}.$$

It is easy to check that C' is an independent subcopula, which in turn shows that H is an independent joint belief function.

4 Bivariate CAR Models

As an application of the joint belief function of random set vector, bivariate CAR models are introduced and its properties are discussed in this section. Also, the computational aspects of CAR model are provided.

Let $(\Omega, \mathcal{A}, P)$ be a probability space, E_1 and E_2 be finite sets. Consider bivariate random vector $(X, Y) : \Omega \to E_1 \times E_2$, and random set vector $(\mathcal{S}_1, \mathcal{S}_2) : \Omega \to 2^{E_1} \times 2^{E_2}$. A **coarsening of** (X, Y) is a non-empty random set vector $(\mathcal{S}_1, \mathcal{S}_2)$ on $E_1 \times E_2$ such that $P(X \in \mathcal{S}_1, Y \in \mathcal{S}_2) = 1$.

Definition 4.1 A coarsening $(\mathcal{S}_1, \mathcal{S}_2)$ is said to be a **bivariate coarsening at random(CAR)** of the bivariate random vector (X, Y), if for all $(A, B) \in (2^{E_1} \setminus \{\phi\}) \times (2^{E_2} \setminus \{\phi\})$,

$$P(\mathcal{S}_1 = A, \mathcal{S}_2 = B | X = x, Y = y) = \pi(A, B) \tag{4.1}$$

for any $x \in A$, $y \in B$. The constant $\pi(A, B)$ is called a **bivariate CAR probability**.

Proposition 4.1 *The condition of bivariate CAR of* (X, Y) *given in (4.1) is equivalent to*

$$P(\mathcal{S}_1 = A, \mathcal{S}_2 = B | X = x, Y = y) = P(\mathcal{S}_1 = A, \mathcal{S}_2 = B | X \in A, Y \in B) \tag{4.2}$$

for any $(A, B) \in (2^{E_1} \setminus \{\phi\}) \times (2^{E_2} \setminus \{\phi\})$ *and* $x \in A$, $y \in B$.

Proof Assume that the condition (4.1) holds. Since $(\mathcal{S}_1, \mathcal{S}_2)$ is a coarsening of (X, Y), we have $(\mathcal{S}_1 = A, \mathcal{S}_2 = B) \subseteq (X \in A, Y \in B)$ so that $P(X \in A, Y \in B | \mathcal{S}_1 = A, \mathcal{S}_2 = B) = 1$. Also,

$$P(X \in A, Y \in B | \mathcal{S}_1 = A, \mathcal{S}_2 = B) = \sum_{x \in A, y \in B} P(X = x, Y = y | \mathcal{S}_1 = A, \mathcal{S}_2 = B)$$

$$= \sum_{x \in A, y \in B} \frac{P(\mathcal{S}_1 = A, \mathcal{S}_2 = B | X = x, Y = y) P(X = x, Y = y)}{P(\mathcal{S}_1 = A, \mathcal{S}_2 = B)}$$

$$= \frac{\pi(A, B)}{P(\mathcal{S}_1 = A, \mathcal{S}_2 = B)} \sum_{x \in A, y \in B} P(X = x, Y = y)$$

$$= \frac{\pi(A, B)}{P(\mathcal{S}_1 = A, \mathcal{S}_2 = B)} P(X \in A, Y \in B) = 1.$$

Therefore, which indicate that (4.2) holds,

$$\pi(A, B) = \frac{P(\mathcal{S}_1 = A, \mathcal{S}_2 = B)}{P(X \in A, Y \in B)} = P(\mathcal{S}_1 = A, \mathcal{S}_2 = B | X \in A, Y \in B).$$

It is easy to show the condition (4.2) implies the condition (4.1). □

Remark 4.1 The characterization of CAR mechanism for univariate random set by utilizing the concept of uniform multi-cover was given in Gill and Grunwald [4] and Li and Wang [9]. Their result can be extended to the bivariate CAR mechanism. Note that

$$\sum_{A \ni x, B \ni y} \pi(A, B) = 1, \text{ for any } x \in E_1, y \in E_2. \tag{4.3}$$

Indeed, we know that $\pi(A, B) = 0$, for any $x \notin A$ or $y \notin B$. For any $x \in E_1$ and $y \in E_2$, we have

$$1 = \sum_{A \subseteq E_1, B \subseteq E_2} P(S_1 = A, S_2 = B | X = x, Y = y)$$

$$= \sum_{A \not\ni x \text{ or } B \not\ni y} P(S_1 = A, S_2 = B | X = x, Y = y) + \sum_{A \ni x, B \ni y} P(S_1 = A, S_2 = B | X = x, Y = y)$$

$$= \sum_{A \ni x, B \ni y} P(S_1 = A, S_2 = B | X = x, Y = y) = \sum_{A \ni x, B \ni y} \pi(A, B).$$

Note that the following Theorem is an extension of the result given in Gill et al. [3] from univariate case to bivariate case.

Theorem 4.1 *Let (S_1, S_2) be a non-empty bivariate random set vector with the joint density $h(A, B)$, $(A, B) \in (2^{E_1} \setminus \{\phi\}) \times (2^{E_2} \setminus \{\phi\})$. Then there exist bivariate CAR probabilities, $\pi(A, B)$, and a joint mass function $p(x, y)$ of a bivariate random vector (X, Y) on $E_1 \times E_2$ such that*

$$h(A, B) = \pi(A, B)p(A, B) \quad \text{for all } (A, B) \in 2^{E_1 \times E_2} \setminus \{\emptyset\},$$

where $p(A, B) = \sum_{y \in B} \sum_{x \in A} p(x, y)$. Furthermore, for each A, B with $h(A, B) > 0$, $\pi(A, B)$ and $p(A, B)$ are uniquely determined by the joint distribution of bivariate random set vector (S_1, S_2).

Proof Consider the maximization problem of

$$L = \sum_{A \subseteq E_1} \sum_{B \subseteq E_2} h(A, B) \log[\pi(A, B)p(A, B)]$$

subject to

(i) $p(A, B) = \sum_{x \in A} \sum_{y \in B} p(x, y)$ with $p(x, y) \geq 0$ and $p(E_1, E_2) = 1$,

and

(ii) $\sum_{A \ni x} \sum_{B \ni y} \pi(A, B) = 1$, for any $x \in E_1, y \in E_2, \pi(A, B) \geq 0$.

Note that L can be rewritten

$$L = \sum_{A \subseteq E_1} \sum_{B \subseteq E_2} h(A, B) \log(\pi(A, B)) + \sum_{A \subseteq E_1} \sum_{B \subseteq E_2} h(A, B) \log(p(A, B))$$

which, as a function of the $p(A, B) \in [0, 1]$ and $\pi(A, B) \in [0, 1]$, is continuous and strictly concave, and takes values in $[-\infty, 0)$. The subset of $p(A, B)$ and $\pi(A, B)$

satisfying constraints (i) and (ii) is convex and compact, so the surpremum is attained uniquely under the given constrains.

Now we study a solution for the $p(A, B)$ separately and consider the maximization of

$$L_2 = \sum_{A \subseteq E_1} \sum_{B \subseteq E_2} h(A, B) \log p(A, B)$$

subject to constraint (i). By concavity of L_2, there exists a Lagrange multiplier λ such that any solution is also solution of the maximization problem:

$$L_2(p, \lambda) = \sum_{A \subseteq E_1} \sum_{B \subseteq E_2} h(A, B) \log p(A, B) - \lambda p(E_1, E_2)$$

over $p(x, y) \geq 0$. For any given solution $P(A, B)$ in $L_2(p, \lambda)$, the partial derivative of $L_2(p, \lambda)$ with respect to $p(x, y)$ satisfies

$$\sum_{A \ni x} \sum_{B \ni y} \frac{h(A, B)}{p(A, B)} - \lambda = 0. \tag{4.4}$$

Note that the maximum of L_2 is attained for all (x, y), $x \in A$, $y \in B$ such that $p(x, y) > 0$. Therefore, for those (x, y) such that $p(x, y) = 0$, we have

$$\sum_{A \ni x, B \ni y} \frac{h(A, B)}{p(A, B)} - \lambda \leq 0. \tag{4.5}$$

If the solution $p(A, B) = 0$, we must have $h(A, B) = 0$ otherwise $h(A, B) \log p(A, B) = -\infty$ and $p(x, y) = 0$ for all $x \in A$, $y \in B$. On the other hand, for any (x, y), such that $p(x, y) = 0$, we have $p(\{x\}, \{y\}) = 0$, and $h(\{x\}, \{y\}) = 0$. So we only need to consider the case where $p(A, B) > 0$, which implies there exist one pair $x \in A$ and $y \in B$ such that $p(x, y) > 0$. Multiplying (4.4) by $p(x, y)$ and adding over all (x, y) such that $p(x, y) > 0$, we obtain

$$0 = \sum_{\substack{x \in E_1, y \in E_2 \\ p(x,y)>0}} p(x, y) \sum_{A \ni x} \sum_{B \ni y} \frac{h(A, B)}{p(A, B)} - \lambda \sum_{\substack{x \in E_1, y \in E_2 \\ p(x,y)>0}} p(x, y)$$

$$= \sum_{\substack{x \in E_1, y \in E_2 \\ p(x,y)>0}} \sum_{A \ni x} \sum_{B \ni y} p(x, y) \frac{h(A, B)}{p(A, B)} - \lambda p(E_1, E_2)$$

$$= \sum_{A \subseteq E_1} \sum_{B \subseteq E_2} \sum_{x \in A} \sum_{y \in B} p(x, y) \frac{h(A, B)}{p(A, B)} - \lambda = \sum_{A \subseteq E_1} \sum_{B \subseteq E_2} p(A, B) \frac{h(A, B)}{p(A, B)} - \lambda$$

$$= \sum_{A \subseteq E_1} \sum_{B \subseteq E_2} h(A, B) - \lambda = 1 - \lambda.$$

So $\lambda = 1$ and

$$\sum_{A \ni x} \sum_{B \ni y} \frac{h(A, B)}{p(A, B)} = 1 \quad \text{if } p(x, y) > 0,$$

$$\sum_{A \ni x} \sum_{B \ni y} \frac{h(A, B)}{p(A, B)} \leq 1 \quad \text{if } p(x, y) = 0.$$

For $x \in A$, $y \in B$, $p(x, y) > 0$, set $\pi(A, B) = \frac{h(A,B)}{p(A,B)}$. For $x \in A$, $y \in B$, $p(x, y) = 0$, set $\pi(\{x\}, \{y\}) = 1 - \sum_{A \ni x} \sum_{B \ni y} \frac{h(A,B)}{p(A,B)}$. It is easy to check

$$\sum_{A \ni x} \sum_{B \ni y} \pi(A, B) = 1 \text{ for all } x \in E_1, y \in E_2.$$

Therefore, $h(A, B) = p(A, B)\pi(A, B)$. □

Remark 4.2 Note that the Theorem 4.1 proves the existence of the values $\pi(A, B)$ and $p(A, B)$, it does not explain how to compute $\pi(A, B)$ and $p(x, y)$ based on the values $h(A, B)$. In the following, a computational method for computing values $p(x, y)$ and $\pi(A, B)$ is given, based on the joint density $h(A, B)$ of (S_1, S_2).

Assume $E_1 = \{x_1, x_2, \ldots, x_{N_1}\}$, $E_2 = \{y_1, y_2, \ldots, y_{N_2}\}$. From Theorem 4.1, we know that

$$\pi(A, B) = \frac{h(A, B)}{p(A, B)} = \frac{h(A, B)}{\sum_{x \in A} \sum_{y \in B} p(x, y)} \quad \text{for all } (A, B) \in 2^{E_1 \times E_2} \setminus \{\phi\}.$$

Note that (4.3) can be rewritten as:

$$1 = \sum_{A \ni x} \sum_{B \ni y} \pi(A, B) = \sum_{A \ni x} \sum_{B \ni y} \frac{h(A, B)}{\sum_{s \in A} \sum_{t \in B} p(s, t)} \quad \text{for any } x \in E_1, y \in E_2. \quad (4.6)$$

The expression (4.6) is a nonlinear system of $N_1 N_2$ equations with unknowns $p(x_i, y_j)$'s $i = 1, 2, \ldots, N_1$, $j = 1, 2, \ldots, N_2$.

We can now use, e.g., BB-package in CRAN-R for solving this system of $N_1 N_2$ nonlinear equations with $N_1 N_2$ unknowns. Once we have determined the $N_1 N_2$ values $p(x_i, y_j)$, we can then compute, for every pair of sets $A \subseteq E_1$, $B \subseteq E_2$, the value $\pi(A, B) = h(A, B)/p(A, B)$.

Table 7 Joint density of $(\mathcal{S}_1, \mathcal{S}_2)$

h_C	{3}	{4}	{5}	{3, 4}	{3, 5}	{4, 5}	{3, 4, 5}
{1}	5/48	13/144	11/144	0	0	0	1/16
{2}	1/12	1/12	1/12	0	0	0	1/12
{1, 2}	1/16	11/144	13/144	0	0	0	5/48

Table 8 Bivariate CAR probabilities

$\pi(\cdot, \cdot)$	{3}	{4}	{5}	{3, 4}	{3, 5}	{4, 5}	{3, 4, 5}
{1}	0.5825	0.5412	0.5001	0	0	0	0.1254
{2}	0.5418	0.5005	0.4595	0	0	0	0.1661
{1, 2}	0.1879	0.2292	0.2702	0	0	0	0.1042

Example 4.1 Suppose the joint density $h_C(A, B)$ is given in Table 7.

By using the above method, we obtain the joint mass function, $p(x, y)$, bivariate random vector (X, Y) on $E_1 \times E_2$,

$$p(1, 3) = 0.1788, \quad p(1, 4) = 0.1668, \quad p(1, 5) = 0.1527,$$
$$p(2, 3) = 0.1538, \quad p(2, 4) = 0.1665, \quad p(2, 5) = 0.1814.$$

The corresponding bivariate CAR probabilities are listed in Table 8.

Acknowledgments The authors would like to thank Professor Hung T. Nguyen for introducing this interesting topic to us.

References

1. D.A. Alvarez, A Monte Carlo-based method for the estimation of lower and upper probabilities of events using infinite random sets of indexable type. Fuzzy Sets Syst. **160**, 384–401 (2009)
2. A.P. Dempster, Upper and lower probabilities induced by a multivalued mapping. Ann. Math. Stat. **28**, 325–339 (1967)
3. R.D. Gill, M.J. Van der Laan, J.M. Robins, Coarsening at random: characterizations, conjectures, counter-examples. Springer Lect. Notes Stat. **123**, 149–170 (1997)
4. R.D. Gill, P.D. Grunwald, An algorithmic and a geometric characterization of coarsening at random. Ann. Stat. **36**(5), 2049–2422 (2008)
5. P. Grunwald, J. Halpern, Updating probabilities. J. Artif. Intell. Res. **19**, 243–278 (2003)
6. J. Harry, *Multivariate Models and Dependence Concepts* (Chapman & Hall, London, 1997)
7. D.F. Heitjan, D.B. Rubin, Ignorability and coarse data. Ann. Stat. **19**(4), 2244–2253 (1991)
8. M. Jaeger, Ignorability for categorical data. Ann. Stat. **33**, 1964–1981 (2005)
9. B. Li, T. Wang, Computational aspects of the coarsening at random model and the Shapley value. Inf. Sci. **177**, 3260–3270 (2007)
10. I.S. Molchanov, *Theory of Random Sets* (Springer, Berlin, 2005)

11. R.B. Nelsen, *An introduction to Copulas*, 2nd edn. (Springer, New York, 2006)
12. H.T. Nguyen, *An Introduction to Random Sets* (CRC Press, Boca Raton, FL, 2006)
13. H.T. Nguyen, T. Wang, Belief functions and random sets, in *The IMA Volumes in Mathematics and Its Applications*, vol. 97 (Springer, New York, 1997), pp. 243–255
14. H.T. Nguyen, Lecture Notes in: Statistics with copulas for applied research, Department of Economics, Chiang Mai University, Chiang Mai, Thailand (2013)
15. B. Schmelzer, Characterizing joint distributions of random sets by multivariate capacities. Intern. J. Approx. Reason. **53**, 1228–1247 (2012)
16. B. Schmelzer, Joint distributions of random sets and their relation to copulas, *Modeling Dependence in Econometrics*. Advances in Intelligent Systems and Computing, vol. 251 (Springer International Publishing, Switzerland, 2014). https://doi.org/10.1007/978-3-319-03395-2-10
17. B. Schmelzer, Joint distributions of random sets and their relation to copulas. Intern. J. Approx. Reason. (2015). In Press
18. G. Shafer, *A Mathematical Theory of Evidence* (Princeton University Press, New Jersey, Princeton, 1976)
19. A. Sklar, Fonctions de répartition án dimensions et leurs marges. Publ. Inst. Statist. Univ. **8**, 229–231 (1959). Paris
20. A.A. Tsiatis, *Semiparametric Theory and Missing Data* (Springer, New York, 2006)
21. A.A. Tsiatis, M. Davidian, W. Cao, Improved doubly robust estimation when data are monotonely carsened, with application to longitudinal studies with dropout. Biometrics **67**(2), 536–545 (2011)
22. Z. Wei, T. Wang, W. Panichkitkosolkul, Dependence and association concepts through copulas, in *Modeling Dependence in Econometrics*. Advances in Intelligent Systems and Computing, vol. 251 (Springer International Publishing, Switzerland, 2014), pp. 113–126. https://doi.org/10.1007/978-3-319-03395-2-7
23. Z. Wei, T. Wang, B. Li, P.A. Nguyen, The joint belief function and Shapley value for the joint cooperative game, in *Econometrics of Risk, Studies in Computational Intelligence*, vol. 583 (Springer International Publishing, Switzerland, 2015), pp. 115–133. https://doi.org/10.1007/978-3-319-13449-9-8
24. Z. Wei, T. Wang, P.A. Nguyen, Multivariate dependence concepts through copulas. Intern. J. Approx. Reason. (2015). In Press

Printed in the United States
by Baker & Taylor Publisher Services